Biotransformations
and Bioprocesses

BIOTECHNOLOGY AND BIOPROCESSING SERIES

Series Editor

Anurag Singh Rathore
Bioprocess Sciences
Pharmacia Corporation
Chesterfield, Missouri

ADDITIONAL VOLUMES IN PREPARATION

Biotransformations and Bioprocesses

Mukesh Doble

Anna University
Chennai, India

Anil Kumar Kruthiventi

Sri Sathya Sai Institute of Higher Learning
Prasanthinilayam, Andhrapradesh, India

Vilas Gajanan Gaikar

Bombay University
Maharashtra, India

CRC Press
Taylor & Francis Group
Boca Raton London New York

CRC Press is an imprint of the
Taylor & Francis Group, an **informa** business

CRC Press
Taylor & Francis Group
6000 Broken Sound Parkway NW, Suite 300
Boca Raton, FL 33487-2742

First issued in paperback 2019

ISBN-13: 978-0-8247-4775-6 (hbk)
ISBN-13: 978-0-367-39443-1 (pbk)

Library of Congress Cataloging-in-Publication Data
A catalog record for this book is available from the Library of Congress.

Visit the Taylor & Francis Web site at
http://www.taylorandfrancis.com

and the CRC Press Web site at
http://www.crcpress.com

We would like to dedicate our contributions to Bhagwan Sri Sathya Sai Baba.
—Mukesh and Anil

I would like to dedicate my contribution to my teachers for all the knowledge imparted to me.
—Vilas

Series Introduction

Biotechnology encompasses all the basic and applied sciences as well as the engineering disciplines required to fully exploit our growing knowledge of living systems and bring new or better products to the marketplace. In the era of biotechnology that began with recombinant DNA and cell fusion techniques, methods and processes have developed mostly in service of protein production. That development is documented in this series, which was originally called Bioprocess Technology. Many protein products that are derived from the technology are already marketed and more are on the way.

With the rapid expansion of genomics, many new biological targets will likely be identified, paving the way for the development of an even wider array of products, mostly proteins. As knowledge of the targets develop, so will rational drug design, which in turn may lead to development of small molecules as healthcare products. Rational genetic manipulation of cells as factories for growing products is also developing. Other examples of the application of genomics in health care include the development of gene therapy by insertion of genes into cells and the blocking of gene expression with antisense nucleotides. In such new directions, nucleotides and other small molecules as well as protein products will evolve. Technologies will develop in parallel.

Transgenic technology, in which the genome of an organism is altered by inclusion of foreign genetic material, is also just beginning to develop. Recombinant protein products can already be made, for example, in the milk of transgenic animals, as an alternative to conventional bioreactors. Newer applications for transgenic technology in agriculture may take time to develop, however. Questions continue to be raised about the long-term environmental consequences of such manipulation.

As technology develops in newer as well as established areas, and as knowledge of it becomes available for publishing, it will be documented in this continuing series under the more general series name of Biotechnology and Bioprocessing.

W. Courtney McGregor

Preface

Biotransformation deals with the use of a biocatalyst for the mediation of a chemical reaction, for the synthesis of an organic chemical or destruction of an unwanted chemical. Bioprocess deals with the application of technology and engineering principles to design, develop, and analyze these processes. The tools of the chemical engineer will be essential to the successful exploitation of bioprocesses. Biotransformation is now playing a key role in many industries, including the arenas of food, chiral drugs and vitamins, specialty chemicals, and animal feed stock. The techniques are also finding their way in the manufacturing of bulk and commodity chemicals. The use of enzyme and microbes for chemical transformation and organic synthesis is expected to grow tremendously since the industries are being forced by the public and nongovernmental organizations (NGOs) to shift toward "green chemistry," which will produce less toxic effluents and also use safer and cleaner chemicals in their manufacturing processes. This interdisciplinary book is well suited to address some of these points.

This book is concise yet comprehensive, covering chemistry and engineering aspects of biotransformation and giving an overview of the various steps involved during the transition from a lab to the plant. Although chemical engineers and organic chemists have worked together during process scale-up related to chemical transformation, together they are entering a completely different field. This book will help them overcome some misconceptions. Organic chemists and chemical engineers differ in their approach to problem solving and this book helps each group see the other's point of view. Other topics covered include molecular structure property, enzyme and microbial kinetics, biotransformation, fermentation, reactors (an in-depth analysis of stirred and tower reactors), separation processes, scale-up issues, and waste treatment with industrial examples.

This book is not intended to be an encyclopedia, but covers the current and relevant matter in a succinct way, addressed to an interdisciplinary audience. The book has illustrations, homework problems, and innovative extensions. This approach will encourage students to obtain a more in-depth understanding of key scientific and engineering concepts. It is designed to be a textbook for undergraduate and graduate-level courses in biotechnology (including fermentation) and other interdisciplinary courses in pharmacy, biosciences, and organic synthesis.

A combination of biotransformations and chemical process engineering (such as kinetics, separations, scale-up) is discussed here, and hence the book will appeal to a diverse audience of chemists, biologists, and chemical technologists/engineers. It will be useful for biologists who would like an overview of chemical and engineering principles and to chemical engineers with no knowledge of biotransformations and biochemical engineering fundamentals. The book assumes that engineers have very little background in synthetic chemistry, and therefore builds up the knowledge from the basics. Similarly, the book assumes that organic chemists have very little knowledge in chemical reaction engineering.

The initial chapters start from the fundamentals of chemistry with an introduction to molecules, structures and their relationships, different types of reactions; from small to supra molecules and extended to enzymes and proteins. Later, an in-depth discussion of the mechanism of the reactions catalyzed by enzymes, whole cells, and microbes is presented. The various experimental and analytical techniques that a bio-organic chemist will employ in the lab are also presented. These techniques are very specific to biocatalytic reactions.

The thermodynamics and kinetics of biocatalytic reactions is dealt with in detail. Selection of a suitable reactor for carrying out the desired trans-formation from a plethora of reactors based on several criteria is discussed, followed by an in-depth design study of two of the most popular reactors (stirred and tower). Since the fermentor has become the workhorse of the biochemical industry, it cannot be ignored. Chapter 9 describes fermentation technology and design of process control strategies.

The underlying reaction engineering and scale-up principles are exam-ined in a detailed discussion that can be viewed as a primer for organic chemists. A manufacturing plant also consists of downstream recovery and purification as well as waste treatment sections. Hence, for the sake of completeness an overview of traditional chemical engineering, special sepa-ration techniques, and waste treatment techniques is also included in this book. One chapter deals with a current biochemical industrial scenario with a few process flow sheets for the manufacture of pharmaceutical intermediates and specialty chemicals.

One chapter very briefly describes the frontier research areas in the area of biotransformation (which includes cross-linked enzymes, designer enzymes, abzymes, site selective modification of enzymes, etc.; all aimed toward improving their stability, activity, and specificity). The book does not cover molecular and cell biology, protein engineering, or metabolic pathways.

This book is based on the lecture courses that all three authors have given to undergraduate and postgraduate students of biotechnology, organic chemistry, and food technology over many years. Dr. Kumar is a conventional synthetic organic chemist and Dr. Gaikar is a solid chemical engineer, while Dr. Doble, who is also a chemical engineer, can be viewed as the bridge between the two disciplines, having worked with chemists for two decades.

Mukesh Doble
Anil Kumar
Vilas Gajanan Gaikar

Contents

1

Introduction and Overview

Biotransformation deals with use of natural and recombinant microorganisms (e.g., yeast, fungi, bacteria), enzymes, whole cells, etc., as catalysts in organic synthesis. Biotransformation plays a key role in the area of foodstuff, chiral drug industry, vitamins, specialty chemicals, and animal feed stock (Fig. 1.1). Scaling up a bioprocess from the lab to a commercial scale is challenging and needs several innovations. Nevertheless, more and more industries are moving toward developing processes based on biocatalysis because of their inherent advantages. In the year 2000 biotechnology stocks traded in the Nasdaq exchange outperformed the overall index by 24% (outperformed the Internet stocks by 17%)! This observed general buoyancy is due to the successful applications of biotransformations in the field of pharmaceuticals, environmental bioremediation, textiles, plastics, and agriculture. Biopolymers made from dextrose and plastics made from corn sugar beet and other biomass compete with polymers made from hydrocarbons. Breakthroughs in the area of optimization, reactor design, separation techniques, and molecular modeling are a few of the underlying reasons for these successes.

The design and operation of industrial reactors for bioprocesses are inherently different from the conventional reactors. This book deals not only with how a biocatalyst could be used for synthesis of an organic molecule but also with the steps involved in the scale-up of a process from the bench scale to

1

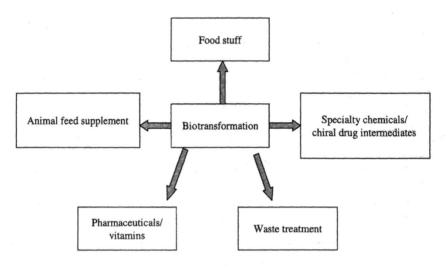

FIGURE 1.1   Role of biotransformation.

the full commercial scale and the reaction engineering aspects of the manufacturing technology, with an in-depth analysis of bioreactor design.

As shown in Fig. 1.2 the field of biotransformation and bioprocess is interdisciplinary in nature. As the process moves from the lab scale to full-scale commercial production, it requires the expertise of biochemist, molecular biologist, synthetic chemist, physical chemist, biotechnologist, and chemical and instrument engineers. All the aspects listed in Fig. 1.2 are dealt in the various chapters of this book. At times the process as it is scaled up may have to go back to the lab because of issues not foreseen earlier.

A biochemical process generally consists of five sections; they are catalyst and raw material preparation, reaction, biocell recovery for reuse or destruction, product recovery and purification, and waste disposal (Fig. 1.3). The book is divided into four parts. The first part deals with the fundamentals, namely chemistry of biotransformation and the associated areas such as synthetic chemistry, enzyme chemistry, frontiers in biotransformations, and enzyme and biocell kinetics. The second part deals with bioreactors selection, types of bioreactors and their design including fermentors, and aspects of biochemical engineering. The third part touches on the downstream separation techniques, and the fourth part, on industrial examples of biotransformations, waste treatment, and scale-up of bioreactions.

The book is written for practicing biochemists and pharmacists who would like to understand the reaction engineering aspects and to chemical engineers who wish to understand the synthetic techniques and organic chem-

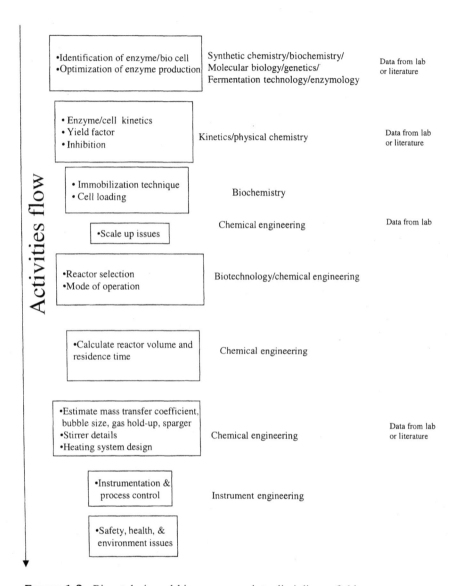

FIGURE 1.2 Biocatalysis and bioprocess: an interdisciplinary field.

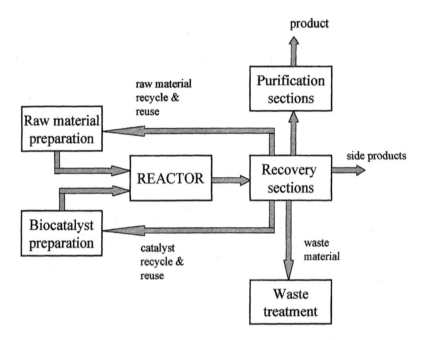

FIGURE 1.3   General flow sheet of a biocatalytic manufacturing process.

istry aspects of this vast field. A large number of problems are given in the end of many of the chapters for students to sharpen their knowledge they would have acquired. This book is not intended to be an encyclopedia for bio-transformations or bioreactors, but a ready reference to the practioners relating the science and engineering.

Chapter 2 gives an introduction to molecules, structures and their relationships, quantum mechanical approach, and different types of reactions starting from small to supra molecules. This explanation is then extended to enzymes and proteins.

Chapter 3 describes the structure and activity of enzymes and proteins, differences between enzymes and conventional heterogeneous catalysts and the thermodynamic aspects of the biocatalytic reaction.

Chapter 4 deals in detail with the reactions catalyzed by enzymes, whole cells, and microbes. Mechanistic aspects of these reactions are also discussed.

Chapter 5 deals with various experimental techniques and analytical techniques a bioorganic chemist will employ in the lab. These techniques are very specific to biocatalytic reactions.

Chapter 6 briefly describes the frontier research areas in the area of biotransformation that includes cross-linked enzymes, designer enzymes, abzymes, site-selective modification of enzymes, etc., all aimed toward improving their stability, activity, and specificity.

Chapter 7 deals with enzyme kinetics, inhibition, Michaleis-Menten approach to modeling biocatalytic reactions, and cell growth. Rate equation for different types of reactions is listed.

Chapter 8 deals with biochemical reactor selection, different types of reactor and their salient features. Basic design equations for various types of reactors and Monod equation are also described here.

Chapter 9 deals with fermentation, namely fermentation classification, issues in fermentation, modeling of molds, and four stages of biocell growth. Details of reactor and process control design are also described here.

Chapter 10 gives an overview of reaction engineering principles such as mass and heat transfers and how they are estimated for design purposes.

Chapter 11 deals with stirred bioreactors in detail, since they are used in general because of simple construction and ease of operation. Several correlations for gas-liquid mass transfer that are needed for design are described here.

Chapter 12 gives a detailed analysis of tower bioreactors including the gas, solid, and liquid mixing and heat transfer issues.

Chapter 13 is an introduction to biochemical separation and downstream processing and purification. The various traditional chemical engineering separation processes such as distillation, extraction, filtration, etc., and separations that are very specific to biochemical processes, namely, chromatography, membrane, electrophoresis, etc. are discussed here.

Chapter 14 deals with industrial examples of where biocatalyst is used successfully including chiral synthesis, pharmaceuticals, specialty chemicals, etc. This chapter is an eye opener to chemists and biologists, giving the industrial world scenario.

Chapter 15 deals with in situ and ex situ waste treatment procedures for solid, liquid, and gas. Different types of reactors used in waste treatment are also discussed. The current chemical methods available in waste treatment are also listed with their advantages and disadvantages over the biochemical approach.

Chapter 16 deals with a large number of scale-up rules which need to be followed for successfully translating a process from bench to commercial scale. They relate to mixing, heat transfer, solid suspension, etc. Several scale-up rules are listed depending upon the criteria one would like to select. A list of innovative techniques reported in literature for scaling up biochemical processes is also tabulated.

2

Chemical Bonding, Structure, and Reaction Dynamics

2.1 INTRODUCTION TO CHEMISTRY

Our unquenchable curiosity about our environs and ourselves is at the root of all the scientific endeavors. We marvel at the way the wings of the butterfly are colored, at the fine smell of a jasmine flower, at the gigantic colossus of the Sal tree trunks, and the list goes on endless. All of this myriad variety, we call *nature*, and in fact our own body is an eternal source of wonder and joy, involving intricate mechanisms (transformations) at the atomic, molecular, supramolecular, and the gross levels. Our journey begins, when we try to understand and imitate this infinite variety. How does all this happen? What are its components? Why is it happening only this way and not by any other way? These are some of the queries, that drive us to explore, analyze, understand, and know.

2.1.1 Origins of Organic Chemistry

As one of the tools that fostered an increased understanding of our world, the science of Chemistry—"the study of matter and the changes they undergo at a molecular level"—developed until near the end of eighteenth century. Initially, there was a single branch, but later with the continued studies of Lavoiser, Berzelius, Wohler, and others the three major branches were recognized. One branch was concerned with the matter obtained from natural or living sources and was called *organic chemistry*. Jons Jacob Berzelius coined this term. The other branch dealt with substances derived from nonliving matter—minerals and the like—it was called *inorganic chemistry*. The dynamic aspects of the changes that occur in both of these compounds had to explained in a more methodical mathematical language and thus started *physical chemistry*. Combustion analysis established that organic chemistry is the study of *carbon compounds*.

Apart, from the water content almost all of the material, except for some trace minerals in living systems, is made up of organic compounds. The food that we take, the fragrances we inhale, the colors we see, the medicines we partake are predominantly organic compounds.

> If we had not had a coating of dipalmitoyl phosphatidyl choline (DPPC) molecules in the lining of our alveoli in our lungs, it would not have been possible for us to breath. These help in maintaining the flexibility of the alveoli.

With the advances in spectroscopic instrumentation and femtosecond laser techniques, organic chemistry has developed into a mature science, and its influence is felt in various other branches of science. A casual look at any modern biology texts or journals makes it amply clear that a good fundamental understanding of the principles of organic chemistry is essential to appreciate those sciences and also to better the advances in them.

During the last decade computational chemistry has joined modern instrumental and chromatographic methods as an important tool in the practice of organic chemistry. The improved performance of microcomputers and the development of high-performance modeling programs with user-friendly graphical input and display capabilities have made the study of three-dimensional (3D) structures and their properties routine. In fact, it is a necessary teaching aid to make the students grasp the nuances of the subject.

In all the types of learning, *understanding* and *visualization* leads to grasping, which furthers to better the application of the knowledge gained. Therefore, an understanding of the molecular structure and the dynamics is essential to appreciate and derive joy there from. Fundamental to understanding is the ability to mentally picturize the statement or object. For example, for an English statement such as *John ate an orange* is appreciated by everyone knowing the English language, because everyone of us was taught initially with visual aids, what an orange meant, what it means to eat, and what is meant by any name; thus, our understanding of the above statement is complete.

Similarly, in order that we appreciate a reaction,

$$CH_3Cl + NaOH \rightarrow CH_3OH + NaCl$$

we need to necessarily develop ability to mentally visualize what we mean by this symbolic language of writing CH_3Cl, NaOH, CH_3OH, and NaCl. Also, what we mean by a reaction (whether catalyzed or uncatalyzed, whether in vivo or in vitro): what is it that happens in a reaction vessel?

2.1.2 Visualization of a Molecule

What do we mean by molecules?
How do they look?

Molecules are obtained by a combination of atoms. Atoms can be visualized as an electron cloud with a (heavy) center—*nucleus*. Almost like a hazy mass of spherical cotton ball with seedlike center nucleus. Therefore, when two electron clouds unite, we get another bigger electron cloud with two nuclei placed at a particular distance from one another. Thus, if more than two atoms combine, we get an electron cloud distributed aross all the nuclei of the atoms. *The shape of this electron cloud is distinct and unique to a molecule.* That is, say, methane—all of methane molecules have the same electron cloud distribution, but the electron cloud distribution of methyl chloride is different from methane, as shown in Fig. 2.1. In fact it is this unique electron cloud distribution over a molecule that determines its reactivity.

2.1.3 Concept of Chemical Bonding

These notes are designed to give a quick guide to bonding—there is a lot more to this subject than is covered here.

The forces that hold atoms together within chemical compounds, the chemical bonds, are electrical in nature. Chemical bonds form to lower the energy of the system, the components of the system becomes more stable through the formation of the bonds. This is true even for noncovalent interactions (hydrogen bond, van der Waals interaction, electrostatic inter-action, etc.), which are far more important among biomolecules and supra-molecular aggregates. These bonding and noncovalent interactions are Nature's way of attaining greater stability. The development of a sound chemical bonding theory started with the idea put forth by G. N. Lewis in 1916, that bonding results from a sharing of electron pair between two atoms.

Two major bonding interactions are common in Organic chemistry: one in which the two atoms constituting the bond share the two electrons, a

(a)

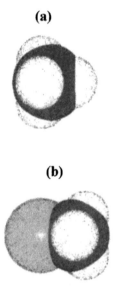

(b)

FIGURE 2.1 Electron cloud distribution over a molecule: Space-filled model of (a) methane molecule and (b) methyl chloride (observe the excess electron density, indicated by size, shift toward the large chlorine atom shown in light grey color).

covalent bond (Fig. 2.2), and the other in which one of the atoms constituting the bond takes both the electrons and attains a negative charge while the other atom loosing an electron attains a positive charge ("sharing" is unequal), an *ionic bond* (Fig. 2.3).

Ionic Bond

From the spectroscopic studies of an atom (hydrogen spectrum) it became clear that electrons reside in shells about the nucleus. The outermost shell of electrons in an atom is called the *valence shell*, and the electrons in this shell are called *valence electrons*. The German Walther Kossel noted (1916) that stable ions tend to form when atoms gain or lose enough electrons that they

FIGURE 2.2 A covalent bond.

$$A\cdot \quad + \quad B\cdot \quad \longrightarrow \quad A^+ \quad {\vcenter{\hbox{$\cdot\kern-0.3em\cdot$}}}B^-$$

FIGURE 2.3 An ionic bond.

have the same number of electrons as the noble gas of closest atomic number. The same observation was restated as the *octet rule*, which states that an atomic species tends to be especially stable when its valence shell contains eight electrons. This is particularly observed with the elements in the A group of the periodic table (while for elements close to helium it is the *duet rule*, two electrons in the valence shell).

Most elements tend to attain this valence electron configuration of a noble gas, either by loosing or gaining an electron. The ability of a neutral atom to loose electrons is measured as ionization potential. The smaller the potential, greater is the ease of removal of the electron. Alkali metals (Li, Na, and K) show the greatest tendency to loose electron. The tendency of an atom to gain an electron (to form a negative ion) is called *electronegativity* and the quantitative measure of this is termed *electron affinity*. The greater the electron affinity, the easier it is to gain an electron. Halogens (F, Cl, Br, and I) are the atoms with high electron affinities.

Thus, when two atoms, one having low ionization potential, another having a high electron affinity combine, an ionic bond results (for example, KCl; see also the box on next page).

Covalent Bond

The type of bonding when electrons are shared equally, between the two bonded atoms, is known as a *covalent bond*. The same octet rule (duet rule for hydrogen), is often obeyed not by the transference of one electron from one atom to the other, as in the ionic bond, but by sharing of the two bonding electrons. This is formed between elements that in general are not readily ionizable, that is, they are neither strongly electropositive nor strongly electronegative. The electron density of the two bonded electrons is distributed between the two atom centers. This electron distribution in a cylindrically symmetric manner about the internuclear axis is termed a *sigma σ bond*. This is represented as a simple line connecting the two atoms in the common notation. That is, C—H, C—C, C—O and so on. Therefore when a line is drawn between two symbols of two elements it symbolizes the presence of two bonding electrons in the region between the two bonded atoms also the presence of an encompassing molecular orbital.

Is there such as thing as an Ionic bond?

When two elements form an ionic compound, is an electron really lost by one atom and transferred to the other one? Consider the data on the ionic solid LiF. The average radius of the neutral Li atom is about 2.52 Å.

Now if this Li atom reacts with an atom of F to form LiF, what is the average distance between the Li nucleus and the electron it has "lost" to the fluorine atom? The answer is 1.56 Å; the electron is now closer to the lithium nucleus than it was in neutral lithium! So the answer to the above question is both yes and no: yes, the electron that was now in the 2s orbital of Li is now within the grasp of a fluorine 2p orbital, but no, the electron is now even closer to the Li nucleus than before, so how can it be "lost"? The one thing that is true about LiF is that there are more electrons closer to positive nuclei than there are in the separated Li and F atoms.

Chemical bonds form when electrons can be simultaneously near two or more nuclei. This is obvious in the covalent bond formation. What is not so obvious (until you look at the numbers such as were quoted for LiF above) is that the "ionic" bond results in the same condition; even in the most highly ionic compounds, both electrons are close to both nuclei, and the resulting mutual attractions bind the nuclei together. This being the case, is there really any fundamental difference between the ionic and covalent bond? The answer, according to modern chemical thinking is probably "no"; in fact, there is some question as to whether it is realistic to consider that these solids consist of "ions" in the usual sense. The preferred picture that seems to be emerging is one in which the electron orbitals of adjacent atom pairs are simply skewed so as to place more electron density around the "negative" element than around the "positive" one.

This being said, it must be reiterated that the ionic model of bonding is a useful one for many purposes, and there is nothing wrong with using the term *ionic bond* to describe the interactions between the atoms in "ionic solids" such as LiF and NaCl.

Hybridization

The probability of finding an electron around the central nucleus is an orbital. To simplify we may even consider it as the region of space occupied by an electron. The outermost shell orbital is termed *valence orbital*. In organic chemistry we come across the spherically symmetrical s orbital and the dumbbell-shaped three p orbitals. This valence orbital may interact with the valence orbital of other atoms without any modifications in themselves, but, many a times these atomic orbital (the s and the p orbital) may mix among themselves. Due to the small energy difference between the s and the p orbital this mixing, also termed *hybridization*, occurs. Such hybridization of the 2s and the $2p$ valence orbital in molecular orbital formation is favorable because it increases the valence-orbital overlap between the bonding nuclei. As a result electron density increases in this internuclear axis region. This not only reduces the overall energy of the system because of more efficient overlap, but also gives a high degree of directionality to the orbital overlap. For example, in water molecule (Fig. 2.4) the bonding arises from the two electrons in an orbital (molecular orbital) formed by the overlap of the spherically symmetrical atomic orbital of each of the two hydrogen atoms, with that of the highly directional two of the sp_3 hybridized orbitals of the oxygen atom.

In the carbon atom the valence $2s$ and the three $2p$-orbitals mix in varying proportions to give the three hybridized carbon entities, namely the sp_3, sp_2, and the sp. Based on the hybridization of the carbon the three-dimensional geometry of the system can be predicted as sp_3, tetrahedral; sp_2, trigonal; and sp, linear (Fig. 2.5).

The occurrences of multiple bonds, i.e., double and triple bonds can also be appreciated by the proper understanding of the hybridization concept. The horizontal overlap of the hybridized orbital yields a sigma bond (already mentioned), while the parallel or the vertical overlap of the unhybridized valence p orbital (in sp_2 and sp hybridized atoms) yields a pi bond. The electron cloud distribution of the pi bond is concentrated above and below the plane of the sigma framework. Thus we observe a sigma bond (sp_3), or a sigma plus a pi bond (sp_2), or a sigma and two pi bonds (sp) in organic chemistry. These are drawn as a single line notation for sigma bond, two lines for one sigma plus one pi, and finally three lines for one sigma and two pi bonds. With the increase in bonding between any two atoms, the internuclear distance also reduces, i.e., the two atoms are held close together. The electron density about the two bonded atoms also increases with increase in the number of bonds. The pi electron cloud is normally localized over the two bonded atoms, in some molecules this pi electron cloud is spread over more than two atoms, i.e., delocalized.

As seen earlier, the overlap of the orbitals makes the electrons get delocalized, that is, more mobile so to say. When this overlap is only between two orbitals of two different atoms, we term it *bonding interaction that is localized*. However, when more than two p orbitals or two p orbitals with a

FIGURE 2.4 Water molecule.

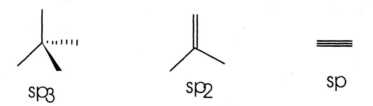

FIGURE 2.5 Three-dimensional geometry of differently hybridized carbons.

nonbonding sp_3 orbital, and so on, overlap, it leads to delocalization of the bonded electrons (normally the pi electrons) over more than two atoms. This is a very common and a stabilizing interaction, normally known as *conjugation*. The occurrence of these various types of conjugative interactions, such as cross-conjugation, hyper-conjugation, skip-conjugation, resonance, and aromaticity, explain in greater detail the electron cloud distribution and hence the reactivity (behavior) of more complex molecules—the discussion of which is beyond the scope of this book.

Polar Covalent Bond

A covalent bond between two atoms of different electronegativities is a polar bond. C−O, C−N, C−Cl, C−F, and C−Mg are a few common examples of polar covalent bonds. As can be observed, due to the differing electronegativities between the two atoms forming the covalent bond, the electrons in the bonding region tend to spend more time on the atom having greater electronegativity, thereby creating a partial charge separation, a dipole along the bond axis. In other words, the sharing of the two bonding electrons is unequal. The measure of this dipole moment is μ, the unit being debye (D). The dipole moment of CH is 0.4. The value for a carbon and halogen bond is around 1.38 to 1.46. It is around 2.3 for C=O.

In fact, it is these polar covalent bonds in an organic molecule, that determine its reactivity. Therefore, these are also termed the *functional groups* of the molecule. These are like the locations through which the molecule expresses itself and responds to the external stimuli (reaction conditions). Like the limbs to the human body, the functional groups are the limbs for the molecule to "express" itself.

2.1.4 Summary: Learning Outcome

Back to visualization of the reaction with which we started:

$$CH_3Cl + NaOH \rightarrow CH_3OH + NaCl$$

Now we can appreciate what we mean by CH_3Cl or CH_3OH. These two are molecules having covalent bonds. In both of these three hydrogens are connected to the central carbon atom to which in one of the molecules a chlorine and in the other a hydroxyl group is attached. Both of these molecules contain polar covalent bonds between the carbon and the heteroatom (Cl, O). Also, from the dipole moment calculations, we can observe that the $C-Cl$ bond is more polar than the $C-O$ bond. Therefore, the electron cloud distribution in both these molecules is *distorted*, creating an inherent dipole (partial electron deficient and electron rich centers) in these molecules. NaCl is an ionic compound, while NaOH is also ionic between the sodium (Na) and the OH ions, but there is a covalent bond between the oxygen and the hydrogen atom.

Molecules, like atoms or even the subatomic particles, are dynamic entities having a variety of movements at the ambient temperature and pressure conditions. Hence, when we mix CH_3Cl and NaOH, the distorted electron cloud of the methyl chloride collides with the electron-rich (almost spherical) hydroxide ion electron cloud in infinite number of ways. The collision in which the electron cloud of the hydroxide ion comes close enough to the electron-deficient center (carbon) of the methyl chloride is the fruitful collision, leading to the product formation.

Learning outcome: Understanding and visualizing the meaning of molecular formulas.

But then the other questions follow:

What does the arrow signify?
Why should only that product form?
What are the interactions between molecules?

We will now attempt to answer these and many more related questions.

2.2 REACTION

People want to change the natural into the useful unnatural, or into the improved man-made or human made. To add value is to profit. So, from the beginning, transformation was essential, whether in the making of metals and alloys, in medicinal preparations, in cooking, in dyeing and coloring, in tanning leather, or in the cosmetics. Reaction is fundamentally *bringing about transformation*. A directed reaction, which is a specific controlled reaction, can also be termed *molecular engineering*. Today, when a chemist thinks of a "*reaction*," he or she sees both the macroscopic transformation as of old and the microscopic molecular change.

Now we know, that in any chemical reaction the motions of the electrons and nuclei of atoms determine how the molecules interact and those interactions in turn create the forces that govern the reaction dynamics.

2.2.1 Basic Types of Reactions in Organic Chemistry

At the molecular level a reaction starts when two molecules come close enough—within the bonding distance, approximately 0.1 nm—then they begin to interact. The electrons get rearranged; they start moving from an electron-rich center to an electron-deficient center. Transitory states are produced which either gets converted to the products or back to the reactants. These transitory states can be a *transition state* or an *intermediate*. Ahmed Zewail, a scientist from Caltech in the United States received the Nobel prize in chemistry (1999) for developing femtosecond laser photography to photograph these transitory states.

The time scales for these transition states range from about 10 to 100 fs. A femtosecond is a very small unit of time. A femtosecond is to a second what a second is to 32 million years. Furthermore, while in 1 s light travels nearly 300,000 km—almost the distance between the earth and moon—in 1 fs light travels only 0.3 μm—about the diameter of a smallest bacterium.

When the movement/rearrangement of these electrons and nuclei is within the same molecule, it is termed an *intramolecular* reaction—similar to clasping of both the hands. When the movement/rearrangement is between two separate molecules it is termed an *intermolecular* reaction—similar to the holding of each others hands to form a human chain. Also, a reaction can be instantaneous/concerted or stepwise. Since, the reactions are basically flow of valence electrons-when an electron-rich center in any molecule seeks to stabilize itself by sharing the excess electrons with an electron-deficient center—then we term these reactions as *nucleophillic* (*nucleo*, positive charge; *phillic*, liking), similar to a very rich individual seeking or wanting to take up charitable works (for obvious tax benefits) or, better still, like a heated object readily loosing its heat to another cooler object or cooler surroundings. On the other hand, when an electron-deficient center goes seeking "begging" to stabilize itself by receiving electrons, we term these reactions *electrophillic* (*electro* electrons; *philic* liking) (But, please bear in mind that in both these instances the electron flow is from a rich to a deficient center). Usually a nucleophile is given the symbol Nu or Nu^-, while the electrophile is shown as E^+. The most common nucleophiles are the lone-pair-containing atoms in compounds such as oxygen in water or alcohol and nitrogen in amines. Apart from these neutral compounds a number of charged species act as nucleophiles, such as the I^-, H_2N^-, CN^-, carbanion (C^-), etc. Most common electrophiles are the nitronium ion (O_2N^+), Cl^+, Br^+, and carbocation intermediates. There are other species which are electrophillic, such as carbonyl carbon ($C=O$) and the carbon attached to a heteroatom ($C-Cl$, $C-Br$, $C-S$, $C-OTs$).

Review

Thus, nucleophillic, electrophillic, intermolecular and intramolecular, concerted, and step-wise are the basic, fundamental types of reactions.

In organic chemistry, we normally come across other transformations, partly involving the above basic concepts. These are addition, elimination, substitution, and rearrangements.

- When molecules/atoms are added to another molecule (across unsaturation), we say it is an *addition reaction.*
- A substitution reaction, is one in which a molecule/atom present on a molecule is replaced by another molecule/atom.
- When molecules/atoms are removed from a molecule, we term it an *elimination reaction.*
- When molecules or atoms in a molecule are not removed, but rearranged the reaction then is called a *rearrangement reaction.*

2.2.2   Arrow Formalism in Explaining the Reactivity

To keep track of the valence electrons during a reaction, a symbolic curved arrow method is used. The direction of the arrow indicates the flow of electrons. We should bear in mind that electrons flow into an empty or partially empty orbital of the partially or completely electron deficient center. In the case of carbon, since tetravalent carbon is the most common form in organic chemistry, when electrons flow towards a partially electron deficient carbon (making its valence temporarily to five), a pair of electrons have to leave with the leaving group. Therefore, two arrows are normally shown, one incoming on to a partially or fully electron deficient carbon and the other leaving from the carbon with the leaving group (Fig. 2.6).   .

2.2.3   Summary: Learning Outcome

Reactions - more so organic reactions are primarily flow of valence electrons from an electron-rich center to an electron-deficient center. This may lead to the formation of either a covalent or a polar covalent bond.

The arrow between the reactants and the products indicates these complex electron flow processes.

FIGURE 2.6 Arrow formalism in depicting the reaction mechanism.

The arrow formalism in showing the mechanism indicates the direction of the flow of the electrons among the reacting groups.

2.3 NONCOVALENT INTERACTIONS

Noncovalent interactions are of paramount importance in biologic systems. These form the basic framework for the three-dimensional shape of the innumerable biomolecules (mostly polymers). These interactions also determine the geometry, rate, and dynamics of interactions between the biomolecules, such as protein–DNA interaction, lipid–protein interaction, and protein–protein interaction, apart from determining the interaction of autocoids and xenobiotics (drug molecules included) with the biomolecules. Noncovalent bonds produce short-lived, reversible interactions. Their small bond energies and limited effective bonding range distinguish these bonds from the covalent type. Major types of noncovalent interactions are shown in Fig. 2.7.

As a result of the variable distance dependencies, multiple bond formations would take place sequentially as the molecules approach each other. Initial bonding attractions would be governed by forces with less critical requirements (ionic and dipolar) followed by those with highly limited effective bonding distances (van der Waals and hydrophobic).

2.3.1 H-Bonding

The hydrogen nucleus is potentially a very effective bonding probe because of its small size and lack of shielding electrons. When hydrogen is involved in strong dipole–dipole interactions, it is known as *hydrogen bonding*. It is generally significant for hydrogen attached to O, F, or N. The other elements are usually not electronegative enough to give rise to hydrogen bonding. Within a given hydrogen bond, $X-H \cdots Y$, hydrogen is viewed as covalently bonded to atom X and ionically bonded to atom Y. Both X and Y have to be electronegative atoms. Hence, hydrogen bonding involves the bonding of hydrogen with a bonding orbital of one atom (here X) and a nonbonding orbital of the other (here Y). The interaction is a special type of dipole–dipole

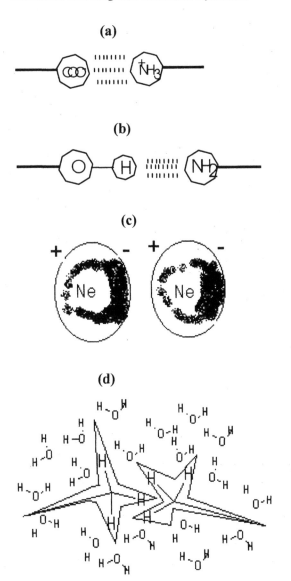

FIGURE 2.7 Noncovalent interactions: (a) ionic, (b) hydrogen bonding, (c) van der Waals interaction (dark region indicates electron density), and (d) hydrophobic interaction.

force, with the positive end of one dipole being the hydrogen atom. The normal distance requirement is about 2.8 Å: in almost all the cases the hydrogen atom lies along the central axis of the two heavy atoms (X, Y). The most frequently observed hydrogen bonds in biologic systems are between the hydroxyl and amino groups and amino (in proteins) and ketone groups (in DNA). Such bonds may occur between molecules (intermolecular), within the same molecule (intramolecular) or as a combination of both of these. In fact it is these interactions that give great stability to the biopolymers (DNA and proteins). The molecules of water are attracted to one another, with the slightly positive hydrogens attracted to the negative "ends" (the oxygens) of other water molecules. This intermolecular attraction is termed *hydrogen bonding* and acts almost like a glue holding the molecules of water together. In the case of water the effect on the physical properties of water are quite astounding: the boiling point of water, for example, is very much greater than would be the case if such bonding did not exist. This fact alone should make the human race (and the rest of life) grateful for hydrogen bonding since water would otherwise be a gas at room temperature.

2.3.2 Van Der Waals Forces

Van der Waals forces are among several types of intermolecular forces that are weak until surfaces get very close. When a large area is in contact, though, they can add up to a strong attraction. Van der Waals forces arise when unbalanced electrical charges around molecules attract one another. Though the charges are always fluctuating and even reversing direction, the net effect is to draw two molecules together. There are two types known, viz., (1) London dispersion forces, which arise due induced dipole–induced dipole or in other words instantaneous dipole caused due to the fluctuating movement of electrons in an atom resulting in a deformation of the electron cloud of the atom approaching close to it, (2) Debye forces, which arise due to dipole–induced dipole interaction. That is, a molecule having a polar covalent bond induces a deformation in the electron cloud (and hence a dipole) in another molecule close to it. The normal distance requirement for both of these types is about 4 to 6 Å between the two atoms or molecules. Since, the dipoles are in alignment by necessity, the interaction is one of attraction. Van der Waals forces are responsible for the attraction between layers of graphite, for example, and the attraction between enzymes and their substrate.

2.3.3 Hydrophobic Interaction

This interaction is to a large extent unique to biologic systems since it involves the attractive influence of two hydrophobic groups in aqueous environment.

It's not glue, suction, or static electricity that keep geckos from falling off the ceiling; it may be van del Waals forces says new study from UC Berkeley, Lewis & Clark, Stanford. Geckos are able to scurry up walls and across ceilings thanks to 2 million microscopic hairs on their toes that glue on to surfaces in a way that has given engineers an idea for a novel synthetic adhesive that is both dry and self-cleaning. The strength of attachment is so strong that a single gecko hair, only one-tenth the diameter of a human hair, could bend the aluminum wire. In fact, a single hair could lift an ant, while a million hairs covering an area the size of a dime could lift a small child of about 45 lb. The key seems to be the hundreds to thousands of tiny pads at the tip of each hair. These pads, called *spatulae*, measure only about ten-millionths of an inch across. Yet, they get so close to the surface that weak interactions between molecules in the pad and molecules in the surface become significant. The combined attraction of a billion pads is a thousand times more than the gecko needs to hang on the wall. Van der Waals forces arise when unbalanced electrical charges around molecules attract one another. Though the charges are always fluctuating and even reversing direction, the net effect is to draw two molecules together, such as molecules in a gecko foot and molecules in a smooth wall.

This type of bonding interaction is observed among two hydrophobic groups in water because by this process the two hydrophobic groups coming close increase the entropy (freedom) of the water molecules. This effect is observed more at ambient temperatures (37°C) or at temperatures below the ambient temperatures, as at these temperatures water molecules exist as clusters. When hydrophobic (nonpolar) molecules come close together in this aqueous environment, they squeeze out the water molecules between them, thereby freeing the water molecules. This phenomenon is perhaps one of the most important interactions involved in maintaining the normal conformation of proteins and also in the substrate-specific enzymatic action.

For an ionic bond the bond strength is 5 kcal/mol, and it varies inversely with the distance. For a hydrogen bond the strength is between 1 to 7 kcal/mol, and the distance varies between $1/r$ to $1/r^2$. For a van de Waals bond the strength is between 0.5 to 1 kcal/mol, and the distance varies as a function of $1/r^4$ to $1/r^6$.

2.3.4 Summary: Learning Outcome

Noncovalent bonds produce short-lived, reversible interactions. Their smallbond energies and limited effective bonding range distinguish these bonds from the covalent type.

These interactions are of paramount importance in biologic systems.

2.4   THREE DIMENSIONAL (3D) STRUCTURE
OF MOLECULES

Molecules are three-dimensional entities. The shape being definite and unique for a given set of molecules, i.e., all of the methane molecules will have the same unique structure, but the structure of either ethane of methyl chloride will be different from that of one another. The distribution of electron density on the whole molecule is determined by a number of factors, viz.,

1. The connectivity of the constituent atoms relative to each other
2. The 3D placement of these atoms relative to each other
3. The polarity differences of these constituent atoms

2.4.1   Internal Coordinates

The internal coordinates of a molecule, such as, bond lengths r, bond angles θ, and torsion angles ω, are helpful in defining the structure of the molecule, though not completely. Thus, for a diatomic molecule A-B, the structure and hence the electron cloud distribution completely are defined by the nature of the nuclei A and B (their polarity difference) and the bond distance or length r between their centers. For a triatomic molecule apart from defining the connectivity (which atom is connected to which other) and bond distances r_1 and r_2, we have to define the bond angle θ. For a tetraatomic molecule apart from the above three, one has to also define the torsion angle ω between the two planes.

Hence the process of completely defining the structure of any molecule can be factorized into constitutional (connectivity and r) and the stereochemical (bond angle and torsion angle) components. Stereochemical component can also be further factorized into the configuration (bond angle θ) and conformation (torsion angle ω). Structure thus encompasses connectivity (r), configuration (θ), and conformation (ω). For example, let us consider an ethane molecule, we need to first define the connectivity, i.e., the connectivity of and the bond lengths of the carbon hydrogen and the carbon carbon bonds. Then we need to define the bond angles of the carbon with the hydrogens and the carbon with the carbon. Finally, we need to also define the angle of one of the carbon hydrogen plane with another carbon hydrogen plane, the torsion angle (Fig. 2.8).

Structural Formulas

Constitution implies the number, kind, and connectivity of atoms in a molecule. Constitution is represented by two dimensional structural formu-

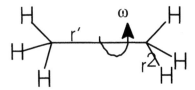

FIGURE **2.8** The three descriptors of the 3D structure of ethane molecule.

las, in a fairly simple straightforward manner. For example, *n*-propane can be drawn as in Fig. 2.9.

Isomerism

When two or more structures can be drawn with the same composition and molecular weight, then these molecules are isomers. Hence, *isomers* connote molecules having the same constituent atoms but arranged differently. This arrangement may differ due to

The difference in connectivity of the constituent atoms, *the constitutional isomers*

The difference in the three-dimensional placement of the constituent atoms having the same connectivity—the stereoisomers—*the configurational is isomers*

The difference in the three-dimensional placement of the constituent atoms having the same connectivity, due to restricted rotation across a bond axis—the stereoisomers—*geometric isomers and the atropisomers*

$$H_3C-CH_2-CH_3$$

or

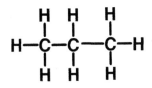

FIGURE **2.9** Structural formula of *n*-propane.

The difference in the three-dimensional placement of the constituent atoms having the same connectivity but vary from one another in having different torsional angles—the stereoisomers—*the conformational isomers*

The constitutional isomers can be conveniently depicted by the standard 2D structural formulas, while for representing the stereoisomers a 3D stereo representation is adopted (Fig. 2.10).

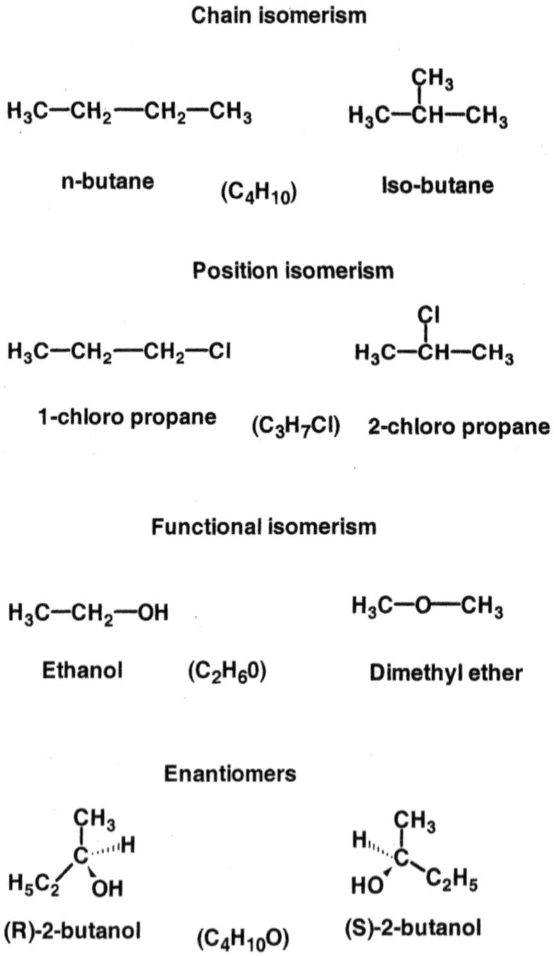

Chain isomerism

$H_3C-CH_2-CH_2-CH_3$

$$\begin{array}{c} CH_3 \\ | \\ H_3C-CH-CH_3 \end{array}$$

n-butane $\quad$ (C_4H_{10}) $\quad$ Iso-butane

Position isomerism

$H_3C-CH_2-CH_2-CI$

$$\begin{array}{c} CI \\ | \\ H_3C-CH-CH_3 \end{array}$$

1-chloro propane $\quad$ (C_3H_7CI) $\quad$ 2-chloro propane

Functional isomerism

H_3C-CH_2-OH $\qquad\qquad$ $H_3C-O-CH_3$

Ethanol $\quad$ (C_2H_6O) $\qquad$ Dimethyl ether

Enantiomers

(R)-2-butanol $\quad$ $(C_4H_{10}O)$ $\quad$ (S)-2-butanol

FIGURE 2.10 Structural formulas of isomers.

2.4.2 Conformation and Structure

From the above brief and succinct overview of the variations in 3D structure of the molecules and their representation, it may be clear that a molecule capable of having torsional rotation can have an infinite number of conformers. For example, if one considers an assembly of four atoms A, B, C, and D. One can envisage different types of connectivity, such as A-B-C-D, A-B-D-C, B-A-D-C, and so on. For any one of these molecules, which are constitutional isomers of A-B-C-D, even if we assume constant bond angles ABC and BCD, one can generate an infinite number of structures changing the torsion angle ABC/BCD. These structures are the different conformers of the molecule A-B-C-D. In spite of this infinite 3D structures possible for molecules, thankfully, since these structures differ in energies, the molecule exists in only a few (sometimes only in one) rapidly interchanging conformers. Thus, it is possible to predict the 3D structure of any molecule by studying the energy of the conformers. The lowest energy conformer is the state in which any molecule will exist in ground state. Thus, for cyclic structures, such as cyclopentane and cyclohexane (which we normally encounter in biologic systems), the probable low-energy conformers are the envelope form for the cyclopentane and the two chair forms for the cyclohexane (as shown in Fig. 2.11). One has to be cautious when applying these above principles because based on the substitutions on theses ring systems the lowest energy conformer may be different. However, these principles being the first principles are of great value in understanding the 3D structure of most of the biomolecules, thereby, their reactivity.

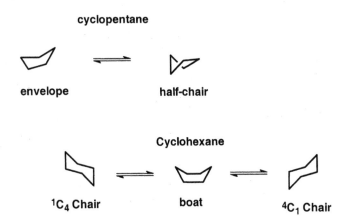

FIGURE 2.11 Stable conformations of cyclopentane and cyclohexane.

2.4.3 Summary: Learning Outcome

The representation of molecules as the structural formula.

Given a molecular formula, there can be more than one compound having the same constituent atoms. All of which are isomers.

The isomers can differ based on connectivity of the constituent atoms (the constitutional isomers) or based on the three-dimensional placement of the atoms (the stereoisomers).

The energies of the various conformers are not the same, and the molecule exists in its lowest enery conformation.

2.5 EXPLANATION OF TERMS ENCOUNTERED IN DESCRIBING STEREOCHEMISTRY

Chiral: This term can be applied to molecules, conformations, macroscopic objects, crystals, etc. When an entity is nonsuperimposable with its mirror image, this term is used.

Achiral: If an entity is superimposable with its mirror image, it is termed *achiral*.

Chiral center: In a tetrahedral (Xabcd) or trigonal pyramidal (Xabc) structure, the atom (X) to which different ligands (atoms or molecules) are attached is termed the *chiral center*. R and S notations based on the Cahn-Ingold-Prelog system (CIP system), which are chirality descriptors can be assigned to this center.

Enantiomers: Mirror-image stereoisomers. These normally have similar physical and chemical properties.

Enantiomer: The pair of molecular species, which are stereoisomers, that are mirror image to each other and are nonsuperimposable.

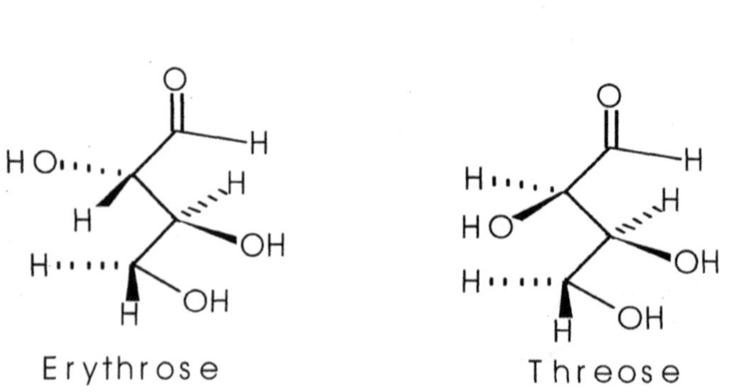

FIGURE 2.12 Diastereoisomers.

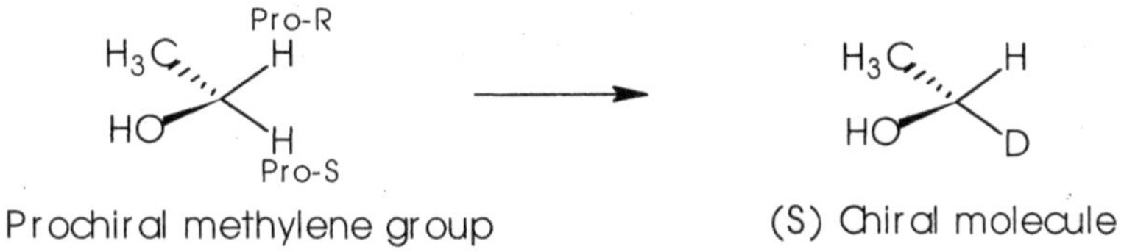

Pro-R

H_3C,,, H

HO H
Pro-S

⟶

H_3C,,, H

HO D

Prochiral methylene group

(S) Chiral molecule

FIGURE 2.13 Enantiotopic atoms in a prochiral group.

Diastereoisomers: Stereoisomers that are not mirror images. These differ in their physical and chemical properties. These can usually be obtained by the presence of more than one chiral center in the molecule, for example, the erythose and threose (tetroses) sugars, as shown in Fig. 2.12.

Prochirality: This is a term used to describe atoms, groups, molecules, or even faces of a particular molecule.

Prochiral atoms and groups: In a tetrahedral carbon (sp_3), when the replacement of one of the two similar ligands leads to a chiral product, the two similar ligands are called as *prochiral atoms* or *ligands*. Pro-R, pro-S descriptors are used to differentiate these two similar ligands. For example, in Fig. 2.13 the two hydrogens are prochiral atoms. They are also termed as *enantiotopic ligands*. This can also be defined as follows: two identical substituents of any kind on an sp_3 hybridized atom (usually carbon) that bear two different additional substituents are *enantiotopic*.

Prochiral faces: In a planar achiral molecule (usually a sp_2 hybridized carbon), when the addition to any one face yields a chiral molecule, then the two faces of the planar molecule are called *prochiral faces, enantiotopic faces*, or *heterotopic faces*. These are differentiated based on the CIP system as the *Re* and *Si* faces (Fig. 2.14).

Re - face

Si - face

Prochiral molecule

FIGURE 2.14 Enantiotopic faces of a planar prochiral molecule.

2.6   DEFINITIONS AND EXPLANATION OF THE TERMS ENCOUNTERED IN STEREOSELECTIVE SYNTHESIS

When there are two functional groups closely related, but slightly differing in their stereo or electronic factors, in a molecule, a given reagent may react preferentially with one rather than the other. Such a reaction is sometimes termed *chemoselective*. When more than one (normally two) products are formed, but one is formed in greater proportion, the term *selective* applies. On the other hand when only one functional group reacts (or only one product forms), the term *specific* applies. Thus, depending on the type of preference, we have

Regioselective: Selectivity with regards to a region in the molecule (commonly observed in the hydroboration of alkenes).

Regiospecific: Reacts only at one center in the molecule.

Chemoselective: When one type of functional group of two or more closely related functional groups react preferentially (Fig. 2.15).

Chemospecific: When only one type of functional group of two or more closely related functional groups react preferentially.

Enanatioselective: When a reaction produces an unequal mixture of enantiomers or when one type of enantiomer reacts preferentially over the other.

Enantiospecific: When only one enantiomer forms or one type of enantiomer reacts over the other.

Enantiomeric excess: It is normally written as ee and is given in percent. In a enantioselective reaction, ee is the percent excess of the enantiomer over the racemate. It is calculated by the equation

$$\%ee = \left(\frac{\text{mole fraction of R} - \text{mole fraction of S}}{\text{mole fractions of R} + \text{S}} \right) \times 100 \qquad (2.1)$$

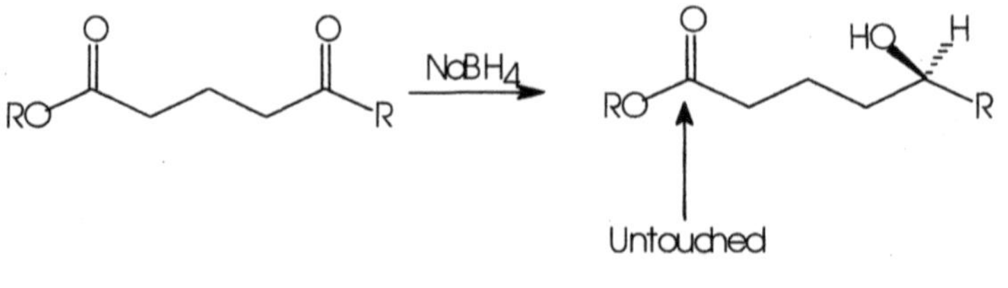

FIGURE 2.15   Selective reduction of ketone in the presence of an ester group.

A 0%ee means a 50:50 (R:S) mixture of R and S enantiomers (racemic mixture), 50%ee means a 75:25 (R:S) mixture, and a 90%ee means a 95:5 (R:S) mixture.

Diastereomeric excess: It is normally written as "de" and is given as percent. In a diastereoselective reaction producing two diastereo-isomers, de is the percent excess of the diastereomer over the mixture. It is calculated by the equation

$$\%\text{de} = \left(\frac{\text{mole fraction of } D_1 - \text{mole fraction of } D_2}{\text{mole fractions of } D_1 + D_2}\right) \times 100 \qquad (2.2)$$

SUGGESTED READING

1. Lewis, G.N. J. Am. Chem. Soc. 1916, *38*, 762.
2. Heitler, W.; London, F. Z Phy 1927, *44*, 455.
3. Smyth, C.P. *Dielectric Behavior and Structure*; McGraw-Hill: New York, 1955; 244 pp.
4. Albert, A. *Selective Toxicity*, 4th Ed.; Barnes and Noble: New York, 1968.

3

Enzyme
Structure and Functions

Enzymes are biological catalysts. They increase the rate of chemical reactions taking place within living cells, without themselves suffering any overall change. In fact, these are central to life, and it may not be an understatement to say that these molecules sustain life on this planet. Most of the enzymes are proteins, to be more specific globular proteins. Since, some RNAs are also known to act as biocatalysts, it is not correct to say that all enzymes are proteins, but for these few exceptions, all enzymes are proteins. More specifically it can be defined as a polypeptide chain or an ensemble of polypeptide chains possessing a catalytic activity in its native form. The reactants of an enzyme-catalyzed reaction are called *substrates*. The catalysis takes place in a specific region of the enzyme, referred to as the *active center* or *catalytic cavity*. Most of the enzymes use a nonprotein component called a *cofactor* for bringing about this catalytic activity.

The outstanding characteristic feature of enzyme catalysis is that the enzyme *specifically* binds its substrates and the reaction takes place in the confines of the enzyme-substrate complex. Thus, to understand how an enzyme works, we not only need to know the structure of the native enzyme, but also the structure of the complex of the enzyme with its substrate. The structure of the native enzyme obviously implies the three-dimensional structure of globular protein, since predominantly all the enzymes are globular proteins. Proteins are macromolecules usually having molecular weight above 10,000 daltons. Before we go into the understanding of the three-dimensional structure of protein, a question may arise as to why should Nature have chosen enzymes to be macromolecules?

This may be understood in two ways:

A substrate molecule and the specific reaction it must undergo must be translated into another structure of higher order whose information content perfectly matches the specifically planned chemical transformation. Only large macromolecules can carry enough molecular information, for substrate recognition.

Coming to the catalytic activity, the active site or the center must be of a defined or highly ordered geometry if it is to contain all the binding and catalytic groups in correct alignment for optimal catalysis. This imposes a heavy entropy demand on the system. This can be compensated at the expense of another already ordered region of the biopolymer losing its organization, as the enzyme binds to a substrate.

For that matter, it is these two functions, namely the molecular recognition and catalytic activity, that necessitate the macromolecule to be highly ordered and organized. Also, the monomers making up this polymer (macromolecule) should have a wide repertoire of functional groups. Hence, we see that proteins have 20 or more different monomers, called *amino acids*, while the information storage device, the DNA macromolecule, has only four nitrogenous bases as its monomers.

3.1 STRUCTURE OF POLYMERS

We have discussed the various parameters/descriptors (r, ω, θ) to define the structure of a molecule, small or big in the earlier chapter. In order to describe a polymer these descriptors have to be complemented with additional descriptors, which specifically describe the elements of the polymer.

What are these additional descriptors of a polymer as against the already mentioned descriptors of a monomer?

Let us consider three monomers A, B, and C forming a polymer. We can obviously have six types of linkages as shown: A-B-C, A-C-B, B-C-A, B-A-C, C-A-B, C-B-A. From among these six, no two polymers will have similar physical and chemical properties. These properties or the behavior of the polymer therefore, is dependent on the sequence of linkage of the monomers. Just as we speak of connectivity [r] as the primary descriptor of a molecule, we need to consider the connectivity of the monomer to form the polymer. This is usually termed the *sequence* of the polymer, also the *primary (1°) structure*. Thus, the primary structure indicates the sequence of connectivity of monomers in the given polymer.

Thus, though we describe the monomers in their totality defining r, ω, and θ for each of the monomer and give all the details of the *sequence* of connectivity of these monomers, the three-dimensional structure of a polymer is still not completely described because the polymer necessarily has a higher degree of structure. How do we understand this?

Consider a thread or a chain. The chain may have individual elements (rings) linked with one another, the sequence of these links in themselves do not describe the three-dimensional structure in detail. We need to necessarily mention the higher level of organization, namely the folding, bending, or clustering of the chain. For example, the chain may be circular, twisted about an axis, or cluttered.

Taking this above example, we can get a better appreciation of structure of proteins since these are also polymeric in nature. These, therefore have a higher level of three-dimensional structure descriptors, namely the secondary, tertiary, and quaternary structures.

3.2 AMINO ACIDS: STRUCTURE AND FUNCTIONS

Let us first take a closer look at the monomers of this sophisticated copolymer termed as *protein*. All proteins consist of amino acid units joined together in series. Amino acid as the very name indicates, is a molecule having both a carboxylic (COOH) group and an amino (NH2 or NH) group. Just as a human chain can be formed only if we use both our arms, the fundamental requirement for any monomer to polymerize is that it should have *more than one* functional group. The term *amino acid* can mean any type of molecule having an amino and carboxylic acid group. Hence, in order to specify, we need to further classify these as α-amino acid, β-amino acid, γ-amino acid, and so on, based on the relative position of the amino group vis-á-vis the carboxyl group in the molecule (Fig. 3.1).

The amino acids that form the constituents of proteins are all α-amino acids. Hence, the general structure can be given as shown in Fig. 3.2.

By the variation in R, the side chain, we have different types of amino acids. The side chains have different sizes, shapes, hydrogen-bonding capabilities, and charge distributions, which enable proteins to display a vast array of biological functions required by the living systems. The 20 α-amino acids, their structures, names, the common three-letter code, and the single-letter code used for abbreviation are given in Fig. 3.3. The amino and the carboxyl groups attached to the α-carbon atom are termed the α-*amino* and α-*carboxyl* groups, to distinguish them from similar groups that may be present as part of the side chain. In all 20 amino acids, except glycine, the α carbon is an asymmetric center. Hence, every amino acid is optically active and will have an enatiomer. The commonly occurring stereoisomer is the L stereoisomer.

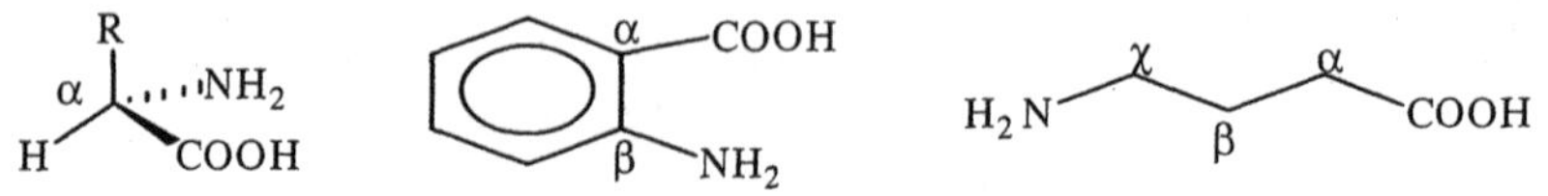

R
α ....ᴵⁱNH₂
H COOH

α-amino acid

α COOH
β NH₂

β-amino acid

H₂N
χ
β
α
COOH

γ-amino acid

FIGURE 3.1 Types of amino acids.

All the proteins synthesized on ribosomes of a living cell use amino acids of L configuration. Two amino acids, isoleucine and threonine, have a second asymmetric center at the β carbon. Its configuration also is constant (as shown in Fig. 3.3). Amino acids with altered β-carbon stereochemistry are called *alloisoleucine* and *allothreonine*.

Based on the nature of the side chain, the amino acids are classified as follows:

Aliphatic: glycine (G), alanine (A), valine (V), leucine (L), isoleucine (I)
Aromatic: phenyl alanine (F), tryptophan (W), tyrosine (Y)
Functional: serine (S), threonine (T), cystine (C), methionine (M), histidine (H)
Acidic: aspartic acid (D), glutamic acid (E)
Basic: lysine (K), arginine (R)
Cyclic: proline (P).(imino acid)

Another useful way of grouping amino acids is based on the polarity of the side chains. Side chains that are uncharged will tend to have low solubility in water–nonpolar amino acids. At the opposite end of the polarity scale are the charged side chains, which have high solubility in water–polar amino acids. Histidine has ambiguous polarity, in a sense it has a dual character due to imidazole ring. Thus, based on this polarity scale, we have

Nonpolar amino acids: Trp, Ile, Tyr, Phe, Leu, Val, Met, Cys, Ala, Gly
Polar amino acids: Pro, Ser, Thr, Asn, Gln, Asp, Glu, Lys, Arg
Ambiguous: His

R COOH
H NH₂

COOH
H₂N ....ᴵⁱR
H

FIGURE 3.2 General structure of alpha-amino acid.

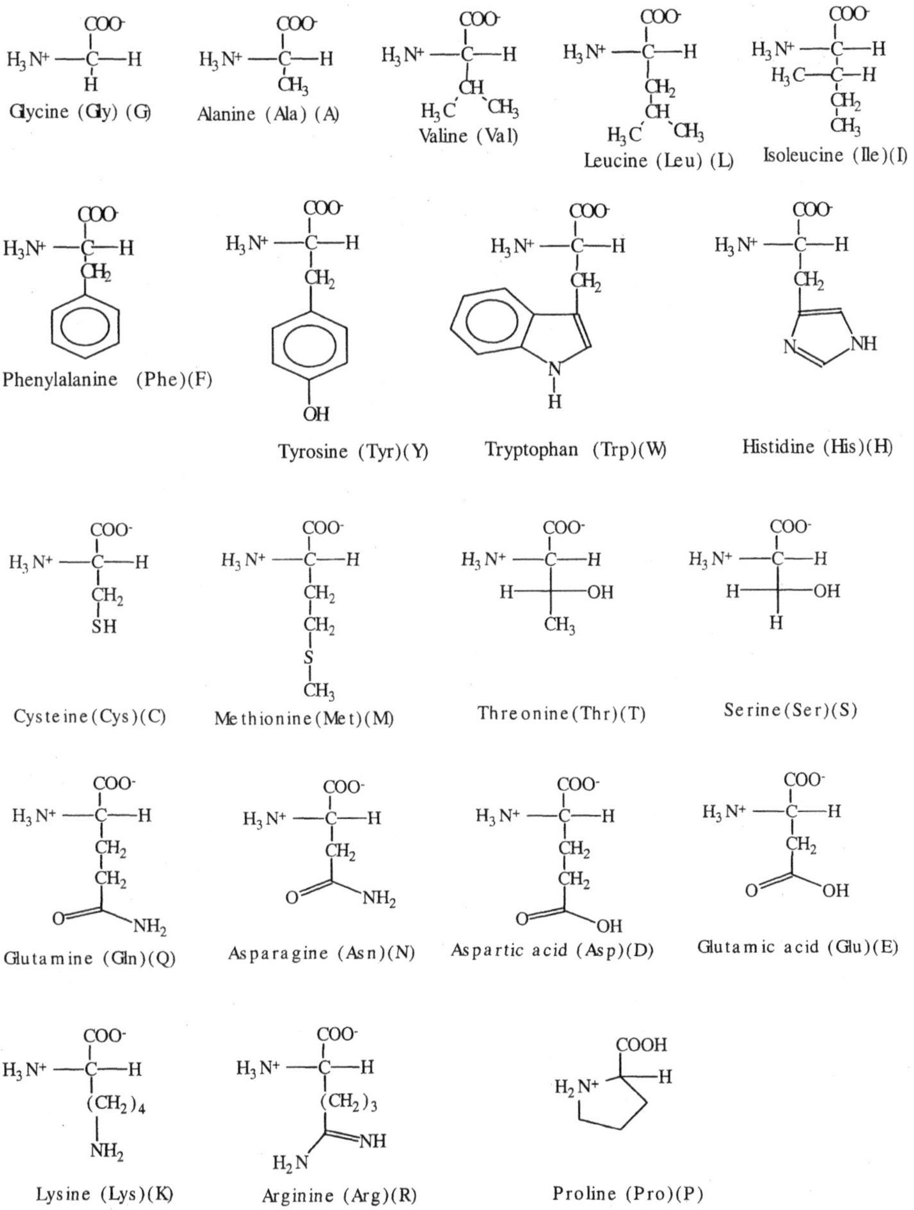

$$H_3N^+ \overset{\overset{\text{COO}^-}{|}}{\underset{\underset{H}{|}}{C}} H$$

Glycine (Gly) (G)

$$H_3N^+ \overset{\overset{\text{COO}^-}{|}}{\underset{\underset{CH_3}{|}}{C}} H$$

Alanine (Ala) (A)

$$H_3N^+ \overset{\overset{\text{COO}^-}{|}}{\underset{\underset{H_3C \quad CH_3}{CH}}{C}} H$$

Valine (Val)

$$H_3N^+ \overset{\overset{\text{COO}^-}{|}}{\underset{\underset{H_3C \quad CH_3}{\underset{CH}{CH_2}}}{C}} H$$

Leucine (Leu) (L)

$$H_3N^+ \overset{\overset{\text{COO}^-}{|}}{\underset{\underset{CH_3}{\underset{CH_2}{H_3C-C-H}}}{C}} H$$

Isoleucine (Ile)(I)

Phenylalanine (Phe)(F)

Tyrosine (Tyr)(Y)

Tryptophan (Trp)(W)

Histidine (His)(H)

Cysteine (Cys)(C)

Methionine (Met)(M)

Threonine (Thr)(T)

Serine (Ser)(S)

Glutamine (Gln)(Q)

Asparagine (Asn)(N)

Aspartic acid (Asp)(D)

Glutamic acid (Glu)(E)

Lysine (Lys)(K)

Arginine (Arg)(R)

Proline (Pro)(P)

FIGURE 3.3 Structures of amino acids.

Grouping this way helps us in predicting the ratio of the amino acids in a protein based on the biological function of the protein. For example,

1. Histones that are relatively small protein molecules that complex with DNA, have a relatively large fraction of positively charged amino acids (Arg, Lys) so as to interact with negatively charged phosphate of DNA molecule.
2. Purple membrane protein, which is a membrane (lipid bilayer), bound protein is rich in nonpolar amino acid residues (Leu, Ala, Gly and Val) so as to be lipophilic.
3. ATPase, which is an enzyme (catalytic subunit), has a combination of polar and nonpolar amino acid residues (Leu, Val, Ala, Gly, His, Ser, Thr, Gln, Phe, Arg) so as to bind to the substrate and bring about the catalytic action.

Hence, from the above examples, which are just indicative, it can be surmised that

Proteins that are membrane bound will have sufficient nonpolar amino acid residues.

Proteins that interact with DNA, RNA, and sugars will have a high degree of polar amino acid residues.

Proteins that act as *enzymes* will have a *judicious* mix of polar and nonpolar amino acid residues.

3.3 LEVELS OF PROTEIN STRUCTURE

The levels of describing the structure of proteins are as mentioned earlier:

Primary structure (1°): the sequence of amino acids (the covalent structure of the polymer)

Secondary structure (2°): regular folding in parts of the polymer

Tertiary structure (3°): further regular folding of the whole molecule so as to be soluble in water

Quaternary structure (4°): interactions of different polypeptide chains

3.3.1 Primary Structure

As we have already mentioned, all proteins contain amino acids joined covalently in a series. The covalent chain structure of these can be written as $R_1R_2R_3R_4R_5...R_i...R_n$, where R_i gives the identity of the ith residue in the chain. A unique feature of proteins that distinguishes it from many synthetic polymers is that the units are arranged in a head-to-tail (directional) fashion.

Thus, the abbreviations R_aR_b and R_bR_a refer to molecules that have different covalent structure and hence will have different chemical properties. Therefore, to define the covalent structure ($1°$ structure), the absolute sequence of residues must be written. Also, there may be more than one polymer chain in the complete covalent structure, and these in turn may contain intrachain or interchain covalent cross-links. The complete covalent structure is the *primary structure*.

To arrive at such a complete description, let us begin with the joining of two monomers, the two amino acids. When the amino group of an amino acid combines with the carboxylic acid group of another amino acid with a concomitant lose of water molecule, a *dipeptide* results (Fig. 3.4).

As can be observed, the dipeptide still has a free amino and carboxylic acid groups, termed the *N-terminal* and *C-terminal ends* respectively. The amino acid residue in a protein having the free α-amino group is called a the *N-terminal amino* acid, and amino acid residue having a free α-carboxylic acid group is called the *C-terminal amino acid*. Also the bond joining the two individual amino acids is an *amide bond*. Extending this understanding, it becomes fairly clear that no matter what the chain length may be, there will still be an N-terminal and a C-terminal ends (except when the chain cyclizes); also there will be $n-1$ number of amide bonds for n amino acid residues in a polymer. Chemically proteins can be described as *polyamides*.

The unique feature of amide bond needs to be appreciated in order to understand the three-dimensional structure of proteins. The lone pair electrons on the nitrogen get delocalized across the $-N-CO-$ bond giving a partial double bond character to the $-N-CO-$ bond, due to the coplanarity of the orbitals, as shown in Fig. 3.5. This makes the whole amide bond planar. Due to the partial double-bond character of the amide bond, there is restriction of rotation (torsion, ω) along the $-N-CO$ bond. This imparts the *rigidness* to the molecule. The dual character (single- and double-bond character) of the amide bond, gives the polyamide a fair degree of rigidity, at the same time allowing conformational flexibility. This dual character is essential, especially for the catalytic activity of these molecules. The conformational flexibility—breathing—allows the enzyme to bind to the substrate (induced

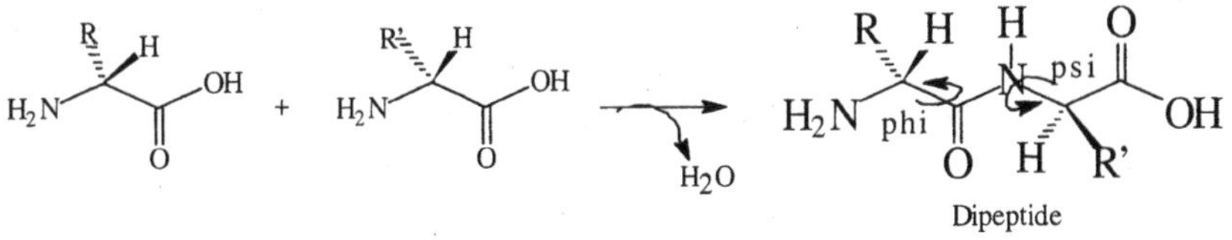

FIGURE 3.4 Condensation of amino acids.

Delocalization of lone pair electrons of nitrogen across the amide bond

coplanarity of the orbitals

FIGURE 3.5 Amide bond (peptide bond).

fit), while the rigidness ensures proper orientation of the reactive groups in the active site.

Just as we speak of geometric isomers (cis and trans) for organic compounds having unsaturation, since the $-N-CO-$ bond also has restriction of rotation, two isomeric forms can exist, viz., cis and trans (Fig. 3.6).

The trans form is the most stable, but whenever proline is condensed, the peptide bond joining the proline residue normally prefers the cis form.

Since the peptide units are effectively rigid groups that are linked into a chain by covalent bonds at the C_α atoms, the only degrees of freedom they have are rotations around the $C_\alpha-C$ and the $N-C_\alpha$ bonds.

A convention has been adopted to call the angle of rotation around $N-C_\alpha$ bond phi (ϕ) and angle around the $C_\alpha-C^1$ bond psi (ψ). In this way each amino acid residue is associated with two conformational angles ϕ and ψ. Since these are the only degrees of freedom, the conformation of the whole main chain of the polypeptide (protein) is completely determined when the ϕ and ψ angles for each amino acid are defined. Most combinations of ϕ and ψ

FIGURE 3.6 Cis and Trans forms of amide bonds.

angles for an amino acid are not allowed because of steric collision between the side chains and the main chain. The angle pairs φ and ψ are usually plotted against each other in a diagram called a *Ramachandran plot* after the Indian biophysicist G. N. Ramachandran, who first made calculations of sterically allowed regions. Figure 3.7 shows the sterically allowed regions. Glycine with only a hydrogen atom as a side chain can adopt a much wider range of conformations than the other residues. Glycine thus plays a structurally very important role; it allows unusual main chain conformations in protein.

Since each of the amino acids have an allowed φ and ψ values, the three-dimensional structure of the protein depends on the composition and the sequence of amino acids. There are a number of ways of sequencing (Edman's method, carboxypeptidase method) the polypeptide chain. Reading the sequence of the polypeptide chain from the sequence of the corresponding gene is by far the most reliable. The discussion of these methods is beyond the scope of the present chapter.

In summary, the function of every protein molecule depends on its three-dimensional structure, which in turn is determined by its amino acid sequence—the primary structure of proteins.

3.3.2 Secondary Structure

Secondary structure (2° structure) refers to regular, repeating patterns formed by backbone of at least part of a polypeptide chain and stabilized by non-

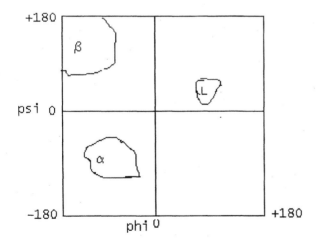

FIGURE 3.7 Ramachandran plot (β in the figure indicates β-sheet region, α in the figure indicates the α-helix region, and the L in the figure indicates left-handed helix region).

covalent interactions. In one of the most impressive feats of structural chemistry, Linus Pauling and Robert Corey postulated the exact 2° structure of the part of polypeptide chain. The postulates they adapted were

> The bond lengths and bond angles of the molecule remain constant in the 2° structure.
>
> The peptide bond adopts a planar trans conformation, except when the bond is with proline.
>
> The secondary structure is held by maximum number of H-bonding interactions between the $-N-H--O=C-$ groups.

Based on these obvious assumptions, they predicted the existence of four main structures, viz, right-handed α-helix, left-handed α-helix, parallel, and antiparallel β-sheets. All the four secondary structures have been found to exist. Of these the α-helix and β-sheet are the most common (Fig. 3.8).

α-Helix. α-Helices in proteins are formed when a stretch of consecutive residues all have phi, psi angle pair approximately −60° and −90°, corresponding to the allowed region in the bottom left quadrant of the Ramachandran plot. The α-helix observed in proteins is right-handed. The α-helix has 3.6 residues per turn with hydrogen bonds between C=O of the residue n and NH of residue n 4. Thus, all NH and C=O groups are joined with hydrogen bonds except the first NH groups and last C=O groups at the end of the α-helix. As a consequence, the ends of α-helices are polar and are almost always at the surface of protein molecule. The α-helix has a pitch of 5.4 Å (1.5 × 3.6) and a diameter of 6 Å. α-Helices vary considerably in length in globular proteins ranging from four or five amino acids to over 40 residues. The average length is around three turns, corresponding to 10 residues. Since the rise per residue is 1.5 Å along the helical axis, this corresponds to about 15 Å from one end to the other of an average α-helix. All the hydrogen bonds in an α-helix point in the same direction. So the peptide fraction has a net dipole moment aligned along the α-helical axis. Different side chains have been found to have weak but definite preferences either for or against being in α-helices. Thus,

> Ala(A), Glu(E), Leu(L) and Met(M) are good α-helix formers.
> Pro(P), Gly(G), Tyr(Y) and Ser(S) are very poor helix formers.

β-Sheet. The second major structural element found in globular proteins is β-sheet. This structure is built from a combination of several regions of the polypeptide chain, in contrast to the α-helix, which is built from one continuous region. The geometry of the peptide backbone in the β-sheets approaches the most extended chair conformations allowed by normal bond lengths and angles. The displacement along the axis is 3.47 Å per residue.

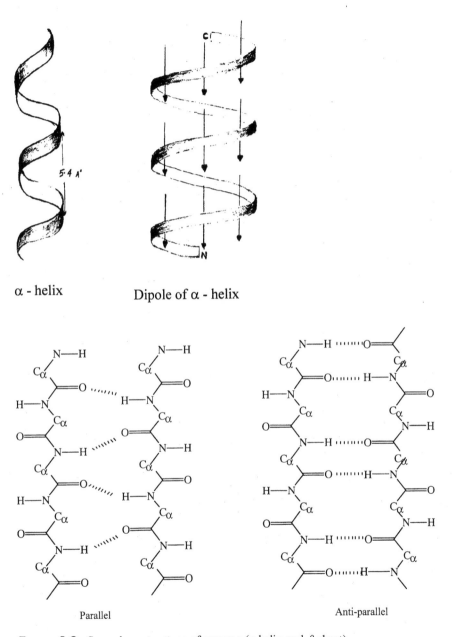

α - helix   Dipole of α - helix

Parallel   Anti-parallel

FIGURE 3.8 Secondary structure of enzyme (α-helix and β-sheet).

These are usually from 5 to 10 residues long. β-Strands can interact in two ways to form a pleated sheet. When the amino terminal of both the strands are in the same direction, it is described as parallel and when the amino terminal of one β-strand is interacting with carboxy terminal of another β-strand, we describe it as antiparallel β-sheet. The antiparallel β-sheet has narrowly spaced H-bond pair that alternate with widely spaced pairs. The parallel β-sheet have evenly spaced hydrogen bonds. β-Strands can also combined into mixed β-sheets, with some β-strand pairs parallel and some antiparallel.

Loop Regions. Most protein structures are built up from combinations of secondary structure elements, α-helices and β-sheets, which are connected by loop regions of various lengths and irregular shapes. Normally, the loop regions are at the surface of the molecule. Loop regions frequently participate in forming binding sites (in antibodies) and enzyme-active sites.

3.3.3 Super Secondary Structure

Simple combinations of a few secondary structure elements with a specific geometric arrangement have been found to occur frequently in protein structures. These units have been called either *supersecondary structures* or *motifs*. Some of these motifs can be associated with a particular biological function. Some of the common motifs with their biological function are

> Helix-loop-helix motif: calcium binding
> Hairpin β-motif: no specific function
> Helix-turn-helix motif: DNA binding
> β-α-β-Motif: no specific function
> Greek key motif: not associated with any biological function

3.3.4 Tertiary Structure

The aqueous environment in the biological systems makes the protein molecule fold, so as to bring the hydrophilic groups to the exterior and the hydrophobic groups (side chains) to the interior of the molecule. This makes the molecule soluble in water and makes the supramolecule attain thermodynamic stability. This further folding is also organized and regular, known as the *tertiary structure*. The folding is of the *whole molecule* unlike in the secondary structure. The noncovalent interactions among the side chains of the amino acid residues and the hydrogen bonding interaction of the side chains of polar amino acid residues with the water molecules reduce the overall energy of the molecule. Several globular proteins also have intra- and intermolecular disulfide linkages, which freeze the three-dimensional conformation. The fundamental unit of a tertiary structure is domain. A *domain* is defined as a polypeptide chain or a part of polypeptide chain that can independently fold into a stable tertiary structure. Domains are also units of function. The domains are held together in a specific arrangement by the noncovalent interactions.

3.3.5 Quaternary Structures

Several identical or nonidentical polypeptide chains may be linked together to form the actual protein. The complete three-dimensional structure, including the interactions between the component polypeptide chains, is termed as *quaternary structure*. It may also be defined, as the arrangement and interactions of different polypeptide chains (subunits) of a single protein. The interaction between these subunits could be noncovalent or they may be chelating to one central group. Most of the oligomeric enzymes have quaternary structures.

Based on these structural parameters all the proteins are classified as

Fibrous proteins, which have only the primary and the secondary structures. In other words the regular, repeating three-dimensional units, known as the α-helix and the β-sheet are uninterrupted through out the molecule. For example, all structural protein components, such as the keratin, fibroin, etc.

Globular proteins, which have the primary, secondary, and tertiary structures. They may also have the quaternary structure. All the functional proteins including enzymes are globular proteins.

The sequence of amino acids in each polypeptide chain constitutes the primary structure of the protein. Regular, repeating three-dimensional features constitute the secondary structure. The overall three-dimensional structure of each polypeptide chain is termed the *tertiary structure*. Proteins may consist of one or more polypeptide chains, and the complete structure is called the *quaternary structure*. For example, the structures of most, larger proteins appear to be composed of independently folded globular units called *domains*. The domains often contain smaller, frequently recurring substructures or folds collectively termed *supersecondary structure*. These are combinations of α and/or β structure. These in turn are formed by the regular rotation of phi and psi angles along the sequence of the polypeptide chain, the primary structure. The primary structure predominantly involves the covalent bonding interactions, while the secondary, tertiary, and quaternary structures predominantly involve the noncovalent interactions.

3.4 ENZYMATIC CATALYSIS

By definition, a catalyst is any substance that alters the speed of a chemical reaction, without itself undergoing change. This is true for enzymes, as they are of the same form before and after the catalytic reaction. According to the current usage, a *catalyst* enhances the velocity of the reaction, and an *inhibitor* decreases the velocity of a reaction. A number of enzymes require a non-protein component, called a *cofactor*, for catalyzing biochemical reactions. These are consumed in the course of the reaction they catalyze but are often restored to their original form by a subsequent reaction.

Those enzymes, which require a cofactor, are inactive in the absence of a cofactor. The inactive form of the enzyme is termed as *apoenzyme*. The apoenzyme along with the cofactor that is the active form of the enzyme is termed as *holoenzyme*.

The cofactors are generally of two types:

1. Small organic molecules bound to the enzyme are called *coenzymes*.
2. Metal ions and organic molecules loosely bound to the enzyme are called as *prosthetic group*.

Catalysis may be classified as heterogenous and homogenous. Heterogenous catalysis involves more than one phase, that is, there is a phase boundary present. Most of the times the catalyst is in the solid phase while the reactants are in the liquid phase. A good example of this type of catalyst is the hydrogenation of ethylene in the presence of palladium. Homogeneous catalysis is limited to one phase. No phase boundary exists. A number of organic reactions are of this type. The acid-catalyzed ester hydrolysis, benzoin condensations by cyanide ions, and hydrogenation of unsaturated compounds by Wilkinson's catalyst are some of the well-known examples.

Enzymes incorporate the features of both the types of catalysis. They may bring reactants together on a protein surface or extract them from aqueous phase into a hydrophobic environment (the active site). For example, the enzyme-catalyzed hydrolysis of amide bond does not proceed merely by the hydrolysis on the protein surface but instead involves chemical interaction of the enzyme to form a susceptible acyl enzyme intermediate, which then undergoes hydrolysis. Homogenous catalysis may be further subdivided into,

Specific acid $\rightarrow$ H^+-catalyzed reaction
Specific base $\rightarrow$ OH^--catalyzed reaction
General acid $\rightarrow$ other acidic compounds
General base $\rightarrow$ other basic compounds
Nucleophillic $\rightarrow$ nitrogenous compounds
Electrophillic $\rightarrow$ metal ions

Specific acid and specific base catalysis is the use of proton and hydroxide ion, respectively, for catalysis. Acid- or base-catalyzed ester hydrolysis are suitable examples for these. Water, acetic acid, phenol, imidazole N-H proton, etc., can act as general acid catalysts. Similarly, water, acetate ion, imidazole ring nitrogen, etc., can act as general base catalysts. For example, the

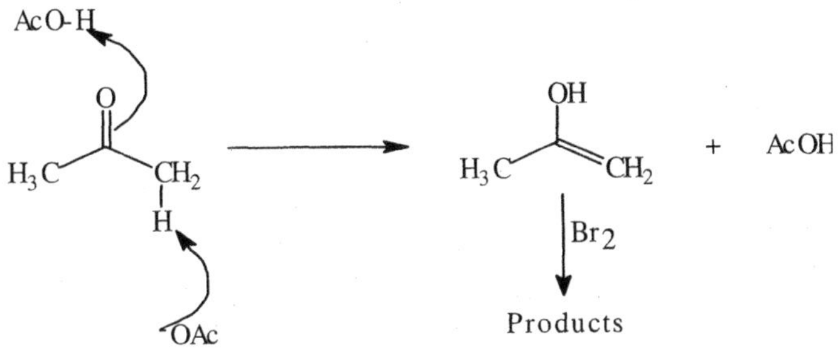

FIGURE 3.9 General acid–general base catalysis.

bromination of acetone in the presence of sodium acetate buffer illustrates the general acid, general base catalysis (Fig. 3.9).

Nuclephilic reaction may be defined as the catalysis of a chemical reaction by a nucleophilic substitution. Thus, if a particular nucleophillic reaction is slow, then the reactant may undergo attack by a nucleophillic catalyst, to give rise to an intermediate that is more susceptible to the desired nucleophillic displacement. A classic example to nucleophillic catalysis is the acetylation of alcohols with active anhydride in the presence of pyridine (Fig. 3.10).

Electrophilic catalysis involves the removal of electrons or electron density from the substrate to the catalyst. Metal ions are the best examples of electrophilic catalysts. The synthesis of 3′,5′-cyclic-guanosine monophosphate from guanosine triphosphate is mediated by divalent metal ions. Similarly hydrolysis of ATP to ADP or ADP to AMP is a classic example of metal ion catalysis (Fig. 3.11).

These above types of catalysis are also observed within a molecule, i.e., intramolecular catalysis. Enzymes may use any of the above-mentioned modes of catalysis in order to catalyze a particular reaction. For example, in the enzyme α-chymotrypsin, the imidazole ring of the histidine residue functions as a general base catalyst, the carboxylic acid group of the aspartic acid residue act as the specific acid, and the hydoxyl group of the serine residue acts as the nucleophillic catalyst (Fig. 3.12).

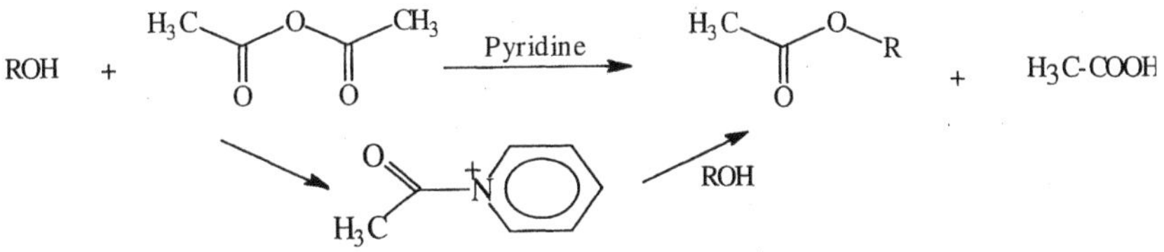

FIGURE 3.10 Nucleophilic catalysis.

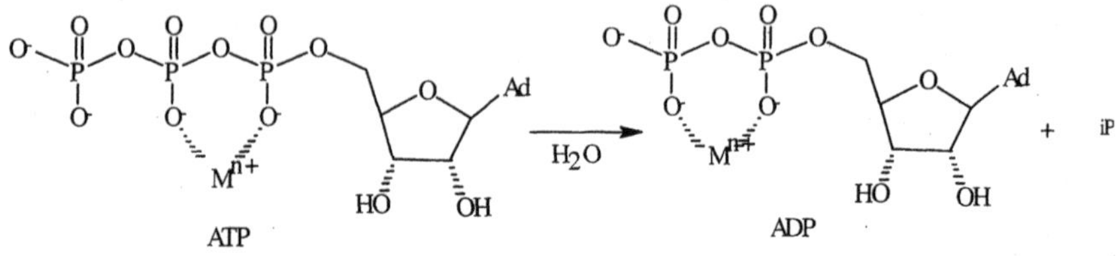

Figure 3.11 Electrophilic catalysis.

Functions of amino acid side chains in catalysis are the following:

The aliphatic and aromatic amino acid residues mainly help in hydrophobic and van der Waals interactions with the substrates, i.e, in the binding and orienting the substrate in the catalytic activity. These interactions, which require short distances, are important during the catalytic cycle.

The acidic and the basic amino acid residues tend to form electrostatic interactions with the incoming substrate. Since electrostatic interactions are long-range interactions, these help in suitably orienting the substrates for binding into the catalytic cavity. These groups also take part in catalytic activity, by providing the necessary acid–base system.

The functional amino acid residues, namely serine, histidine, cysteine, methionine, and threonine, are the most valuable in catalytic activity.

     Histidine: The imidazole group of the histidine takes part in the reversible exchange of proton. The nitrogen atom is used to bind to the metal ions.

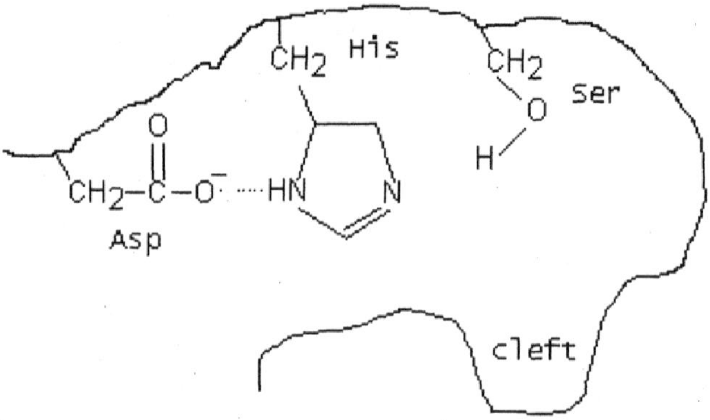

Figure 3.12 Side chains in active site of α-chymotrypsin.

Serine: The primary hydroxyl group of serine provides the nucleophilic oxygen, useful for a variety of reactions.

Methionine: The $-S-CH3$ group of methionine is the methyl donor.

Enzymes employ more than one catalytic parameter during the course of their action. It is by this successful integration of a combination of individual catalytic process that a rate enhancement as high as 10^{14} is achieved. By the use of hydrophobic (nonpolar) amino acid residues or even the charged amino acid residues, the enzyme stereospecifically binds to the substrate and thus brings the groups reacting in proximity and also orients them appropriately. By these noncovalent interactions with the substrates, the enzymes achieve stereospecificity. Thus, enzymes bring about efficient and effective catalysis—efficient in the sense of rate enhancement and effective in the sense of specificity.

Based on the number of polypeptide chains constituting the enzyme, enzymes are classified as

1. Monomeric enzymes consist only a single polypeptide chain (example: carboxypeptidase, proteases).
2. Oligomeric enzymes consist of two or more polypeptide chains (subunits) linked by noncovalent interactions.

Oligomeric enzymes consisting of two identical sub units are called as protomers. Depending on the number of subunits, oligomeric enzymes are called *dimeric* (2), *trimeric* (3), and *tetrameric* (4). Monomeric enzymes are normally biosynthesized in their inactive form, known as *zymogens*. These are activated by a modification in the enzyme structure. For example, trypsinogen gets converted to active trypsin by the cleavage of internal peptide bonds in trypsinogen.

3.5 BIOCATALYST AND HETEROGENEOUS CATALYST

Enzymes and biocells are biological catalysts, which facilitate and/or speed up the reaction without them being consumed, similar to heterogeneous or homogeneous catalysts. Metals like Pt, Pd, Ni, Fe, Co, etc. and oxides like alumina, zeolites, clay, and silica are a few examples of heterogeneous catalysts. Examples of biocatalyst are plant or animal cells, lipase, yeast, etc. For example, urease catalyze the hydrolysis of urea at a rate 10^{14} faster and carbonic anhydrase catalyze the hydrolysis of carbonic acid 10^7 faster than the corresponding uncatalyzed reactions. There are several similarities and differences between these and heterogeneous catalysts, which are listed below.

3.5.1 Reaction Conditions

The reaction conditions are more severe in the heterogeneous than enzyme catalyst systems. The operating temperature in the case of the former is very wide, generally in the range of 50° to 300°C. In contrast, enzyme catalyzed reactions are operated between 30° to 90°C. Also, heterogeneous catalyst reactions are carried out at higher pressure, whereas reactions in the later case are generally carried out at atmospheric pressure.

The pH of the reaction medium plays an important role in the enzyme catalysis, and each enzyme has a characteristic pH optimum and is active over a relatively small pH range.

3.5.2 Active Site

It is known that the substrate binds to a specific region of the enzyme called the *active site*, where reaction occurs and products are released. The interactions between the substrate and the enzyme are due to hydrogen bonding and van der Waal interaction. In heterogeneous catalysis the substrate is chemisorbed on the catalyst site due to pi-bond interaction between the substrate and the metal atoms. The size of the active metal sites are of the order of microns whereas that of enzymes is in the range of 50-100 Å.

3.5.3 Activation Energy

The activation energy for native enzyme-catalyzed reactions are of the order of 20–100 kJ/mol, whereas the activation energy for heterogeneous catalyzed reactions are generally larger than that of the former (75–250 kJ/mol), which means that enzymes are able to reduce the energy barrier more easily than the other catalyst.

3.5.4 Turnover Number

At ambient temperatures where enzymes are most active, they are able to catalyze reactions faster than the majority of the heterogeneous catalysts (0.1–1000 molecules per site per second for enzyme-catalyzed reactions as against < 0.01 for solid-catalyzed reactions at 25°C).

3.5.5 Inhibition

Large amounts of substrate or product are known to inhibit the biocatalytic reactions. Such substrate and product inhibition effects are not normally observed in heterogeneous catalyzed systems. Palladium catalyst is known to get deactivated by amines formed during the hydrogenation of nitro compounds.

3.5.6 Deactivation

Enzymes are in general far more fragile than their counterparts. The activation energy for enzyme denaturation is of the order of 160–280 kJ/mol, which is attributed to removal of water surrounding the enzyme, which keeps it flexible. Two irreversible deactivations, namely sintering and poisoning, are known in heterogeneous catalysis.

3.5.7 Reaction Mechanism and Kinetic Equation

Both heterogeneous and enzyme catalysis operates through an intermediate species formed between the substrate and the active catalyst species (enzyme-substrate complex in the case of biocatalytic reactions or chemisorbed species in the case of heterogeneous catalyst), which leads to similar reaction mechanistic steps and kinetic equations. The forms of the Michaelis-Menten equation for biocatalytic mechanism and Langmuir-Hinshelwood equation for heterogeneous catalytic reactions are similar. More discussion are given on Michaelis-Menten kinetics in the subsequent sections.

3.5.8 Catalyst Support

The metal catalysts such as nickel, platinum, palladium, etc., are generally dispersed on inert supports such as carbon, alumina, silica magnesia, etc., to enhance its activity and stability. Immobilization of the enzyme is achieved by adsorption or binding onto various supports such as glass, polymers, silica, alginate, etc., or entrapment in membranes or porous materials. Immobilization of enzyme or supporting of metal catalyst leads to mass transfer resistance, which leads to the slowing down of the reaction. Generally, greater overall rates can be obtained by subdividing the same amount of solid catalyst material into smaller pieces. This is also valid in the context of biological catalyst, for it indicates how immobilized enzyme pellet size could be controlled to achieve maximum efficiency. A fascinating evolutionary concept is that biological structures ranging from cells to organelles to multienzyme complexes have developed in a manner that minimizes diffusional effects on overall rates. Smaller structures are used in nature to process low-concentration intermediates at high rates. Yeast cells typically range from 5 to 50μm while enzymes range from 50–100 Å.

3.5.9 Reaction Engineering

Slurry, packed, fluidized, and trickle bed reactors are used for both catalyst systems. The basic mass transfer equations and correlations for diffusion coefficients are the same for both catalyst systems, which leads to similar basic reactor design for both systems.

3.5.10 Preparation

Soluble enzymes are extracted from animal or plant sources by the disruption of the membranes, followed by extraction with solvent such as *n*-butanol. Techniques for the purification of the extracted enzymes include electrophoresis, ion exchange chromatography, gel filtration, and affinity chromatography and are described in detail in Chapter 13.

Traditional catalysts are prepared chemically from the corresponding metal salt by precipitation/coprecipitation, by impregnation of the metal salt solution onto a support, viz., palladium, platinum, and ruthenium on silica, alumina, or carbon, by digestion of metal alloy to prepare finely divided metal as in the case of Raney catalyst, or by fusing several metal ingredients. The catalyst is activated before use, which consists of calcination at high temperature followed by reduction.

3.5.11 Steric Effect and Specificity

Steric bulk of the substrate has generally very little effect on the reaction kinetics and product selectivity in heterogeneous catalytic systems, except in zeolites. Enzymes exhibit chemical and stereochemical specificity to both substrate and products. The matching of the size and shape of the substrate with that of the enzyme active site is a necessary condition for the reaction to take place.

3.5.12 Safety Hazards Handling

Enzyme-catalyzed reactions are generally milder and inherently safer than heterogeneous catalyzed reactions. Since the latter requires higher temperature and pressure, the reactor design has to consider hot spots and temperature runaway conditions. Also finely dispersed metal catalysts or activated nickel or Raney-nickel catalysts are pyrophoric and will require inert atmospheres or wet conditions during transportation and charging into the reactor. On the contrary, since most of the enzymes are of plant or animal origin, they are safe and can be transported and charged with very little safety precautions.

3.5.13 Economics

The main drawback of enzyme catalyst is that it is more expensive than metal catalysts, contributing almost 15% to the total cost of the product, since they are produced from natural resources and hence require abundant raw material. Enzyme catalyst may not be able to replace heterogeneous catalyst for the production of bulk chemicals but can be used for the manufacture of value-added chemicals, which cannot be manufactured by traditional methods.

3.5.14 Environmental Issues

Enzyme-catalyzed reactions are ecofriendly and produce very little effluent. The reactions are also being termed as *green chemistry*. On the other hand, heterogeneous catalyzed reactions produce harsher effluents. Also, the catalyst used in the process have to be fully recovered to avoid heavy metal pollution in the liquid effluent. Enzymes are produced from natural products; so used solid enzymes can be disposed in the soil without any extra precaution.

3.6 CLASSIFICATION AND NAMING OF ENZYMES

Enzymes are known and are being used in industry for more than a century now. In order to remove any ambiguity and systematically name and classify this growing number of enzymes, an Enzyme Commission was appointed by the International Union of Biochemistry. Its report and the later updates form the basis of the present accepted system of classification and naming of enzymes.

The enzymes are divided into six main classes on the basis of the total reaction catalyzed. Each enzyme was assigned a code number, consisting of four elements, separated by dots. The first digit shows to which of the main classes the enzymes belongs. The main classes are as follows:

1. Oxidoreductases: oxidation–reduction reactions
2. Transferases: transfer of atom or group between two molecules
3. Hydrolases: hydrolysis reactions
4. Lyases: removal of a group from the substrate (not hydrolysis)
5. Isomerases: isomerization reactions
6. Ligases: the joining of two molecules/atoms, with the breakdown of a pyrophosphate bond in a nucleoside triphosphate (ATP)

The second and third digits in the code further indicate the kind of reaction being catalyzed. The fourth digit describes the actual substrate being transformed. For example, lactate dehydrogenase (trivial name) catalyzes as follows:

$$CH_3CH(OH)COO^- + NAD^+ \longleftrightarrow CH_3COCOO^- + NADH + H^+$$

$$L - Lactate \qquad\qquad\qquad pyruvate$$

This enzyme is given the EC number E.C.1.1.1.27.

The first digit indicates that this enzyme is an oxidoreductase.
The second digit indicates that acts on a secondary hydroxyl (CHOH).

The third digit indicates that it uses $NAD^+/NADP^+$ as cofactor, and finally.

The fourth digit indicates the substrate, here L- lactate.

The commission also gave a systematic nomenclature. This systematic name includes the name of the substrate or substrates in full and a word ending in -*ase* indicating the nature of the process used. For example, the above enzyme is given the systematic name, L-lactate: NAD^+ oxidoreductase. Thus, this enzyme gets the complete name L-lactate: NAD^+ oxidoreductase: E.C.1.1.1.27.

3.7 THERMODYNAMICS

Bioenergetics is the quantitative study that deals with how biological systems gain and use energy, and it is a branch of thermodynamics. Bioenergetics is essential for understanding how metabolic processes provide energy for the cell structure of macromolecules and how membrane transport occurs. Although bioenergetics tells whether a process will occur spontaneously, it does not mention how fast the reaction proceeds (rate of reaction).

The first law of thermodynamics states that "energy can neither be created nor destroyed," which means that it can be converted from one form to another to perform work. This means a biocell can convert food to heat or work. The second law states "entropy, or degree of disorder, of the universe is always increasing." Life, a state of high order (or low entropy), is maintained by food, but ultimately every organisms reaches the state of equilibrium, namely death and decay.

Gibbs Free energy is an important thermodynamic function that is useful in biochemistry, and it is related to other parameters such as enthalpy at constant pressure (H), temperature (T), and entropy (S) by the equation $G = H - TS$. For any process to be favorable, change in enthalpy (ΔH) has to be negative (heat is released) and change in entropy (ΔS) has to be positive (increase in randomness of the system). If the change in free energy ΔG, namely the difference between the free energy of the product(s) and that of the reactant(s), is negative, then the process is spontaneous. If $\Delta G = 0$, the process is at equilibrium, and if ΔG is positive, the process is not spontaneous (in fact, the reverse process is spontaneous). The free-energy change is related to enthalpy and entropy by

$$\Delta G = \Delta H - T\Delta S \tag{3.1}$$

If $\Delta S > 0$, then the system is less ordered, and if $\Delta S < 0$, then the system is more ordered. If protein denaturation occurs spontaneously at high temperature ($>60\,^{\circ}C$), where $T\,\Delta S$ dominates, then the right-hand side of the equation is negative.

For an equilibrium reaction of the form

$$A + B \rightleftharpoons C + D$$
$$\Delta G = -RT\log_e K_{eq} \tag{3.2}$$

The equilibrium constant $K_{eq} = ([C_{eq}][D_{eq}])/([A_{eq}][B_{eq}])$. When the initial concentrations of all reactants and products $= 1$ mol/l, then free energy is called *standard free-energy change*.

If the equilibrium constant is estimated at different temperatures, then the slope of the plot between $\log_e K_{eq}$ and $1/T$ will give ΔH°, the enthalpy change in the standard state. Fermentation of glucose to yield ethanol and CO_2 has a $\Delta G = -218$ kJ/mol. In this reaction, both ΔH and $-T\,\Delta S$ are negative. Combustion of ethanol is an enthalpy-driven reaction since it has both ΔG and ΔH negative, but $T\,\Delta S$ is positive. Decomposition of solid nitrogen pentoxide to NO_2 and O_2 is an entropy-driven reaction where both ΔG and $-T\,\Delta S$ are negative and ΔH positive.

An endergonic reaction is one in which energy is input, and it has positive free-energy change and is unfavorable. Processes that characterize life are endergonic. An exergonic reaction has negative free-energy change. Endergonic reaction can be made to proceed in the desired direction if it is coupled to an exergonic reaction. Phosphorylation of glucose to produce glucose-6-phosphate is a very important reaction in the cell and is given by

$$\text{Glucose} + \text{Pi} \rightleftharpoons \text{glucose-6-phosphate} + H_2O \qquad \Delta G^{\circ} = +14 \text{ kJ/mol}$$

$$\text{ATP} + H_2O \rightleftharpoons \text{ADP} + \text{Pi} \qquad \Delta G^{\circ} = -31 \text{ kJ/mol}$$

The first reaction is not favorable, whereas the second reaction is favorable. By coupling the two one gets a favorable reaction as shown below:

$$\text{Glucose} + \text{ATP} \rightleftharpoons \text{glucose-6-phosphate} + \text{ADP} \qquad \Delta G^{\circ} = -17\text{kJ/mol}$$

This is also known as the *principle of common intermediate*. ATP (Adenosine triphosphate) is used by organisms to transfer energy within the cell and gain this energy during the transfer of electrons from an electron donor to an electron acceptor. ATP, also known as *high-energy compound*, has a large negative free energy of hydrolysis and is used by cells to drive many energy consuming reactions. ATP is present in cells at 1 to 10 mM, it is anionic, carrying four negative charges at pH 7.0, and it is neutralized by complexing with Mg^{2+}. In bacteria such as *E. coli*, energy is provided by ATP for most biosynthetic reactions and some transport systems.

Apart from photosynthesis, all organisms derive their energy by means of oxidation–reduction (redox) reactions. In heterotrophic organisms, respiration reactions provide the energy required for cell metabolism and growth. The respiration reaction is of the general form,

Electron donor(reduced) + electron acceptor(oxidized)

→ reduced electron acceptor + oxidized electron donor

where electron donors could be organic substrates (e.g., the biochemical constituents such as carbohydrates, proteins, fats, etc.), which oxidize to products such as CO_2. Electron acceptors for aerobic processes are oxygen (whose reduced product is H_2O) and for anaerobic processes could be CO_2, SO_4, NO_3, and partially oxidized organic compounds.

Bacterial synthesis occurs when a reduced electron donor (generated from the previous reaction), supplemented with a nitrogen source, is used and incorporated in bacterial cell biomass:

Reduced electron donor + NH_4^+ → bacterial cells

In an aerobic process the sugar gets oxidized as shown below:

$$\frac{1}{24}C_6H_{12}O_6 + \frac{1}{4}O_2 \rightleftharpoons \frac{1}{4}H_2O + \frac{1}{4}CO_2 \qquad \Delta G_r = -28.675$$

In an anaerobic process the reaction is as follows:

$$\frac{1}{24}C_6H_{12}O_6 \rightleftharpoons \frac{1}{8}CH_4 + \frac{1}{8}CO_2 \qquad \Delta G_r = -4.237$$

where ΔG_r is the standard free energy for respiration.

Metabolism is the sum of all the chemical reactions that occur within an organism and is a sum of catabolism and anabolism. Catabolic reactions are those metabolic reactions that are involved in the "generation" of cellularly useful energy and are involved in the breaking down of more complex molecules to simpler ones. On the other hand, anabolism involves the net use of energy to build more complex molecules and structures from simpler ones. The energy "generated" by catabolic processes is harnessed by cells to perform anabolic processes and is known as *energy coupling*. While catabolic processes represent a chemical movement toward equilibrium, anabolic process represent the reverse, the movement away from equilibrium. Hydrolysis is an example of a catabolic reaction, and dehydration synthesis is an example of an anabolic reaction.

3.7.1 Dissociation

In biological systems one or more small molecules bind to a protein or enzyme through noncovalent bond and this biding determines the biological

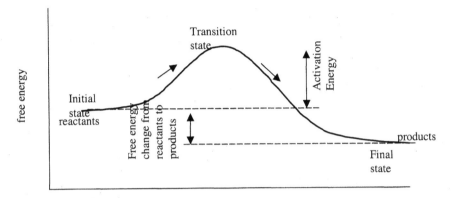

a

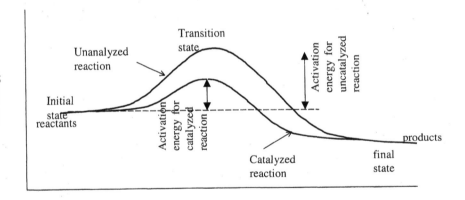

b

FIGURE 3.13 Free-energy change during a chemical reaction (a) via the formation of transition state, (b) effect of catalyst.

activity. The dissociation of the ligand (L) from the protein (P) is given by the equation

$$P + L \rightleftharpoons PL$$

the dissociation constant K_d for the reaction is $[P][L]/[PL]$.

3.7.2 Transition-State Theory

Transition-state theory states that all reactants should raise from their initial state to a more energetic state, called the *transition state*, before the reaction can take place. This energy barrier, which the molecules have to overcome, is called the *energy of activation* and is shown in the Fig. 3.13. The transition state is unstable, and the molecule quickly breaks down to form the product. The final state of the product is lower than the initial state. Activation energy barrier could be overcome by increasing reaction temperature, pressure, etc. Some reactions have high activation energies and some have low, which means the latter are facile reactions and the former need more energy input. Energy input can be in the form of heat or pressure.

A catalyst accelerates the reaction without affecting the thermodynamics. The activation energy or the barrier decreases in the presence of a catalyst as seen in Fig. 3.13b, without altering the free enegies of the reactants or the products.

3.8 SUMMARY

Enzymes are globular proteins that catalyze in a highly specific way the chemical reactions taking place within a living cell. The binding and the catalytic activity occur in active site of the enzyme, wherein all the necessary side chains of the amino acid residues are arranged. The active-site geometry is determined by the three-dimensional structure (primary, secondary, tertiary, and quaternary structures) of the molecule and is maintained by the noncovalent interactions within the enzyme (protein) molecule. Enzymes use all the forms of catalysis for attaining a high degree of efficiency. Because of the lack of consistency in the names of the enzymes, the Enzyme Commission has developed a systematic method of classifying and naming the enzymes.

SUGGESTED READING

1. Fersht, Alan. Enzyme Structure and Mechanism. W. H. Freeman: New York, 1985.
2. Dugas, Hermannn; Christopher, Penney. Bioorganic Chemistry. Springer-Verlag: New York, 1981.
3. Branden, Carl; Tooze, John. Introduction to Protein Structure. Garland Publishing: New York, 1991.

4

Biotransformations

Microorganisms for their growth, sustenance, and reproduction need a suitable form of reduced carbon source (chemical energy), which under normal conditions of culture broth are the common sugars. Amazingly, to a very large extent, there is no limit to the adaptability of these microorganisms to metabolize various other forms of carbon sources, other than the common sugars. For example, biphenyl (I) and monohydroxy biphenyl (II) are toxic to the fungi, but when strains of *Asperigillus parasiticus* adopted to monohydroxy biphenyl (II) were used, biphenyl was converted to the 4,4'-dihydroxy biphenyl (III) in good yields (Fig 4.1).

Similarly for the currently used resolution of racemic 2-chloropropanoic acid (CPA) to L-CPA by the dehalogenase from *Pseudomonas putida*, the necessary strain, AJ1, was isolated from the environment with high-chlorine-containing compounds—the road tanker off-loading point (Fig 4.2).

The same can also be done by the soil enrichment methodology. This helps in developing strains of microorganisms, which adapt to high substrate concentration (which thereby develop enzymes to metabolize that form of reduced carbon—the substrate). Since all the reactions in the biological systems, be it in plant, animal, or microorganisms, are enzyme catalyzed, the metabolism of the compounds by microorganisms, the in vitro enzymatic reactions, the biosynthetic pathways in the plants, the metabolic processes in the animal systems, and the drug metabolism together can be termed as

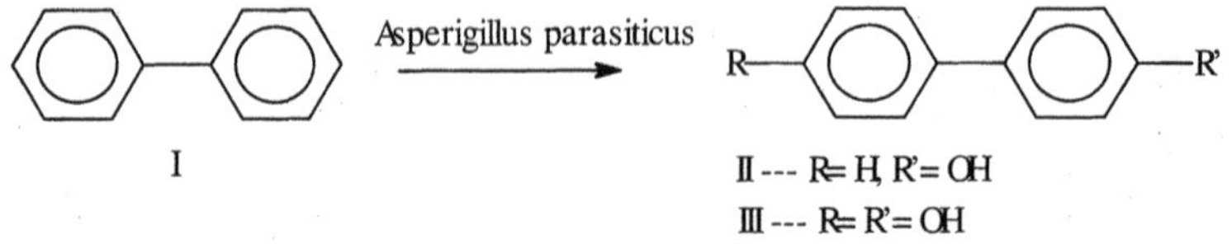

$$\text{I} \xrightarrow{\textit{Asperigillus parasiticus}} \text{R} - \bigcirc - \bigcirc - \text{R}'$$

II --- R= H, R'= OH

III --- R= R'= OH

FIGURE 4.1 Dihydroxy biphenyl from biphenyl.

biotransformations. Hence, biotransformations cover a broad spectrum of interdisciplinary fields and have application in diverse ways. The results in a simple laboratory experiment can be of use in directing the course of research in food technology, cosmetic industry, industrial catalysis (in developing ecofriendly methods), cleaning up an oil slick, medicinal chemistry and drug metabolism, to name a few. With the ever-increasing problem of waste disposal and recycling taking on gigantic proportions, biodegradation, or enzyme/microbe-mediated degradation, is one of the mainstay in the search for a suitable waste disposal technology. Though enzymes are specific, many enzymes do accept a wide range of substrates. Although the reaction of unnatural substrates may be slower than that of the natural substrate, the specificity of many enzymes is broad enough to allow a wide range of compounds to be utilized at an acceptable rate. It is the degree to which enzymes discriminate between structural and stereochemical features of their substrates that determines their synthetic utility. As with all reactions, it is the thermodynamically preferred products that accumulate in enzyme-catalyzed reactions. However, by suitably modifying experimental conditions (continuous flow method), thermodynamically less preferred kinetic products could be isolated. Biocatalysis is widely employed in industry. For, bringing about a transformation or reaction by biocatalytic means, one can use an entire microorganism or an isolated enzyme and, more recently, cross-linked enzymes. Each of these has their share of advantages and disadvantages. Some of the transformations in vogue in industry being:

Production of glucose from starch by *Asperigillus niger*

Esterification of myristic acid to isopropyl myristate by lipases

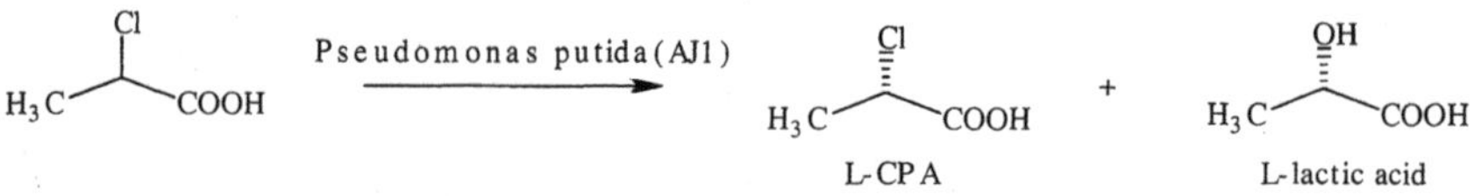

FIGURE 4.2 Resolution of racemic 2-chloropropionoic acid.

Racemic resolution of amino acids via the acylase method

Production of acetic acid from ethanol using immobilized *Acetobacter strain* (immobilized on beech wood shavings)

Steroid hydroxylations using microorganisms, *Rhizopus arrhizus*

Synthesis of glucose-6-phosphate using glucokinase/acetate kinase

Production of high fructose corn syrup from glucose by glucose isomerase

Fermentation of sugar to ethanol by yeast (*Saccharomyces cerevisiae*)

Production of citric acid from sucrose using *Asperigillus niger*

4.1 BIOTRANSFORMATIONS

Extensive and thorough review articles and books dealing with the various types of transformations effected by the enzymes and whole cell systems have appeared in literature from time to time. The present chapter is by no means an exhaustive up-to-date compilation of the work, but an attempt is made here to direct the attention on the mechanistic aspects of the enzymes, in conjunction with an overview of the multifarious transformations these enzymes have effected. Many a times, unlike in the chemical catalysis, where the chemo-, regio-, and stereoselectivities and outcome of a reaction are predictable, in the biocatalysis (brought by the microorganism and the enzymatic conversions) the transformations are deciphered only after the reaction (i.e., after the isolation and purification of the products). This lacuna is due to the lack of adequate attention on the mechanism of enzymatic action. The central aspect in utilizing this vast resource of ecofriendly catalysts is the understanding and appreciation of the mechanism of transformations brought about by these enzymes. This equips us to choose the right enzyme/microorganism in which the enzyme of choice is most active (either naturally occurring or made to the requirement by genetic engineering or the soil enrichment technique). Any chemical transformation (reaction) or molecular transformation involves two factors, viz., (1) the steric factor, the three-dimensional structure (the gross molecular size), and (2) the electronic factor, the electron density contour over the whole molecule (the charge density distribution).

These two factors play a more prominent role in the enzymatic trans-formations. To a large extent the substrate specificity and the enantioselec-tivity of an enzymatic transformation is determined by the steric requirement of the active site. While, the regio- and chemoselectivity of the transformation is determined by the "electronic requirement," i.e., the mechanistic detail of the transformation. Hence, a sound understanding of the mechanism is necessary to forehand expect a transformation along with its stereo- and regiochemical implications. The understanding of the mechanistic aspects helps in controlling the chemo-, regio-, and stereoselectivity in the course of the reaction. Also, an appreciation of the active site geometry (in the enzyme) or even the knowledge of substrate transformed, will equip us with the ability to suitably modify the substrate to either enhance the reaction or inhibit the reaction. Therefore, a brief note on isostere modification and its effect on the overall reaction is necessary to broaden the range of substrates that can be utilized for the transformation.

4.1.1 The Concept of Isosterism

The main criterion for isosterism is that two molecules must present similar, if not identical, volumes and shapes. We arrive at isosteres by the replacement of an atom or a group of atoms by another presenting a comparable electronic and steric arrangement. That is, molecules A-X and A-Y are said to be isosteric to each other if X and Y have comparable electronic and steric arrangement. In a sense X and Y also are isosteric to each other. Since the hydroxyl group (OH) and the fluorine group (F) are isosteric, phenol (C_8H_5OH) and fluorobenzne (C_8H_5F) are isosters (Fig 4.3).

Langmuir (1) in 1919 defined this concept, which was later modified and improved by Grimm (2). Erlenmeyer (3) in 1932 extended this concept to biological problems.

4.1.2 Bioisosterism

Friedman (4), recognizing the usefulness of isosterism in the design of biologically active molecules, proposed to call bioisosteres compounds "which fit the broadest definition of isosteres and have the same type of biological activity." Thomber (5) redefined *bioisoteres* as "molecules which have chemical and physical similarities producing similar biological effects," to encompass a wider group of compounds. Some of the groups which are isosteric/bioisosteric to each other and which are currently in practice are given in Table 4.1.

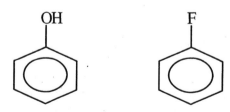

FIGURE 4.3 Isosteres.

TABLE 4.1 Molecules and Groups That Are Bioisosteric to Each Other

Type of groups	Isosteric groups
Univalent	-H -F -OH -NH$_2$ -CH$_3$ -Cl -CF$_3$ -CN -SCN -COOH -CONHOH -SO$_3$ -H$_2$PO$_3$ -SO$_2$NHR
Divalent	-O -NH -CH$_2$ -S
Trivalent	
Rings	
Rings	
Rings	
Urea	
Urea	

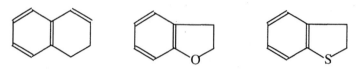

FIGURE 4.4 Isosteres–ring equivalents.

In general, the isosteric replacement, even though it represents a subtle structural change, results in a modified profile, specifically, some properties of the parent molecule remain unaltered, others will be changed. The similar shape and polarity within a series of substrates of different reactivity (bio-isosteres) eliminates effects due to differences between enzyme-substrate binding (ES) and *hence is a good method of extending the range of substrates that can be chosen for the transformation.*

A number of instances can be sited form literature wherein the isosteres had similar transformations. The isosteres, 1,2-dihydro naphthalene, 2,3-dihydro benzothiophene, and 2,3-dihydro benzofuran gave the similar corresponding diol products (Fig 4.4) on incubation with *Pseudomonas putida* UV4.

Microbes that possess the metabolic pathways to metabolize benzene, substituted benzenes, and phenols were found to metabolize fluorinated benzenes (isosteres) in a similar manner (Fig 4.5).

Similarly, the activity and enantioselectivity of α-chymotrypsin and subtilisin C in transesterification was studied on an isosterically varied N-acetyl-alanine and phenyl alanine series (Fig 4.6).

A word of caution, replacement of groups with isosteres does not always give similar transformations. In certain cases it leads to inhibition of the enzymatic action. The bicyclic haloketone underwent a facile Baeyer–Villiger oxidation with *Acinetobacter* sp. (Fig 4.7), while the isosteric hydroxy or methoxy were inert under the conditions.

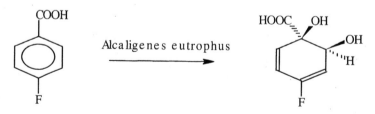

FIGURE 4.5 Biotransformation of *p*-hydroxy benzoic acid isostere.

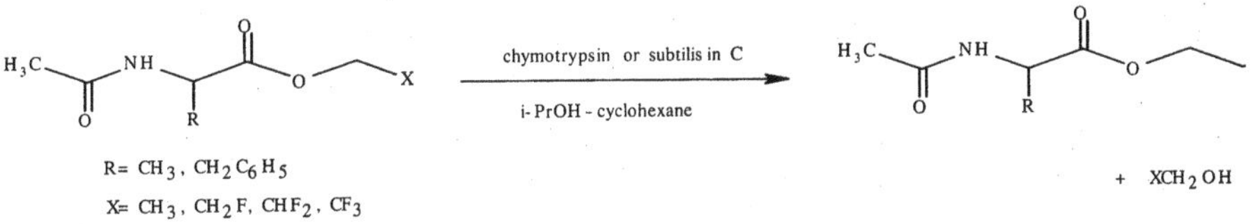

R= CH$_3$, CH$_2$C$_6$H$_5$

X= CH$_3$, CH$_2$F, CHF$_2$, CF$_3$

+ XCH$_2$OH

FIGURE 4.6 Chymotrypsin activity on isosterically varied phenyl alanine.

4.2 BIOTRANSFORMATIONS: A MECHANISTIC APPROACH

The following account of the enzymatic/microbial transformations is tailored only for a mechanistic appreciation of these transformations:

Enzymes are categorized into six groups:

1. Oxidoreductases catalyze oxidation–reduction reactions involving removal or addition of hydrogen atom equivalents and oxygenations.
2. Transferases mediate the transfer of groups such as acyl, sugar, phosphoryl, and aldehyde/ketone moieties from one molecule to another.
3. Hydrolases catalyze the hydrolysis of esters, lactones, amides, nitriles, anhydrides, glycosides, and peptides.
4. Lyases catalyze the addition of HX to double bonds (CC, CN, and CO) and removal of HX (dehydrohalogenation and the reverse process).
5. Isomerases bring about isomerization and racemization.
6. Ligases catalyze the formation of CC, CO, CN, CS and phosphate ester bonds.

4.3 OXIDOREDUCTASES

Microorganisms/enzymes have the ability to effect chemical oxidation on a wide variety of substrates, many of which occur with chemo-, regio- or

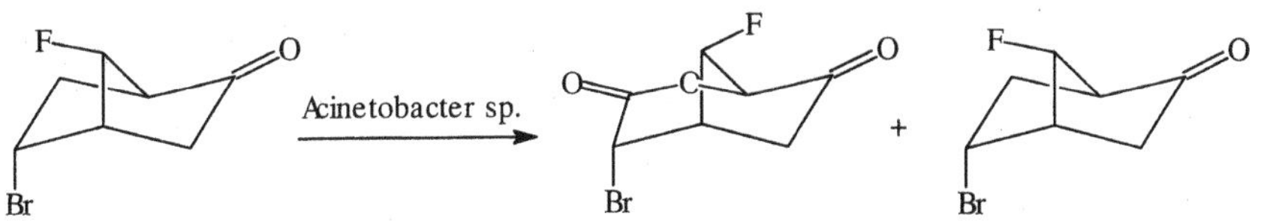

FIGURE 4.7 Baeyer–Villiger oxidation.

stereoselectivity, unattainable by conventional chemical methods. In fact, hydroxylations of the alicyclic, aliphatic, and aromatic systems are the single most important reactions effectively catalyzed by the enzyme/microorganisms. Regio- and stereoselective hydroxylations in steroids by the biocatalysts led to the unprecedented application of microorganisms as reagents in organic synthesis. Similar encouraging results were observed in the reactions incorporating a hydroxyl group on an unactivated carbon in sesquiterpenes, which are of value to the perfumery and pharmaceutical industry. Though these reactions are not high yielding, they provide valuable intermediates in the synthesis of complex organic compounds. These biooxidation compounds are also valuable in assessing the quantitative structure activity relations of the pharmacophore. A number of reports on regiospecific hydroxylation of monoterpene have appeared in the literature. Biocatalytic oxygenation reactions are becoming increasingly important since conventional methodology is either not feasible or makes use of hypervalent metal oxides, which are ecologically undesirable when used on a large scale. As the use of isolated oxygenases will be always hampered due to their requirement of NAD(P)H recycling, many useful oxygenation reactions such as mono- and di-hydroxylations, epoxidation, sulphoxidation, and Baeyer–Villiger reactions will continue to be performed using whole cell systems. The study of the stereo- and regioselectivity of these reactions is very valuable to the understanding of the mechanism of the enzymatic transformations, also providing viable alternatives to the general reagents.

4.3.1 Oxidoreductases: Salient Features:

1. This class of enzymes is the largest one, which catalyze oxidation or reduction of the substrate. The general form of the equation being

$$AH_2 + B \rightarrow A + BH_2$$

2. The oxidation–reduction is brought about by the transfer of two reducing equivalents (electrons followed by proton uptake or hydrogen molecule uptake) from one substrate to another.
3. These operate via redox cofactors (NADH, NADPH, FAD, etc.)
4. The transfer of these two reducing equivalents (two electrons) may be stepwise or in a single step. When flavin (FAD) groups are present as the prosthetic groups (flavoproteins, cytochrome oxidases, ferrodoxins, etc.), one equivalent oxidoreductions are effected.
5. These enzymes form the backbone for most of the catabolic processes in the microorganism and the animal systems.
6. These transformations are commonly observed in the catabolism of the endogenous and exogenous compounds (xenobiotics and drugs)

in the blood stream by the enzymes of the liver microsomes. In fact, oxidative transformations are by far the most important in drug metabolism. For example, tetrahydro cannabinol (THC) is oxidized to THC-7-oic acid by the cytochrome P-450 enzymes in the liver microsomes (Fig. 4.8).

7. Since, these reactions involve the transfer of electrons followed by proton uptake, the overall reaction will vary with pH.

8. This class of enzymes includes

   a. Dehydrogenases: Those that remove a molecule of hydrogen from the substrate. Since these reactions are reversible, these enzymes can be also made to add a molecule of hydrogen to the substrate. These are also referred to as oxidases.

   b. Oxidases: Those that remove a molecule of hydrogen from the substrate, the acceptor being oxygen (O_2).

   c. Peroxidases: Those that use hydrogen peroxide (H_2O_2) as the oxidant.

   d. Oxygenases: Those that incorporate molecular oxygen into the substrate. These are two types, viz., monooxygenases and dioxygenases.

   e. Hydroxylases: Those that incorporate oxygen atom, sometimes a hydroxyl group.

9. Since these require cofactors for their activity, when isolated enzymatic transformation is planned, cofactor recycling or supply be-

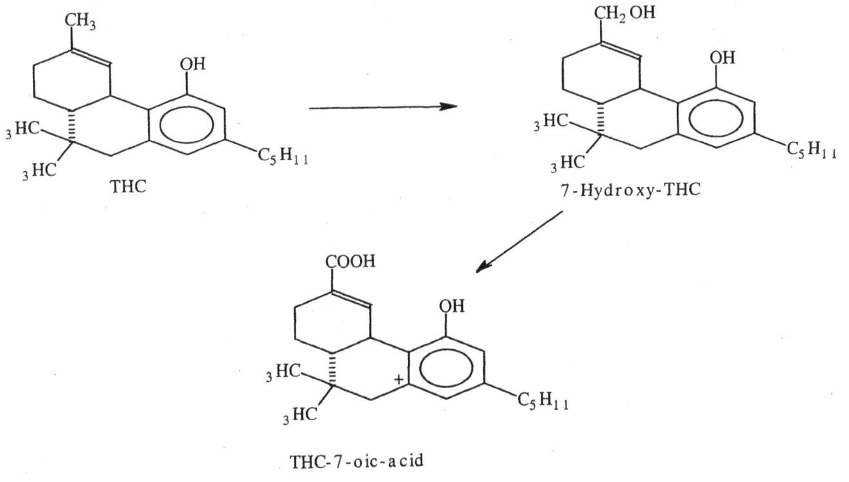

FIGURE 4.8 Metabolism of tetrahydro cannabinol.

comes another important component of experimental design. To overcome this additional cumbersome procedure, it is usually common that whole cell systems (yeasts, bacteria, and fungi) are used for these reactions, albeit, a small trade-off with the enantiomeric purity in chiral reduction. All the same, it should be borne in mind that an astonishing variety of optically pure compounds are synthesized by this whole cell biocatalytic approach.

4.3.2 Oxygenases–Aromatic Hydroxylation

The two main enzyme systems capable of oxygen insertion into organic molecules are the mono oxygenases (EC.1.13.12) and dioxygenases (EC.1.13.11). These are versatile, not highly specific enzymes (because a lot of isozymes are known). These catalyze the oxidation/hydroxylation of a variety of substrates.

The ubiquitous monooxygenases are the cytochrome-P-450 (cytochrome-b) haemoproteins dependent on NADH or NADPH, in the Eukaryotic organisms. The catalytic cycle is as shown in Scheme 4.1. The crucial step is the transfer of oxygen from the highly electron deficient $[FeO]^{4+}$ complex to the substrate to yield the corresponding hydroxylated substrate. Instead of molecular oxygen, other oxygen atom donors such as alkyl hydroperoxides, peracids, iodosobenzene, amine oxides, hydrogen peroxide, sodium periodate, or sodium perchlorate can also bring about the hydroxylation. Note:

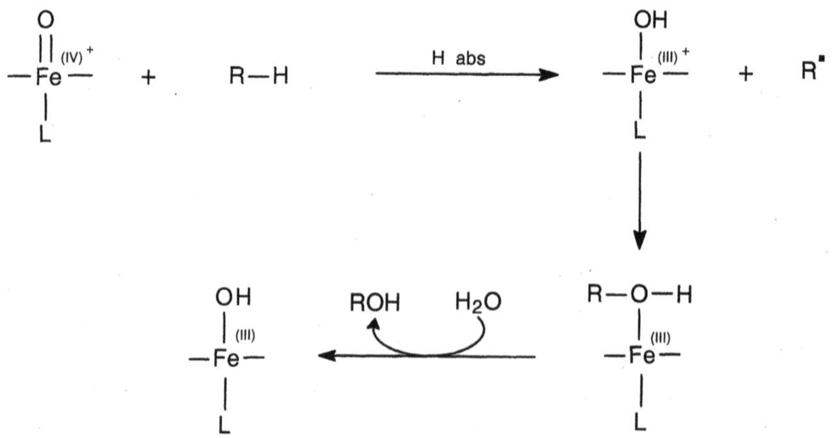

SCHEME 4.1   Catalytic cycle of monooxygenases.

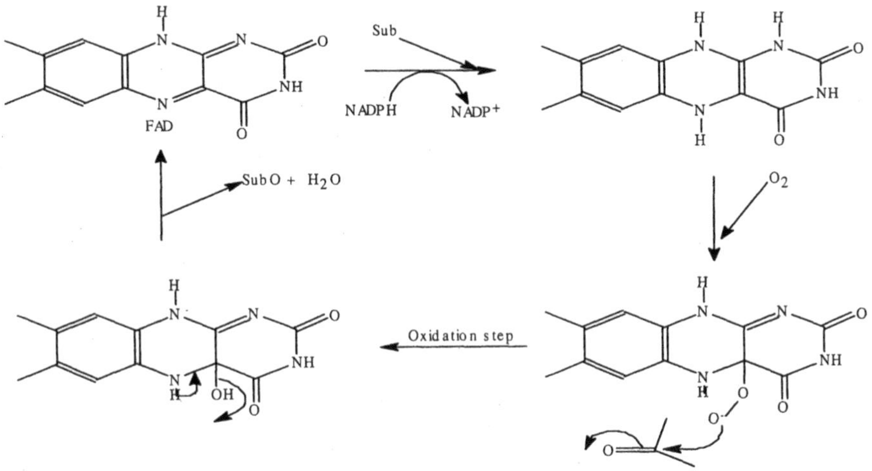

FIGURE 4.9 Catalytic cycle of FAD.

the site of the reaction on the substrate is the high electron density moiety.

The flavin-dependent monooxygenases use flavin as the cofactor. The catalytic cycle is as shown in the Fig 4.9. The crucial step is the transfer of oxygen to the substrate by the nucleophillic attack of the peroxide anion intermediate of the oxidized enzyme. Hence, the site of reaction on the substrate is the electron deficient center, such as the carbonyl carbon, or an electronegative heteroatom.

These monooxygenases (CytP-450) are known to react with a variety of organic substrates. Some of the widely known being (1) arenes to arenols, (2) alkene to epoxide, (3) alkane C-H to alcohols (hydroxylation of saturated carbon), (4) ethers to alcohols, (5) N-alkyl moieties to amines, (6) N-oxide formation of secondary amines, (7) thio ethers to thiols, and (8) S and N oxides from thio ethers and secondary amines, respectively. A graphical account of some of the common transformations is as given in Fig 4.10.

The aromatic hydroxylation by the monooxygenases (CytP-450) is well studied. It has been shown that almost all aromatic hydroxylation reactions proceed through an epoxide intermediate, an "arene oxide." This rearranges in two ways as shown in Fig 4.11.

The arene oxide intermediate formed spontaneously rearranges to a carbocation intermediate. The carbocation formed is the most *resonance stabilized*

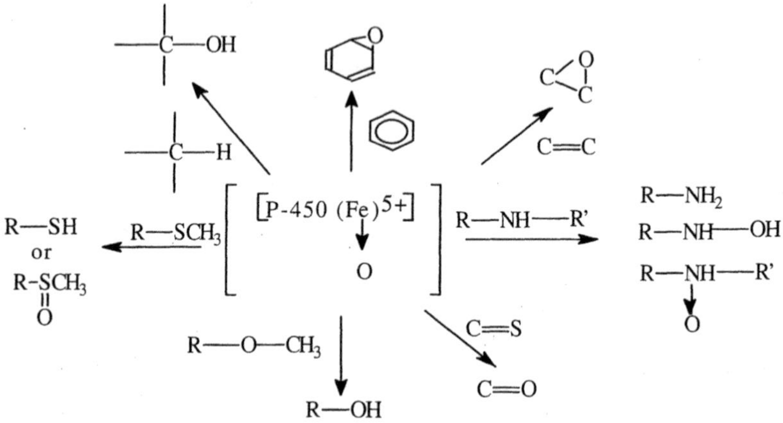

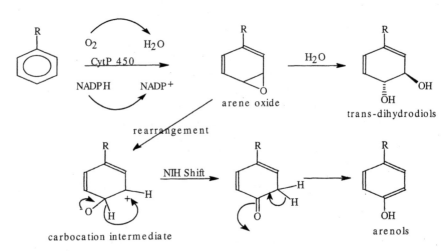

FIGURE 4.10 Reactions catalyzed my monooxygenases.

carbocation. (See Box 4.1, for the mesomeric stabilization of the carbocation.) This then undergoes an NIH shift to give the arenols. Note that the hydroxylation has occurred para to the ring substituent. Alternatively, the arene oxide also opens up in the presence of water to give *trans*-dihydrodiols, which may also further get oxidized in the presence of cofactors (NADP+) to yield a catechol derivative (Fig 4.12).

FIGURE 4.11 Mechanism of aromatic hydroxylation by monooxygenases.

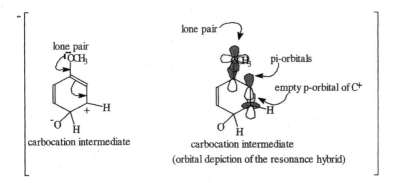

Box 4.1 Mesomeric stabilization of the carbocation.

Phenols and substituted phenols are hydroxylated to give catechol derivatives by polyphenol oxidase enzymes (copper containing monophenol monooxygenases, EC.1.10.3.1).

In the prokaryotic organisms (bacteria), the hydroxylation is brought about by the cycloaddition of molecular oxygen with a double bond to yield a dioxetan. This is then reduced to cis-dihydro diol. Rearomatization by dihydro diol dehydrogenase yields catechol derivative (Fig 4.13).

It is reported that toluene and its derivatives were first oxidized to benzoic acid, which then yielded the catechol via the cis-1,2-diol intermediate (Fig 4.14).

Studies on bacterial metabolism of aromatic compounds indicated benzylic hydroxylation. Dioxygenases are also known to induce a radical formation. Boyd et al. (6). have observed the transformation of 1,2-dihydronapthalene to (1R,2S)-cis-1,2-dihydro napthalene-1,2-diol via the triol intermediate (believed to have arisen by the benzylic hydroxylation), using the growing cell cultures of *Pseudomonas*

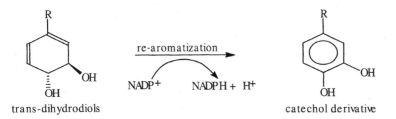

Figure 4.12 Catechol formation from dihydrodiols.

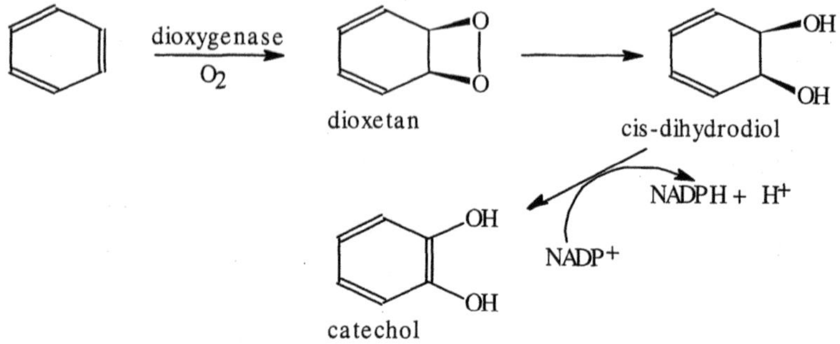

FIGURE 4.13   Mechanism of aromatic hydroxylations by dioxygenases.

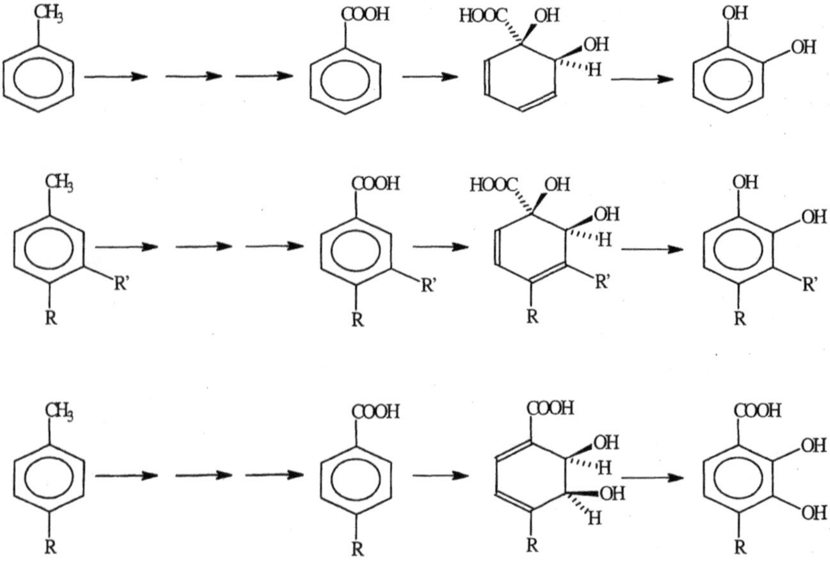

FIGURE 4.14   Biotransformation of toluene and its derivatives.

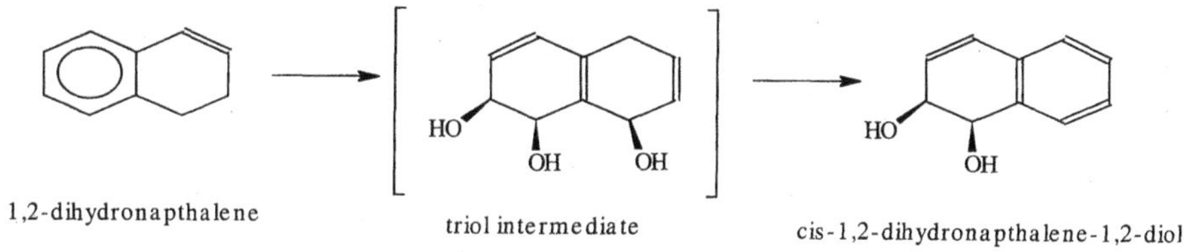

1,2-dihydronapthalene   triol intermediate   cis-1,2-dihydronapthalene-1,2-diol

FIGURE 4.15 Transformation of 1,2-dihydro naphthalene by *P. putida*.

putida UV4 (Fig 4.15). A schematic presentation of the mechanism of dioxyegenase action is given in Scheme 4.2.

4.3.3 Summary of the Oxygenation of Aromatic Rings by Mono- and Dioxygenases

1. Electron-rich unsaturation (unsaturation with electron pumping substituents, i.e., OH, OR, NH_2, NHR, alkyl) is susceptible to oxidation either by a mono- or dioxygenase more readily than the electron-deficient unsaturation (unsaturation with electron-withdrawing substituents, i.e., NO_2, NHCOR, halogens, SO_2R groups)

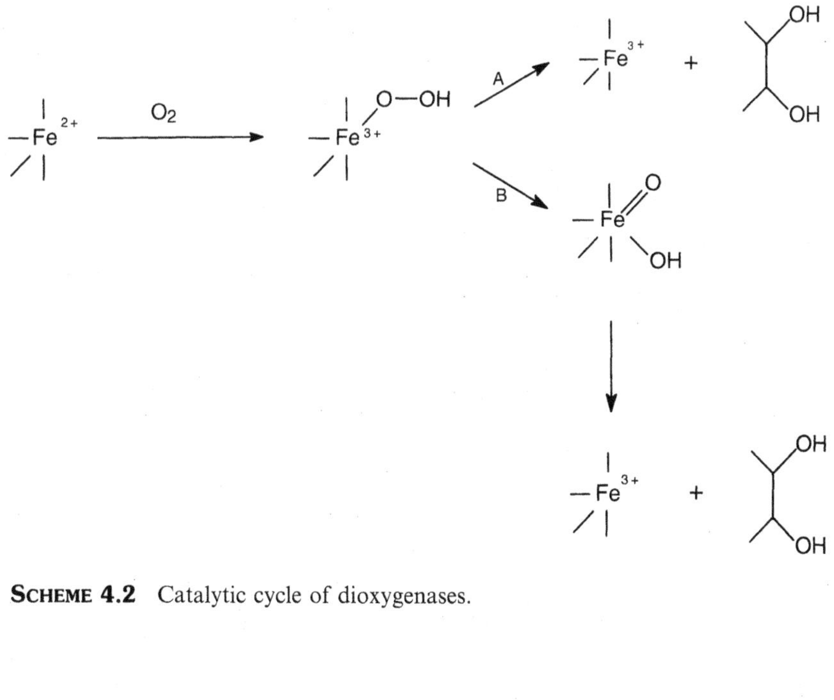

SCHEME 4.2 Catalytic cycle of dioxygenases.

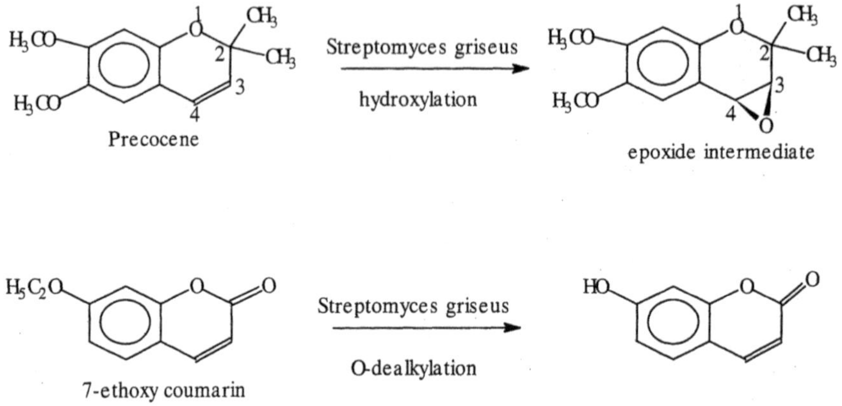

FIGURE 4.16 Epoxidation of unsaturation having higher electron density.

because it is from the highly electron deficient $[FeO]^{4+}$ complex that the transfer of oxygen to the substrate occurs. Therefore, the substrate should be able to readily donate the electrons. This can be appreciated better when we consider the conversion of precocene by *Streptomyces griseus* to three hydroxylated metabolites, all of which are believed to arise from the epoxide intermediate. However, the same organism could only bring about o-dealkylation of 7-ethoxy coumarin. Note that the C3, C4 unsaturation is being withdrawn due to the lactone carbonyl group in 7-ethoxy coumarin while it is more or less unconjugated in precocene hence having high electron density (Fig 4.16).

2. Aromatic hydroxylation reactions by NADH-dependent monooxygenases or dioxygenases appear to proceed more readily in activated (electron rich) rings, whereas deactivated aromatic rings are generally slow or resistant to hydroxylation. For example, when a substituted chlorobenzene (deactivated ring) was incubated with

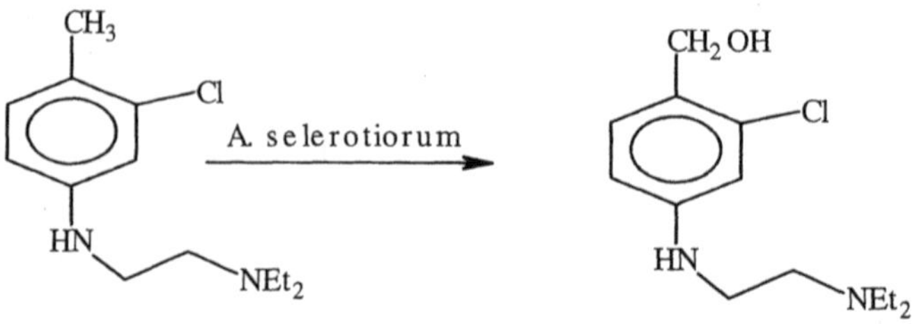

FIGURE 4.17 Benzylic hydroxylation.

Asperigillus selerotiorum, only the benzyl alcohol derivative was obtained (Fig 4.17).

4.3.4 Dioxygenases

1. Unlike the NADH-dependent monooxygenases, dioxygenases selectivity for either electron-rich or electron-deficient unsaturation is not as much.
2. Benzylic hydroxylation is also observed.

4.3.5 Epoxidation

Stereoselective epoxidations even on an unactivated alkene are routinely achieved by biocatalytic means.

The enantiospecific epoxidation of di-substituted (terminal) and tri-substituted alkenes have been reported.

Different organisms have varying regioslectivity of alkene epoxidation. Cyclic and internal olefins, aromatic compounds, and alkene units, which are conjugated to an aromatic system, are not epoxidized by hydroxylase of *Psuedomonas oleovorans*, while the monooxygenases of *Cunnighamella blaksleena* or *Curvularia lunata* were able to epoxidize endocyclic alkenes.

When more than one alkene group is present in the substrate molecule, the rate of epoxidation of the alkene with greater electron density is faster. For instance, 3-oxo-4,8-diene steroid derivatives on incubation with *Cunnighamella blakslena* gave only the 8,11-epoxide (Fig 4.18). The electron density of the 4,10-alkene being conjugated to the 3-keto group is less as compared to the 8,11-alkene, which is isolated and hence electron dense.

Though epoxidation is highly stereoface selective, when whole cell systems are used, the further transformations of the epoxide to a diol are a competing side reaction.

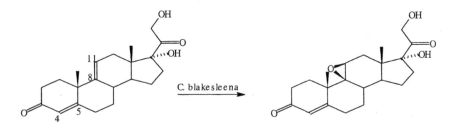

FIGURE 4.18 Selective epoxidation of unsaturation with higher electron density.

Epoxides formed during the transformation are normally toxic to the cell. Hence, *cosolvent* (organic solvents) procedures, wherein the epoxide formed is selectively partitioned into the organic layer, are adopted. This helps in limiting the concentration of the toxic metabolite in the reaction medium (aqueous), thus increasing the yields in the whole cell epoxidations.

(R) epoxides are normally formed. The formation of (R) epoxides is accounted for by preliminary hydrophobic bonding of the alkene to the active site in a such a way as to ensure that the epoxide closure must be from the *Si* face of C-2.

Chiral epoxides of predictable absolute configurations can also be obtained from halohydrins produced by chloroperoxidase-catalyzed addition of hypohalous acids to double bonds. The enzyme utilizes iodide, bromide, and chloride (Fig 4.19).

4.3.6 Hydroxylation at Allylic and Benzylic Centers

Biocatalytic procedures for the stereo- and regiospecific hydroxylation at activated or unactivated carbon centers are irreplaceable by chemical means, be it in the hydroxylation of steroids or the hydroxylation of benzyllic/allylic centers. In the case of hydroxylation of allylic centers, epoxidation and the corresponding diol formation are significant side reactions.

It is believed that at some stage the hydrocarbon residue is converted into a short-lived ($<10^9$ s) free radical.

Since, benzyllic and allylic radicals are resonance stabilized (hence more easily formed), hydroxylations at the benzyllic and allylic carbon atoms is very common. In fact, it is one of the major ways of drug

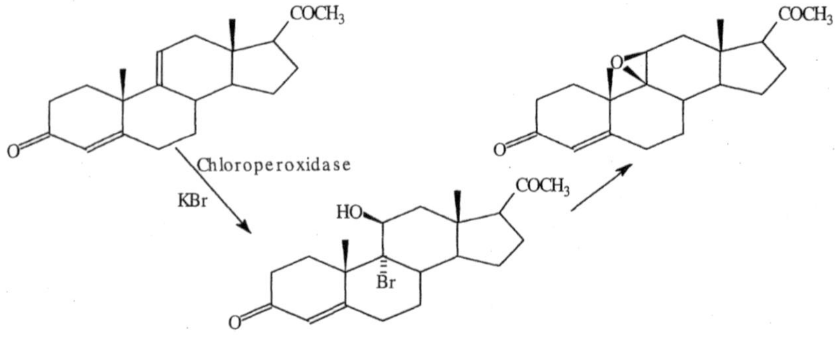

FIGURE 4.19 Chloroperoxidase catalyzed epoxidation.

detoxification pathways in the liver microsomes by the mixed function monooxygenases present there (Fig 4.20).

The stability of the radical formed (though for a short period) is of paramount importance in deciding the direction of the reaction.

Geraniol and nerol did not undergo any notable bioconversions when incubated with the cultures of *Asperigillus niger*, while the corresponding acetates led to exclusive hydroxylation at allylic position (C-8). This may be partly because the hydroxyl is not lipophillic and the monooxygenases being membrane bound demand a certain hydrophilic/lipophillic balance for the approach and concominant transformation (Fig 4.21).

As discussed earlier both the steric and the electronic factors are to be weighed to predict the outcome of a reaction. Therefore, just the presence of an allylic CH does not vouchsafe allylic hydroxylation. Grindelic acid on incubation with cultures of *Asperigillus niger* and

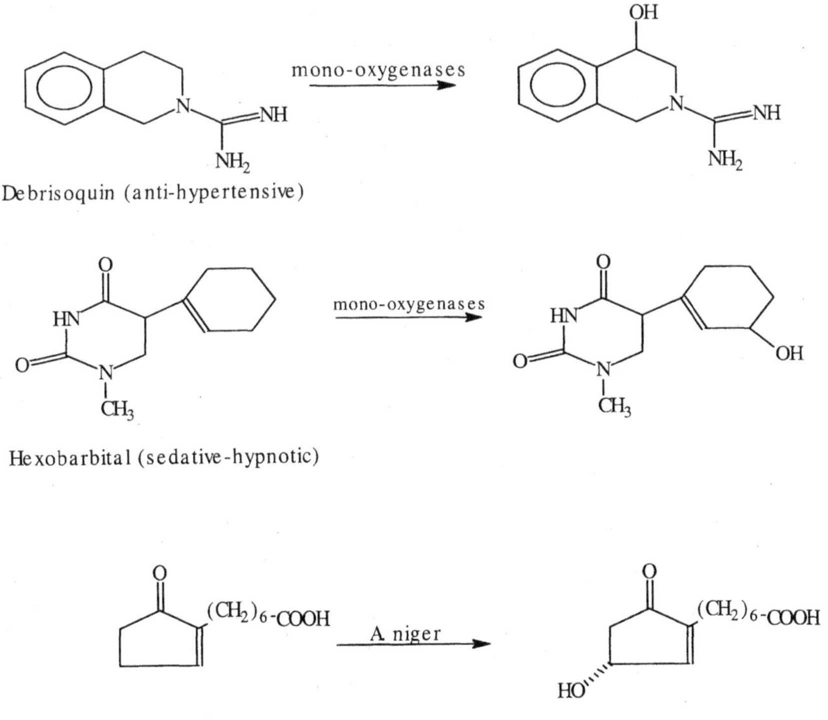

Debrisoquin (anti-hypertensive)

Hexobarbital (sedative-hypnotic)

FIGURE 4.20 Benzylic and allylic hydroxylations.

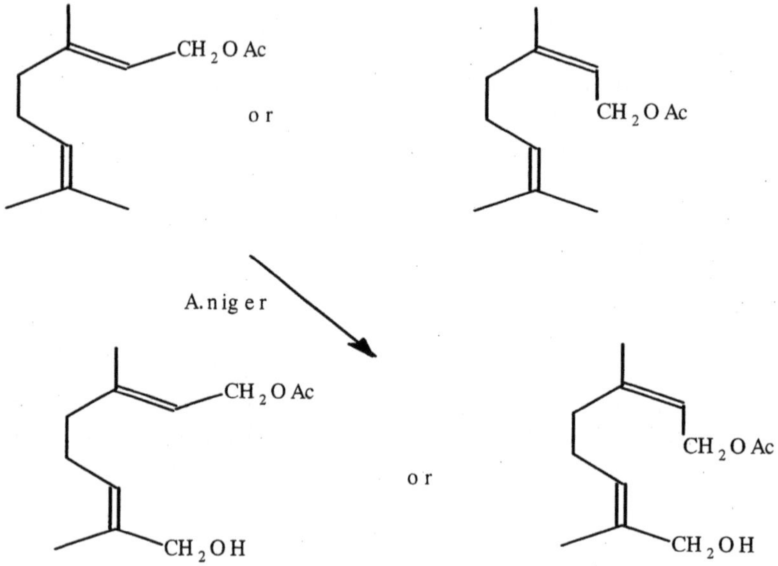

FIGURE 4.21 Enhancement in transformation by lipophillic derivatization.

Sordaria bombioidee, did not yield any allylic hydroxylation, but instead gave the 3-hydroxy grindelic acid (Fig 4.22).

4.3.7 Hydroxylation at Unactivated C–H

In the field of steroid hydroxylations virtually any carbon center in the steroid nucleus can be hydroxylated stereospecifically using a range of microorganisms.

To a very large extent, fungi are of greatest use in the hydroxylations of steroids, and five genera in particular have been found to be

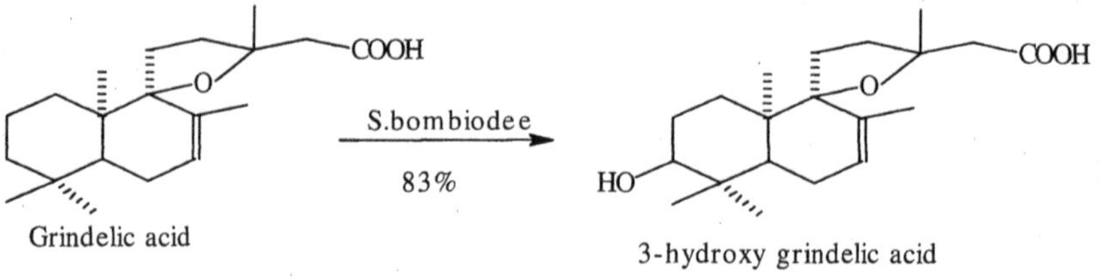

FIGURE 4.22 Transformation of Grindelic acid.

TABLE 4.2 Hydroxylation of Steroids

Microorganism	Product
Regioselectivity	
Asperigillus ochraceus	11α-OH
Asperigillus ochraceus spores	11α-OH
Asperigillus phoenicis	11α-OH
Rhizopus nigricans	11α-OH
Curvularia lunata	11β-OH
Calonectria decora	12β, 15 α- dihydroxylation
Rhizopus arrhizus	6β-OH
Cunninghanella elegans	7β/7β-OH
Stereoselectivity	
Calonectria decora	Equatorial −OH
Curvularia lunata	Axial −OH
Rhizopus nigricans	Equatorial −OH
Cunninghanella elegans	Axial and equatorial −OH

extremely useful, viz., *Rhizopus, Calonectria, Asperigillus, Curvularia*, and *Cunninghamella*.

A number of microorganisms show a tendency to hydroxylate in certain positions of the steroid, irrespective of substituent patterns. The regioselectivity and the stereoselectivity of some of the well-known transformations are as given in the Table 4.2.

Functional groups, especially polar groups help in increasing the yield of the hydroxylated products. Probably, these groups enhance the water solubility of the steroid substrates, thus increasing the catalytic efficiency.

The unique selectivity of the organisms can be used effectively to bring about multiple transformations on the substrate. For example, the fermentation of the steroid (shown in Fig 4.23) with *Pellicularia*

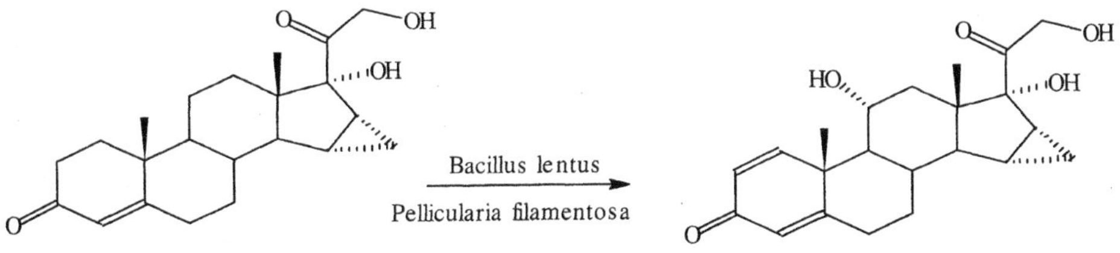

Bacillus lentus
Pellicularia filamentosa

FIGURE 4.23 Multiple transformations.

filamentosa and *Bacillus lentus*, brought about 11-β-hydroxylation along with a Δ'-dehydrogenation.

Most of the mixed function oxygenase-catalyzed transformations have a radical intermediate.

The stability of carbon radicals is tertiary > secondary > primary. But, this is slightly modified in the case of bicyclic and fused alicyclic systems: since tertiary carbon radical now is more strained, the stability of carbon radicals in such systems is secondary >> primary > tertiary. We observe the similar trend in steroidal hydroxylations, indicating that a radical intermediate may be formed.

In cases where a tertiary radical can be formed, it is preferred. 1,4-Cineole gives 8-hydroxy cineole on incubation with *Streptomyces grieseus* as the major product along with 2-exo- and 2-endo-cineoles as minor products (Fig 4.24).

Dioxygenases add molecular oxygen to the substrate. There is probably no formation of a radical intermediate. Hence, streroidal and un-activated carbon hydroxylations by prokaryotic organisms (having dioxygenases) are rare.

Microbial oxidation of alkanes can take place at terminal carbon, in which case an alcohol is the product, or at subterminal position to give either a secondary alcohol or a ketone. Most of these undergo further oxidation and get degraded (metabolized). In order to isolate the primary oxidative products, mutant strains of the microorganisms have to be used.

Monooxygenases are known to bring about hydroxylation at CH position with inversion of stereochemistry.

β-Hydroxylation of short-chain aliphatic carboxylic acids can be accomplished by a number of microorganisms such as *Endomyces reesii*, *Trchosporum fermentans*, *Torulopsis candida*, *Micrococus flavus*, *Candida rugosa*, baker's yeast (*Saccharomyces cerevisiae*), *Rhodococus* spp., and *Pseudomonas putida*.

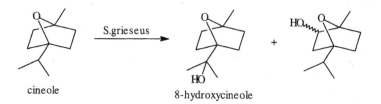

cineole   8-hydroxycineole

FIGURE 4.24 Transformation of Cineole.

The enantiotopic discrimination of hydrogens during oxidation of un-activated C-H bonds by microorganisms is synthetically very useful. For example, (R)-3-hydroxy butanoic acid obtained on incubation of the cultures of *Candida rugosa* with n-butanoic acid is a versatile homochiral synthon (Fig 4.25).

4.3.8 Reductions of Carbonyl Groups

Enzymes can operate stereospecifically on one of the two enantiotopic or diastereotopic faces of planar groups, such as C=C, C=N, or C=O. With a few exceptions, such as conjugated carbonyl functional groups, stereospecific reduction of an aldehyde or ketone in virtually any molecule can be effected either enzymatically or microbially. The asymmetric reduction of carbonyl compounds by microorganisms, a method outside the traditional arena of chemical synthesis, is now well recognized as an invaluable tool for the preparation of chiral alcohols. Because a single microorganism contains a number of oxidoreductases, it can mediate the reduction of a variety of arti-ficial ketones to produce chiral alcohols of remarkable optical purity. Most of the oxidoreductases use NADH or NADPH as cofactor: hence, use of iso-lated pure enzyme system necessitates the regeneration of this cofactor, while the use of whole cell systems are devoid of this component. Since the whole cell systems have more than one oxidoreductase with varying enantiospeci-ficities, racemic mixtures are obtained. Be that as it may, various methods such as the following may be adopted to circumvent the problem of the mul-tiple transformations and still use whole cell transformations:

Extensive screening to identify the micoorganism with a single, most active enzyme

Extractive biocatalysis (use of hydrophobic polymer resin for product adsorption)

Recombinant DNA techniques to create a set of strains in which the enzyme of choice is overexpressed or the other enzymes are suppressed (absent)

Expressing individual reductases in a heterologous host, such as *Escherichia coli*

The Prelog school (7) has investigated the purified NADPH specific dihy-droxyacetone reductase of *Mucor javanicus*, extensively. This enzyme is

FIGURE 4.25 Synthesis of (R)-3-hydroxy butanoic acid.

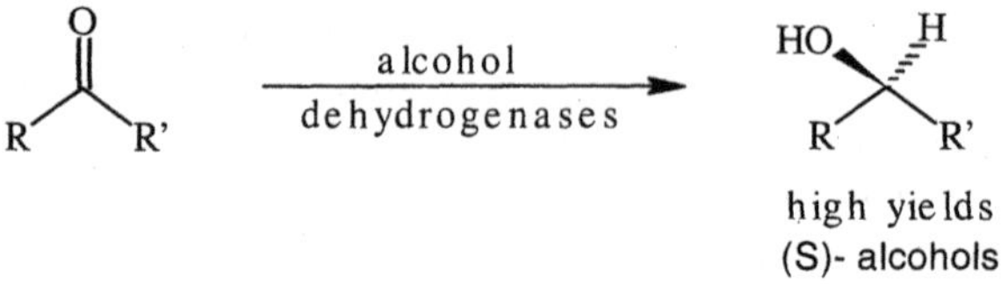

FIGURE 4.26 Prelog rule [R=CH$_2$COOC$_2$H$_5$, (CH$_2$)$_2$CH=CH$_2$, aryl, pyridyl; R′=CH$_3$, CH$_2$COOH, CF$_3$, CH$_3$].

known to transfer hydrogen exclusively to the *Si* face of the carbonyl group of most substrates giving (R) alcohols. From the product analysis of the reduction of 2-alkanones, it was concluded that the reaction rate and specificity increased with the increase in alkyl chain length. Also, it was observed that a sufficiently large hydrophobic group is needed for higher stereoselectivity, that is, for getting higher ee values. The stereochemical course of reduction of acyclic ketones by baker's yeast (*Saccharomyces cerevisiae*), on the other hand, proceeds via hydrogen transfer to the *Re* face of the prochiral ketone, giving an (S) alcohol.

The stereospecificities of yeast alcohol dehydrogenases-catalyzed reductions of ketones can be predicted by the Prelog rule. This was initially postulated for reductions of decalones by *Curvularia lunata*. This states that when groups R and R′ (the two substituents on either side of carbonyl group) are sterically larger (L) and smaller (S), respectively, as given in Fig 4.26, the hydride equivalent is delivered to the *Re* face of the carbonyl group as defined by the Cahn-Ingold-Prelog (CIP) priority sequence of oxygen > L > S. This rule also applies to other oxidoreductases, such as horse liver alcohol dehydrogenase.

As indicated earlier, other alcohol dehydrogenases having opposite stereospecificities also are common. Hence, either enantiomer of an alcohol

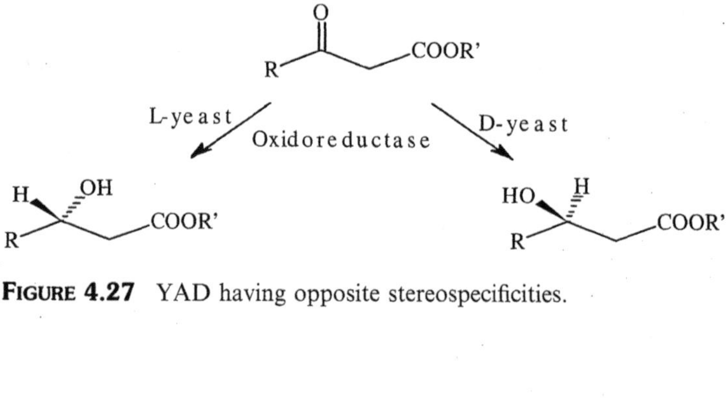

FIGURE 4.27 YAD having opposite stereospecificities.

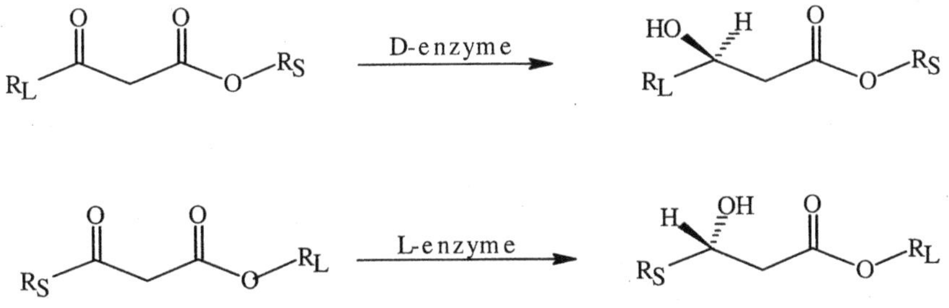

FIGURE 4.28 Reductions of β-keto acids.

can be produced at will by selecting enzymes with opposite enantiotopic face specificity for the same carbonyl substrate. Yeast contains two fatty acid synthetases with such properties. For example, keto esters are reduced to the R alcohols by the D enzyme and to the S alcohols by the L enzyme (Fig 4.27). The chiralities of the hydroxy products can be predicted for both the enzymes using a rule based on steric size distinctions between R and R'.

The stereochemical course of the reductions of β-keto carboxylic acid derivatives by yeast is influenced by substituents at both the ends of the molecule. While the enantioselective reductions are dependent on the affinity of the competing enzymes of opposite chirality, the enzymes affording (S) alcohols appear to prefer large hydrophobic substituents at the carboxy end, whereas (D) enzymes appear to prefer those at the hydrocarbon end (Fig 4.28).

Substituted cyclic ketones have also been reduced stereospecifically. Brooks et al. (8), in their study designed to establish a relationship between enantioselectivity and size differences of the alkyl substituents at position 2 of cyclic ,3-diketones, found that the enantioselectivity of the reaction depends

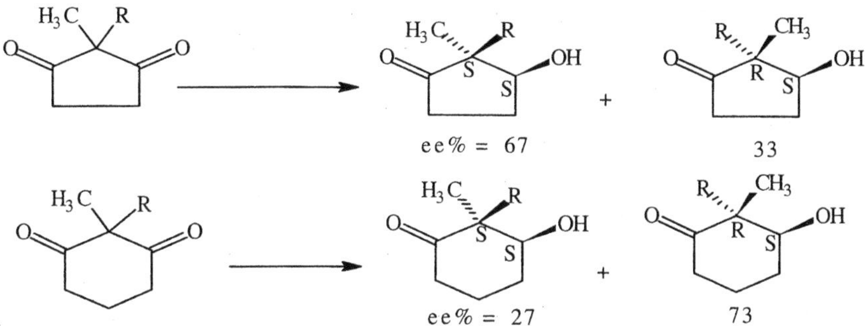

FIGURE 4.29 Reduction of cyclic ketones.

not only on the alkyl substituents but also on the ring size. In cyclopentane series, the (2S,3S) isomers were the major species, whereas the (2R,3S) isomers predominated in the cyclohexane series (Fig 4.29).

4.3.9 Baeyer–Villiger Oxidations

Enzymatic Baeyer–Villiger oxidation of ketones is usually catalyzed by flavin-dependent monooxygenases. The oxidized flavin cofactor acts as the nucleophile. This is mechanistically similar to the peracid catalyzed chemical Baeyer–Villiger oxidations. The strength of enzyme-catalyzed Baeyer–Villiger oxidation resides in the recognition of chirality, unthinkable by chemical means. Enzymatic Baeyer–Villiger oxidations are relatively common and can sometimes provide access to lactone products that are either sterically or electronically unfavored with peracid reagents. For example, the microbial conversion of fenchone yields a 9:1 mixture of the products as shown in Fig 4.30, in contrast to the 3:2 mixture afford by chemical oxidation.

4.4 PYRIDOXAL PHOSPHATE-DEPENDENT ENZYMES

Pyridoxine (vitamin B_6) is a very important coenzyme catalyzing a large variety of reactions. The aldehyde form is called *pyridoxal*, and its phoshate ester is implicated in many enzyme-catalyzed reactions of amino acids and amines.

4.4.1 Mechanism of the Coenzyme

The aldehyde group of the pyridoxal phosphate condenses with amino group of the substrate to form an imine (schiff base) (Fig 4.31). The pyridine ring in the schiff base acts as an "electron sink", which very effectively stabilizes the negative charge. Hence protonation of pyridine ring is essential for catalysis. It is the direction of this delocalization that dictates the reaction type and in model systems more than one reaction pathway is often observed. Thus the

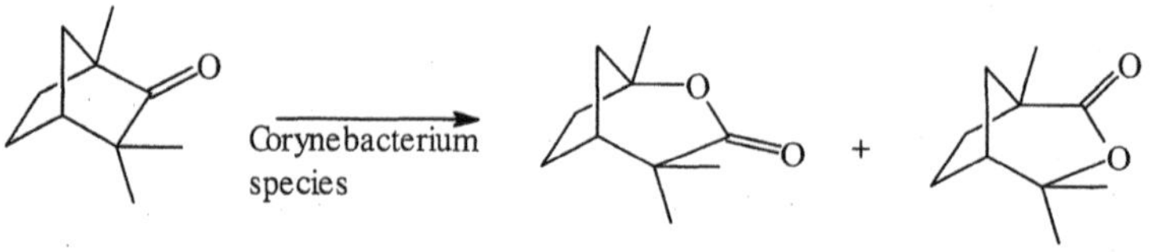

FIGURE 4.30 Baeyer–Villiger oxidation of Fenchone.

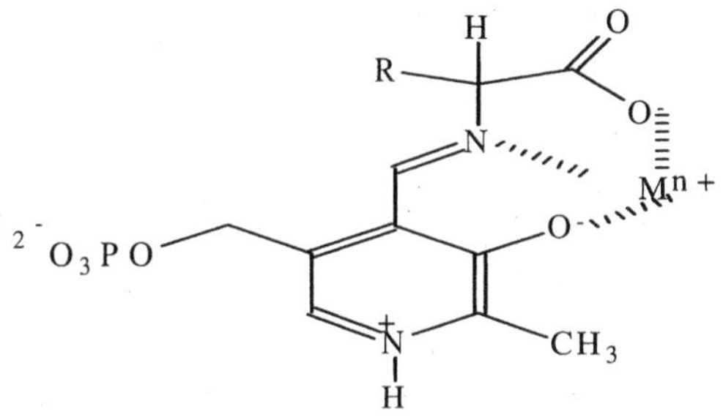

FIGURE 4.31 Schiff base of an amine with pyridoxal phosphate.

enzyme both enhances the rate of the reaction and also gives direction to the reaction.

Pyridoxal phosphate catalyses very different reactions, such as

1. Transamination
2. Elimination
3. Decarboxylation
4. Reverse condensation
5. Racemization

In the schiff base-[I] the removal of α-hydrogen gives a key intermediate [II] that may react in several different ways (Fig 4.32).

1. Racemization: Addition of proton back to the amino acid will lead to racemization (Fig 4.33).
2. Transamination: The pyridoxamine formed by the addition of proton to carbonyl carbon of pyridoxal and hydrolysis may react with a α-keto acid to yield pyridoxal and amino acid (Fig 4.34).

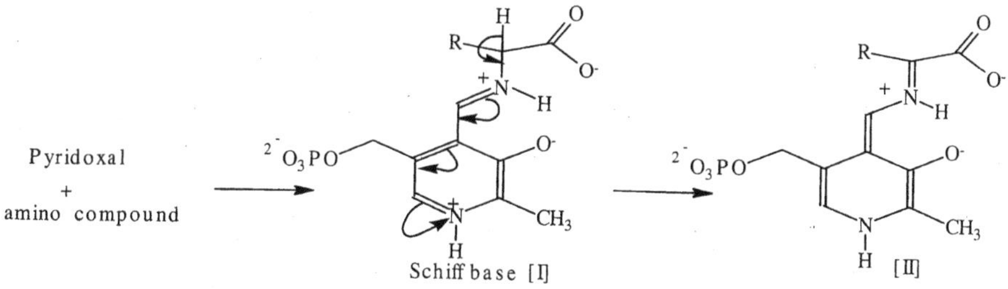

FIGURE 4.32 Formation of key intermediate [II].

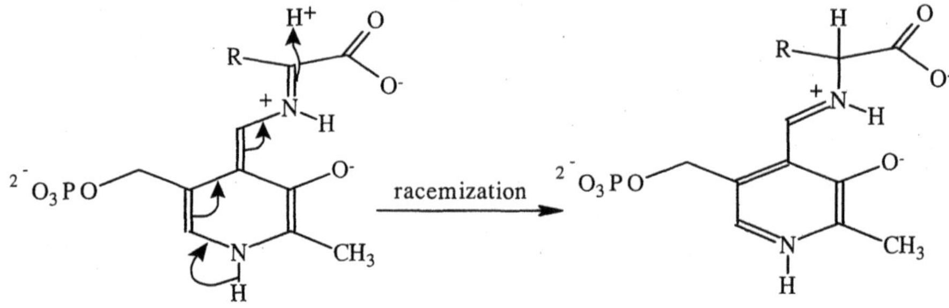

FIGURE 4.33 Racemization reaction.

3. β-Decarboxylation: When the amino compound reacting with pyridoxal to form the schiff base has a β-carboxyl group, facile decarboxylation is possible. For example, aspartic acid can be converted to alanine by pyridoxal (Fig 4.35).

4. Side-chain cleavage: When the amino compound reacting with pyridoxal to form the schiff base has a good leaving group (−SH, −OH, indole, etc.) on the β-carbon, the leaving group may get eliminated (Fig 4.36)

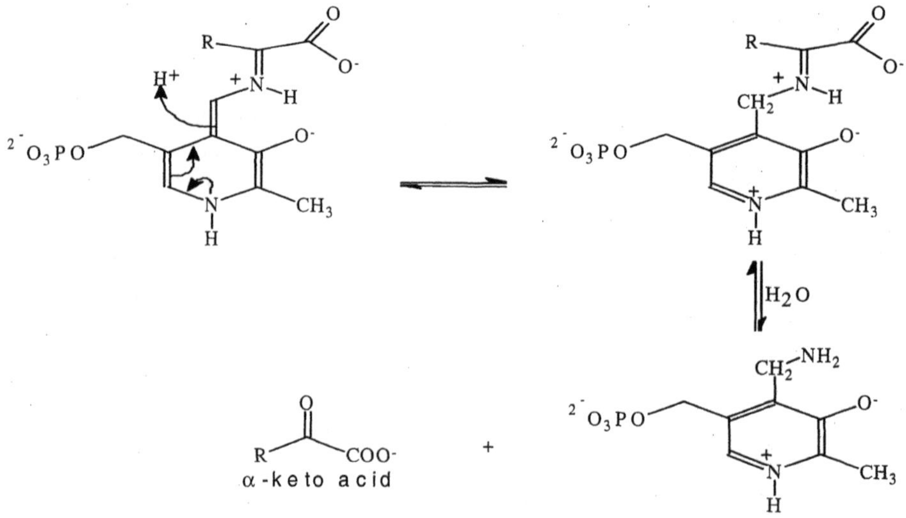

α-keto acid

FIGURE 4.34 Transamination reaction.

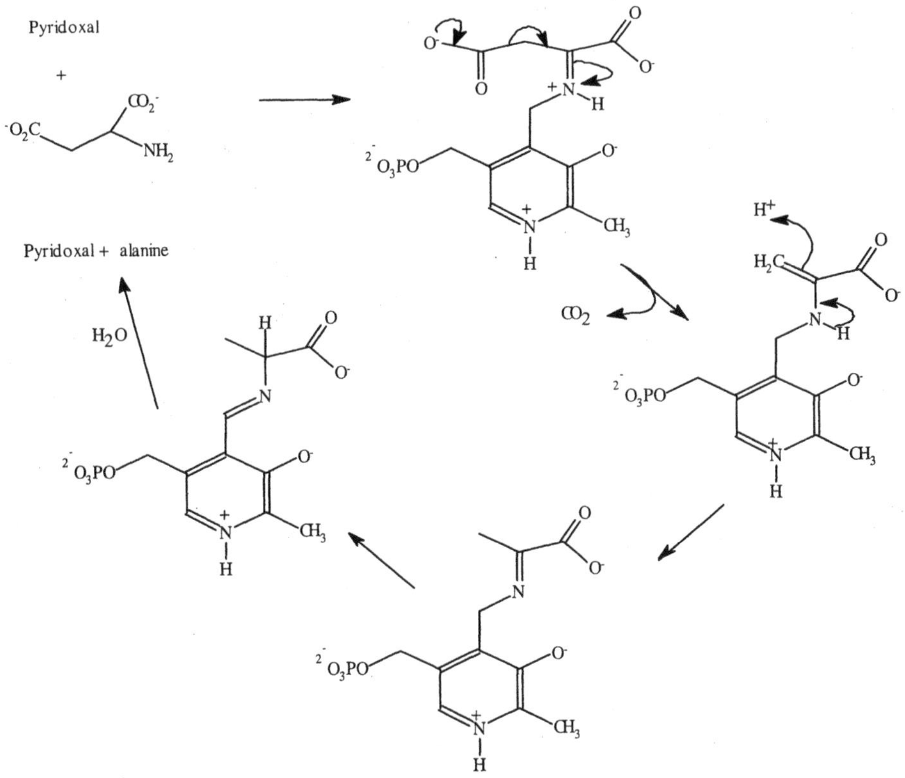

FIGURE 4.35 β-Decarboxylation reaction.

5. α-Decarboxylation: When the amino compound reacting with pyridoxal to form the schiff base has a α-carbonyl group, decarboxylation might occur because of the electrophilic character of the pyridine ring. Further modification will lead to amine formation (Fig 4.37).

4.4.2 Transaminases in Synthesis

Of all the pyridoxal dependent enzymes, transaminases are the most important. The pivotal role played by amines in stereoselective organic synthesis makes the transaminases, which convert a prochiral carbonyl group to optically active pure amine, an important addition to the repertoire of organic reagents. L-Tyrosine, L-DOPA, L-tryptophan, [^{15}N]- or [^{13}C]-labeled L-tyrosine, and similar amino acid derivatives are prepared using transaminases,

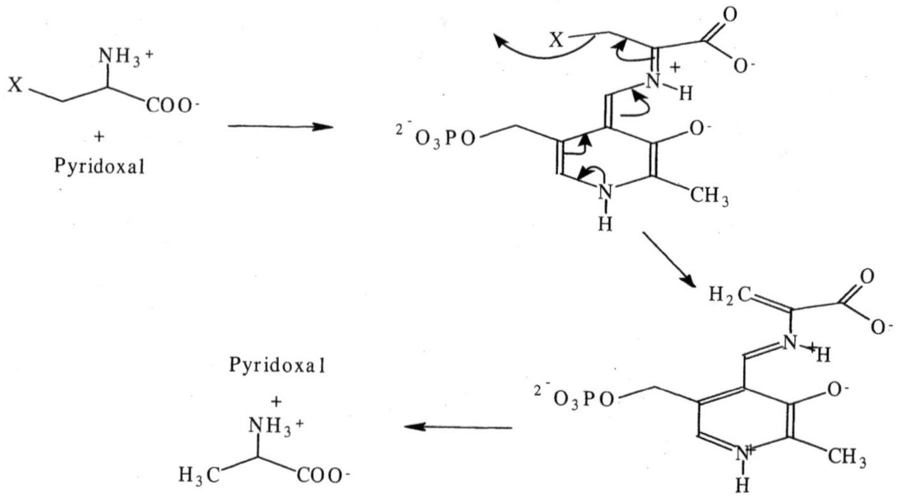

FIGURE 4.36 Side-chain cleavage reaction.

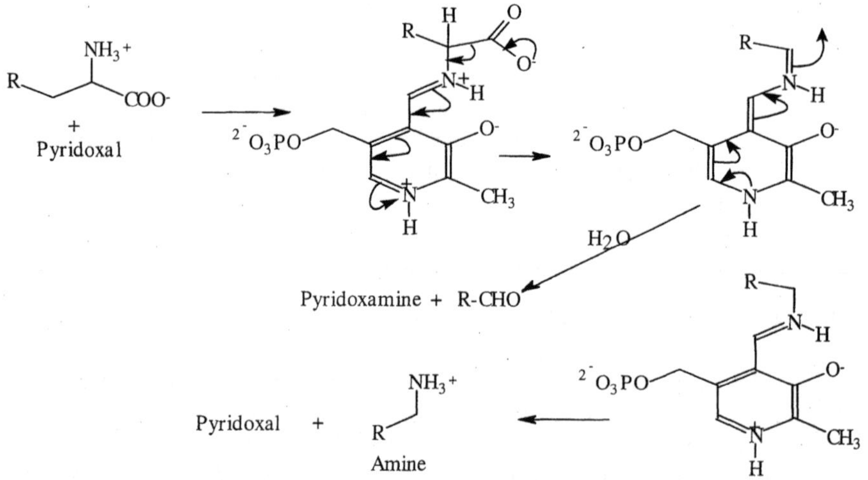

FIGURE 4.37 α-Decarboxylation reaction.

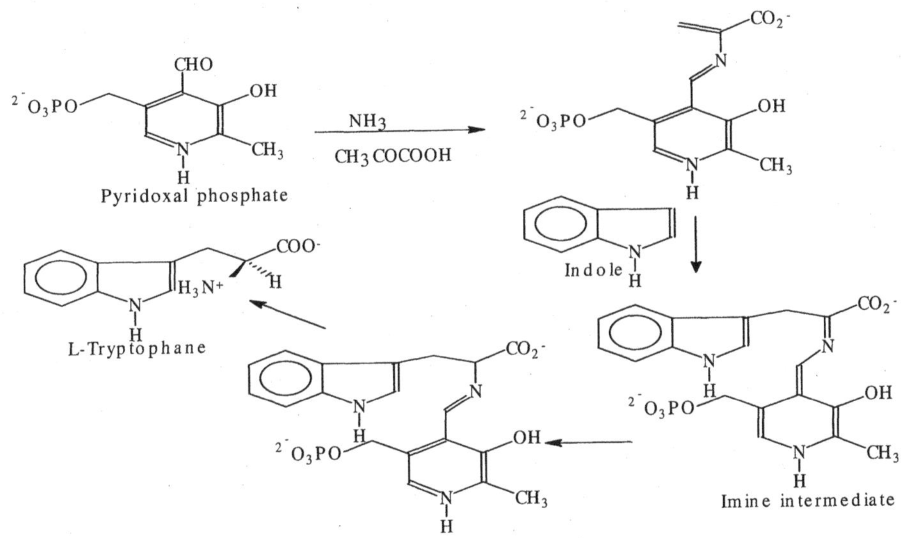

FIGURE 4.38 Transaminase in organic synthesis.

starting from unassuming achiral precursors. For example, from pyruvic acid, ammonia, and indole, L-tryptophane was synthesized (Fig 4.38).

The imine intermediate cleaves hydrolytically under the enzyme's influence such that the addition of proton at the α carbon of the target amino acid occurs with Re-face specificity. Similarly, this approach has been extended to the synthesis of unnatural amino acids, amino alcohols, and amines. Shin and Kim (9), used a ω-transaminase from *Vibrio fluvialis* JS17 to convert acetophenone to (S)-α-methyl benzyl amine, using L-alanine as the amino donor. In order to overcome the product inhibition by pyruvate, in situ

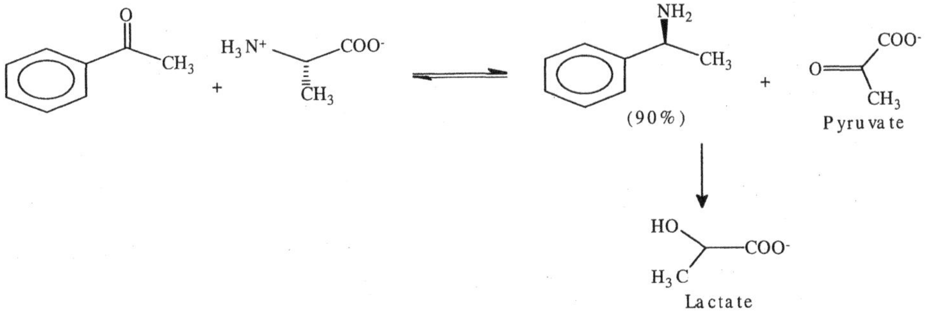

FIGURE 4.39 Transaminase-catalyzed synthesis of unnatural amines.

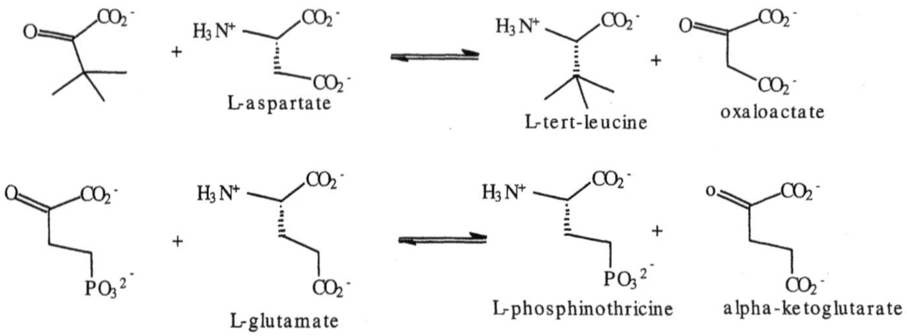

FIGURE 4.40 Bacterial transaminases in the synthesis of amino acids.

removal of pyruvate by reduction to lactate was carried out. A 90% yield of (S)-α-methyl benzylamine with an optical purity of >99% was achieved within 24 h (Fig 4.39).

Then and co-workers (10) have described the use of bacterial transaminases to produce L-tert-leucine and L-phosphinothricine using aspartate and glutamate, respectively, as the amino donors. Molar excesses of amino donors with *E. Coli* strains over expressing an *E. Coli* transaminase were used (Fig. 4.40).

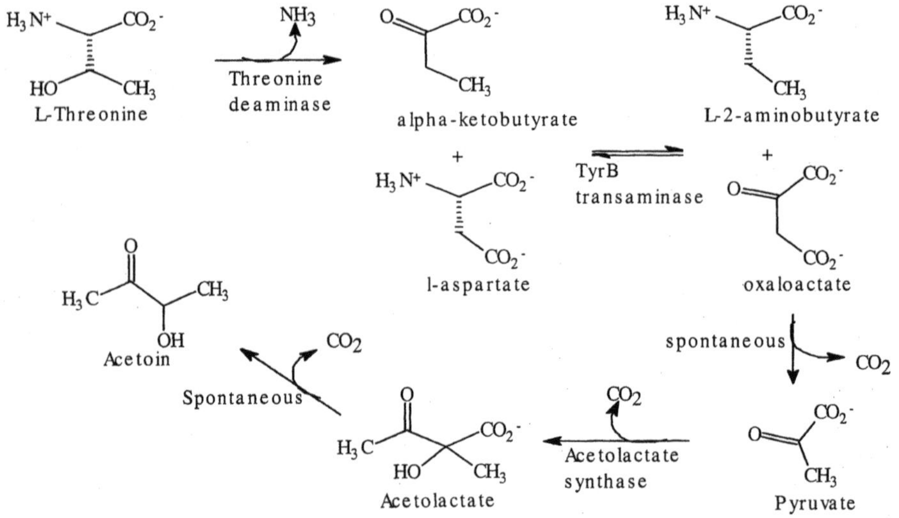

FIGURE 4.41 Synthesis of L-2-aminobutyrate by the removal of keto acid byproduct.

One of the limitations in the use of transaminases in the synthesis of chiral amines is that the enzymes do not accept amines of synthetic interest; also the equilibrium constants are near unity. The use of multistep, irreversible pathways to remove the undesired α-keto acid by-product from the reaction is one methodology of choice to ensure forward reaction. In an interesting report, on the synthesis of L-2-aminobutyrate the above methodology is effectively used. Three *E. Coli* strains, each overexpressing one enzyme, were mixed in appropriate amounts to carry the whole multistep transformation. L-Threonine was deaminated to yield α-ketobutyrate by *E. Coli* overexpressing threonine deaminase. *Escherichia Coli* overexpressing Tyr B transaminase converted α-ketobutyrate and L-aspartate (added externally) to L-2-aminobutyrate and oxaloacetate. Oxaloacetate being an inhibitor had to be in situ removed. This was accomplished by *E. Coli* strains overexpressing acetolactate synthase, which transformed oxaloacetate formed into volatile acetoin (Fig 4.41).

4.5 HYDROLASES

The natural ability of enzymes to discriminate even subtle distinctions among chemical substances to catalyze reactions in a chemo- and stereoselective manner is even more pronounced in this group of enzymes. These catalyze the hydrolysis of esters, amides, phosphate esters, glycosidic bonds, epoxides, lactones, lactams, and nitriles. These enzymes are ubiquitous in both prokaryotic and eukaryotic organisms. In fact, hydrolysis is one of the common pathways for the metabolism of the endogenous and exogenous (inclusive of drugs) substances in an organism. A classic example of ester hydrolysis in mammals is the metabolic conversion of aspirin (I) to salicylic acid (II) (Fig 4.42).

The ability of bacteria to develop enzymes to cleave the *β*-lactam ring (the pharmacophore) of penicillins makes them become resistant to penicillins. The hydrolysis of the amide bond of penicillin-G to yield 6-aminopenicillnic acid (6-APA) is economically valuable since this acid is the principal intermediate in the preparation and manufacture of semisynthetic penicillins (Fig 4.43).

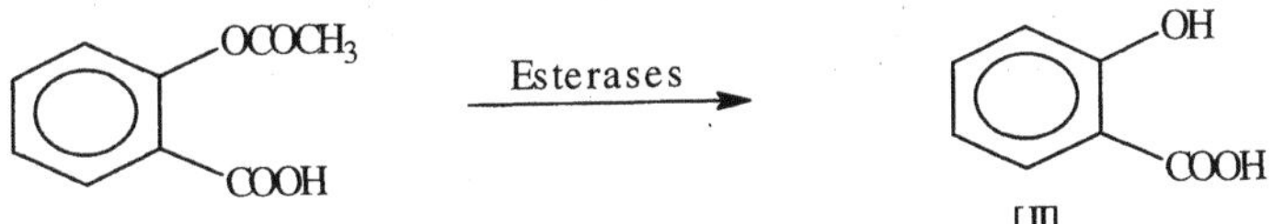

FIGURE 4.42 Asprin metabolism.

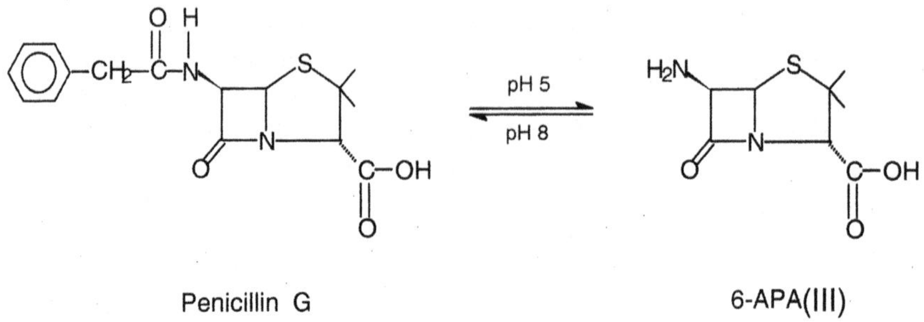

Penicillin G   6-APA(III)

FIGURE 4.43   Synthesis of APA.

The greatest single factor that has made this group of enzymes unique is their ability to bring about highly chemo-, regio-, and stereoselective transformation inspite of having low substrate specificity. Also, these enzymes do not require any cofactor and are easily available. In a sense, these are Nature's primary "gift" to organic chemists.

The simple esterases and lipases are concerned with the hydrolysis of uncharged substrates, and the main factors influencing the specificity are the length and shapes of the hydrophobic groups on either side of the ester/amide linkage.

This group of enzymes consists of

1. Esterases: These catalyze the hydrolysis of esters and lactones (lactonases).
2. Lipases: These are also known as triacylglycerol hydrolases. These also bring about the hydrolysis of ester, but the chain length of the acyl component is usually large.
3. Peptidases/Amidases: These are proteolytic enzymes. These catalyze the hydrolysis of the specific peptide bond in a peptide (endo peptidases) or cleave the peptide bond from a terminal (exopeptidases). These also catalyze the hydrolysis of normal amide/nitrile bonds. The peptidases also include

    a. Serine proteases, which have a serine and histidine in their active center
    b. Thiol proteases, with cysteine in the active site and
    c. Acid protease, which contain a bound metal which is essential for the catalytic activity

4. Glycosidases: Includes enzymes that catalyze the hydrolysis of glycosidic bond of simple glycosides and polysaccharides. This

group also includes enzymes that catalyze the hydrolysis/formation of N-glycosidic and S-glycosidic bonds.

5. Phosphatases: These enzymes can be regrouped as phosphate esterases and phosphodiesterases. Phosphate estreases sometimes also known as phosphate mono esterases, catalyze the hydrolysis of phosphate esters—these have low substrate specificity. Phosphodiesterases, also known as nucleases, catalyze the hydrolysis of specific phosphodiester bonds in polynucleotides. Depending on whether they cleave a phophoester bond within a polynuclotide chain or a terminal phosphoester bond, they are classified as endo on exo nucleases, respectively.

6. Pyrophosphatases: This includes enzymes that catalyze the hydrolysis of acid anhydrides and phosphate anhydrides.

Hydrolases are generally used in organic synthesis for

1. Cleavage of racemic esters to afford an optically active acid
2. Removal of acyl group from racemic acylate to produce optically active alcohol
3. Bringing about transesterification
4. Bringing about regiospecific acylation (normally of carbohydrate substrates)

4.5.1 Mechanism of Enzymatic Hydrolysis

The serine proteases are families of closely related enzymes that contain a uniquely reactive serine residue at the active site. The family includes α-chymotrypsin, trypsin, subtilisin, elastase, thrombin, and lipase. Extensive work has resulted in a fairly clear understanding of the structure and mechanism of these enzymes. Though there may be variation in the sequence of amino acid residues among the different enzymes of this class, the sequence of amino acids at the active site is more or less preserved. Generally, the amino acid residues that form the active site of an enzyme can be divided into those forming the binding site and those forming the catalytic site. The three-dimensional structure of bacterial lipases were determined to understand the catalytic mechanism of lipase reaction. Structural characteristics include

An α/β hydrolase fold.

A "catalytic triad" consisting of a nucleophillic serine located in a highly conserved Gly-X-Ser-X-Gly (X is any other amino acid) pentapeptide.

A aspartate or glutamate residue that is hydrogen bonded to histidine, very similar to α-chymotrypsin.

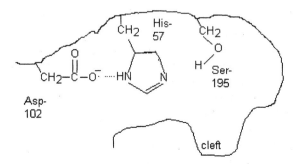

FIGURE 4.44 Sketch of active site of α-chymotyrpsin.

Four substrate binding pockets which were identified for triglycerides: an oxyanion hole and three pockets accommodating the three fatty acids of the triglycerides. The differences in the size and the hydrophillicity/hydrophobicity of these pockets determine the enantiopreference of a lipase.

A simplified sketch of the active site of α-chymotrypsin is shown in Fig 4.44. This essentially contains all the necessary details. The cleft adjacent to the catalytic site, having predominantly the nonpolar amino acid residues, is a hydrophobic pocket, which (extends into the hydrophobic interior of the enzyme) helps in binding and orienting the substrate. Hence, the nonpolar noncovalent interactions—the van der Waals and hydrophobic interactions—are the key to the affinity of the substrate to the active site in α-chymotrypsin. This holds the substrate in place and facilitates a reaction. The specificity of an enzyme is also determined by the geometry and the hydrophillic/lipophillic

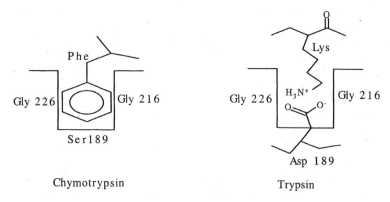

FIGURE 4.45 Specificity due to cleft geometry.

nature of this pocket. For example, the cleft in trypsin contains aspartic acid (Asp) residue, which can form an electrostatic interaction with the protonated e-amino group in lysine (Lys) or the protonated guanido group in arginine (Arg) (Fig 4.45). Similarly, the cleft in elastase has valine (Val) and theronine (Thr) residues. The catalytic site includes side groups of three amino acid residues, viz., serine (Ser), histidine (His), and aspartic acid (Asp). These three act together as a "catalytic triad" forming a charge-relay system via hydrogen bonds. The "buried" aspartic acid (Asp-102) group polarizes the imidazole ring of the histidine (His-57), which in turn picks up a proton from the serine (Ser-195). Thus, generating a highly reactive nucleophillic serine residue "CH$_2$-O⁻." This process gets triggered only after the binding of the substrate in the active site. Because the partial positive character of carbonyl carbon induces the polarization of the electron density of the alcoholic oxygen of serine (Ser-195) residue, the alcoholic proton is made acidic enough to be picked up by the nitrogen of the histidine (His-57) residue.

A nucelophillic attack of this nucleophillic side group on the carbonyl carbon of the amide/ester, suitably positioned by the binding sites, leads to the tetrahedral intermediate, the oxyanion. The −NH$_2$ protons of glycine (Gly-193) residue again stabilize the charge on the oxygen atom in the oxyanion intermediate. The rearrangment of this oxyanion leads to the acyl enzyme intermediate. Histidine-57 once again deprotonates the new nucleophile (normally water) with concomitant release of proton to aspartic acid (Asp-102). The activated nucleophile thus formed concertedly attacks the carbonyl carbon of the acyl enzyme, again generating an oxyanion intermediate, which

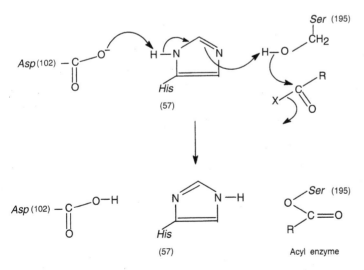

FIGURE 4.46 Catalytic cycle of chymotrypsin.

collapses to yield the acid or ester (depending on the new nucleophile being either water or another alcohol molecule) and the serine-OH. Note that, at the end of one transformation, the catalytic triad goes through one cycle of protonation/deprotonation to get back to their normal condition. The catalytic cycle can start again. Schmatic description of this sequence is shown in Fig 4.46.

From the studies on the hydrolysis of *p*-nitro phenyl acetate with α-chymotrypsin, it was found that the acylation step is usually rapid; hence it is termed as *burst phase*. The deacylation step where the enzyme and the acid are released is slow and is called the *steady-state phase*. Thus, the mechanism of serine proteases can be divided into four steps as shown in Scheme 4.3.

The active site of herpes virus proteases consist of a novel Ser-His-His catalytic triad instead of the classical Ser-His-Asp triad. Though a simplified overview is presented, there has been much debate over the finer details of the mechanism. In some cases, it has been indicated that the synchronization of ester hydrolysis and substrate release may take place. Be that as it may, the overall electronic and the steric aspects are the same. *Note the similarity of the reaction with the classical general base-catalyzed hydrolysis of esters or amides.* The fundamental difference is in the chiral

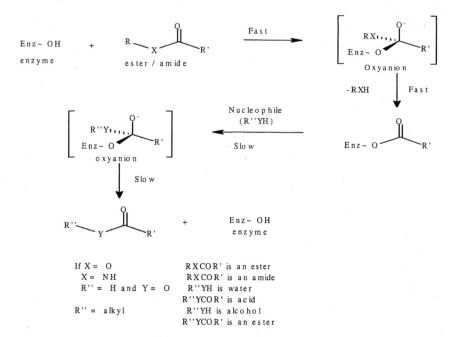

SCHEME 4.3 Mechanism of serine proteases.

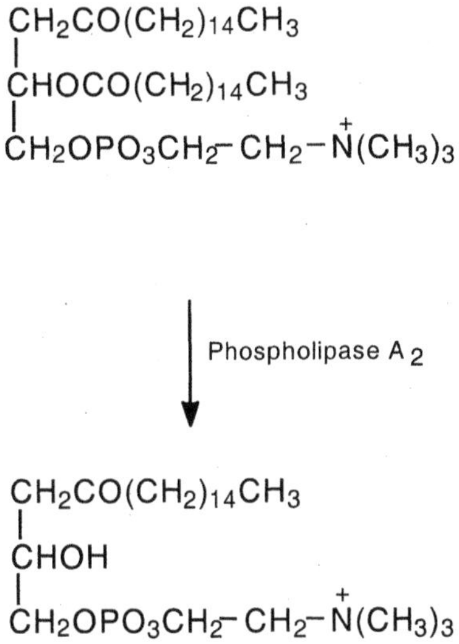

$CH_2CO(CH_2)_{14}CH_3$
|
$CHOCO(CH_2)_{14}CH_3$
|
$CH_2OPO_3CH_2^-CH_2-\overset{+}{N}(CH_3)_3$

Phospholipase A$_2$

$CH_2CO(CH_2)_{14}CH_3$
|
$CHOH$
|
$CH_2OPO_3CH_2^-CH_2-\overset{+}{N}(CH_3)_3$

FIGURE 4.47 Regiospecific hydrolysis.

environment that the enzyme provides. This imparts the enzyme with an ability to discriminate two chemically similar groups and perform highly regio- and enantioselective transformations. For example, regiospecific hydrolysis of polyesters is achievable, as illustrated by the phospholipase A$_2$-catalyzed hydrolysis of the phospholipid. Cleavage occurs at the secondary alcohol center (Fig 4.47).

Hydrolysis of prochiral diesters with either α-chymotrypsin or pig liver esterase (PLE) yields optically active monoester. When C-3-substituted

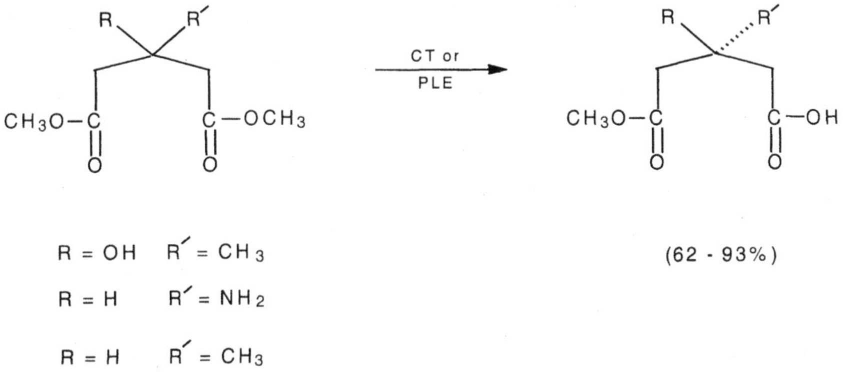

R = OH   R′ = CH$_3$ (62 - 93%)

R = H   R′ = NH$_2$

R = H   R′ = CH$_3$

FIGURE 4.48 Hydrolysis of prochiral diesters.

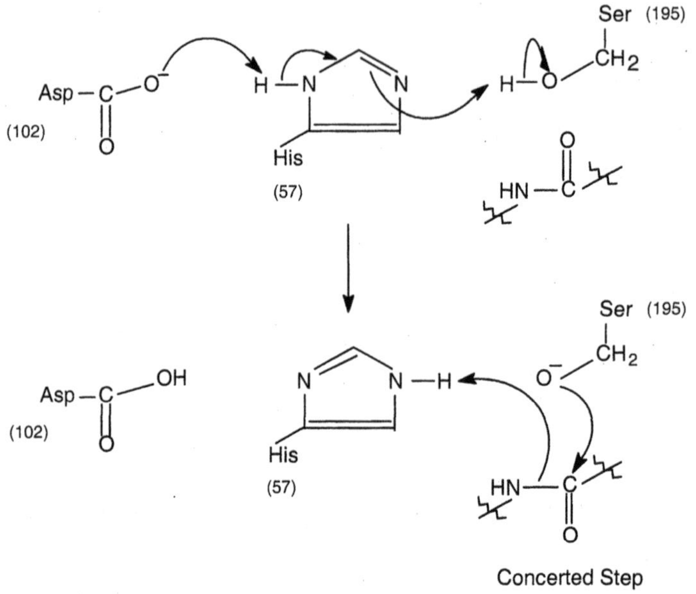

FIGURE 4.49 Amide hydrolysis mechanism.

glutarate diesters were hydrolyzed with α-chymotrypsin or PLE, (S) half-ester products were obtained (Fig 4.48).

4.5.2 Mechanism of Enzymatic Hydrolysis of Amides

The mechanism of the enzymatic hydrolysis of anilides and esters proceed through a discrete tetrahedral intermediate. But in the enzymatic hydrolysis of amides, the proton abstracted from the hydroxyl groups of the serine residue (Ser) by the imidazolyl group of the histidine residue (His) is donated to the nitrogen atom of the leaving group of the amide before the bond between the carbonyl carbon atom of the amide and the attacking serine oxygen atom is completed. Thus, it is believed that amide hydrolysis proceeds through a S_N2-like reaction, as shown in Fig 4.49.

Amide hydrolysis is used extensively in the preparation of L-amino acids. Amino acids and their acyl derivatives are obtained chemically as racemic mixtures of their D and L enantiomers. These racemic mixtures are

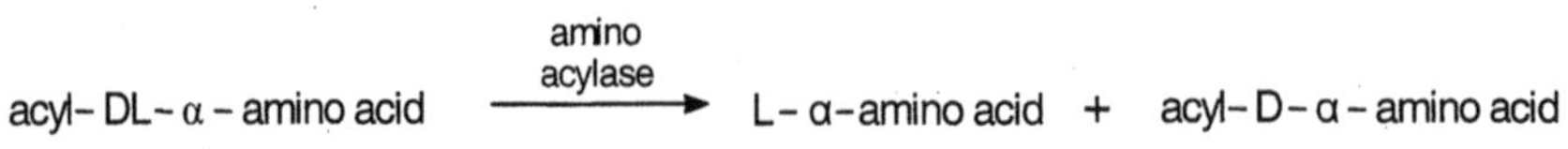

SCHEME 4.4 Resolution of racemic mixture of amino acids.

resolved using amino acylase from *Asperigillus oryzae*, which specifically hydrolyzes L enantiomers, as shown in Scheme 4.4.

4.5.3 Hydrolytic Reactions in Organic Synthesis

As with all reactions, it is the thermodynamically preferred products that accumulate in enzyme-catalyzed processes. However, by rapid removal of unstable product by flow systems with immobilized enzymes or using organic solvents or precipitation, thermodynamically less preferred products could be synthesized. For example, thermolysin-catalyzed coupling of DL-phenylalanine methyl ester [I] with N-carbobenzoxy-L-aspartic acid (II) gives excellent yields of the kinetic product, the aspartame precursor [III]. The unreactive D-phenyl alanine methyl ester [IV] is recovered as shown in Fig. 4.50. The reaction is driven in the peptide-bond-forming direction by using an immobilized enzyme system and by exploiting the fact that the product formed [III] is insoluble, and therefore precipitates. The specificity of the enzyme precludes any need to selectively protect the side chain carboxyl group of [II].

The specificity of enzymes can be effectively used to bring about chemoselective hydrolysis of esters. This is used extensively in protecting group chemistry. For example, acetate and dihydrocinnamate are hydrolyzed at almost identical rates with hydroxide ion, wheres α-chymotrypsin has been shown to be completely specific in removing dihydrocinnamate protection

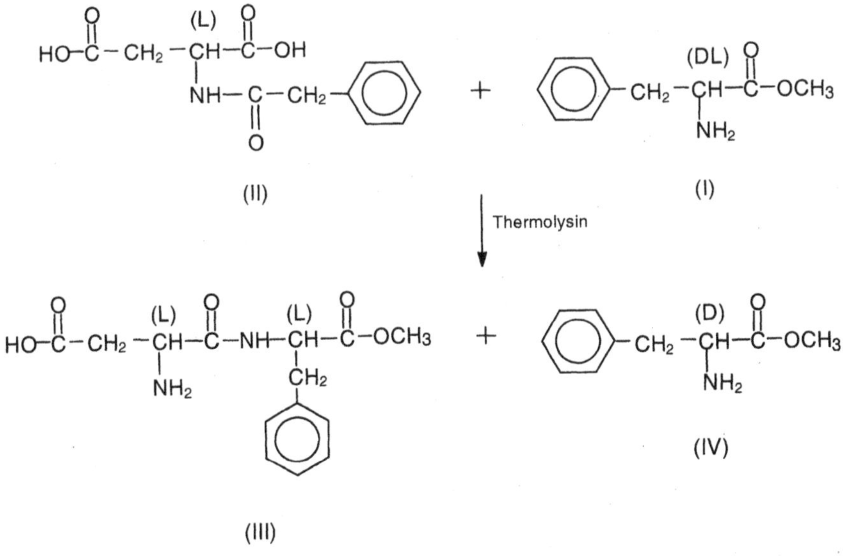

FIGURE 4.50 Isolation of kinetic product by immobilized enzyme.

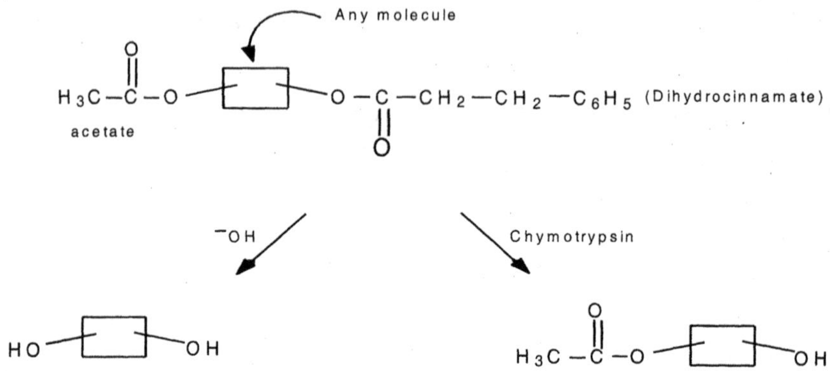

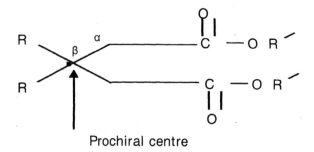

SCHEME 4.5 Selective deprotection by α-chymotrypsin.

from simple mixed acetate-dihydrocinnamate diesters. This is schematically depicted in Scheme 4.5.

As already mentioned, due to the chiral environment of the active site, a high degree of stereoselectivity can be achieved in the hydrolysis of prochiral diesters (such as the diesters of glutaric acid) to half-esters. The following rules may be useful in predicting the selectivity:

1. In order to obtain high stereoselectivity, the ester group must not be further away from the induced prochiral center than the β position (Fig. 4.51).
2. Diesters bearing a higher substitution pattern are hydrolyzed with higher stereoselectivity than their less substituted counter parts (Fig. 4.52).
3. Cyclic diesters show higher stereoselectivities than the corresponding acyclic analogues (Fig. 4.53).

FIGURE 4.51 Stereoselective hydrolysis distance parameter.

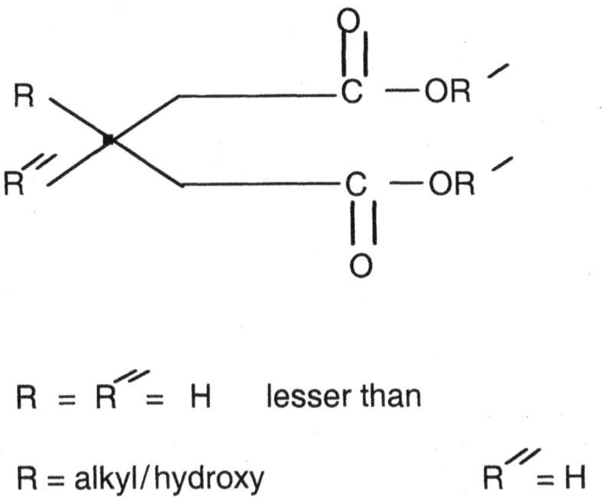

R = R″ = H   lesser than

R = alkyl/hydroxy   R″ = H

FIGURE 4.52 Stereoselective hydrolysis substitution parameter.

4. Substituents of different polarity and different size show opposite effects on selectivity of the enzymic hydrolysis (Fig. 4.54).

5. The stereoselectivities of the same enzymes from different sources (microorganisms) varies. For example, the meso diester can be selectively hydrolyzed with *Acinetobater lowfii* to give (R) acid, whereas both *Corynebacterium equi* and *Arthrobacter sp.* exhibit opposite enantioselectivity and produce the (S) acid (Fig. 4.55).

6. The structural features of the majority of the substrates that can be transformed by esterases, lipases, and proteases are the following:

    a. Esters of the form $R_2R_1CHCOOR$, where R_1, R_2, and R may be either alkyl or aryl. Both the acyclic and the cyclic esters are hydrolyzed.

    b. Acyl alcohols of the general form $R_2R_1CHOCOR$, where R_1, R_2, and R may be either alkyl or aryl. Both the acyclic and the cyclic esters are hydrolyzed.

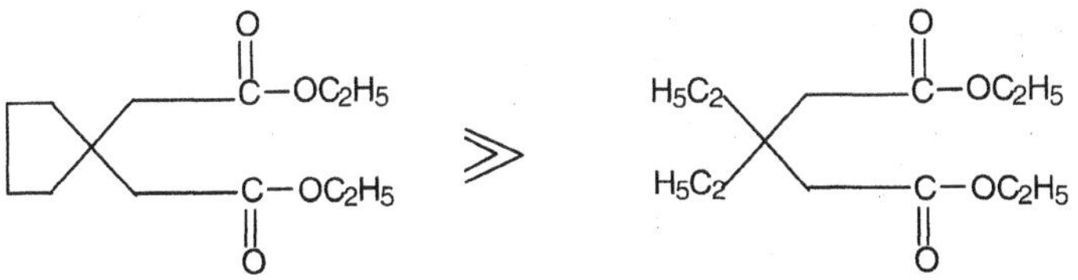

FIGURE 4.53 Stereoselective hydrolysis steric parameter.

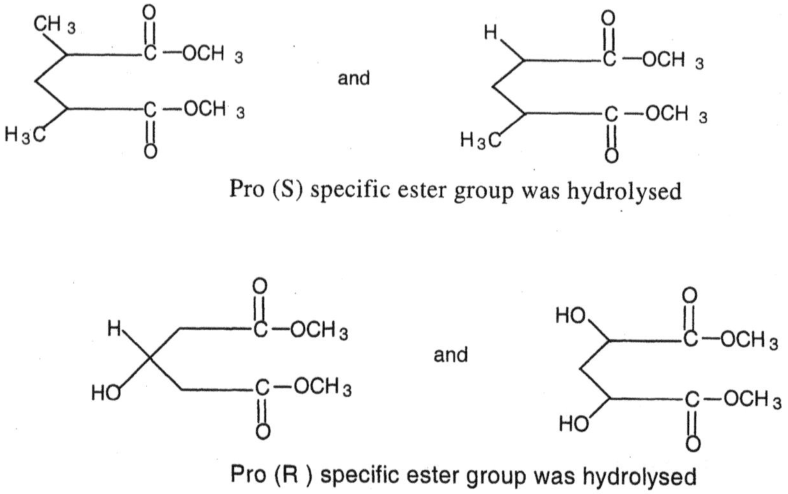

Pro (S) specific ester group was hydrolysed

Pro (R) specific ester group was hydrolysed

FIGURE 4.54 Stereoselective hydrolysis stereoelectronic parameter.

c. Prochiral diesters of the general form R_2R_1CH [$(CH_2)_n$ COOR]$_2$, where R_1, R_2 and R are usually alkyl groups and n should not be more than 2. Both the acyclic and the cyclic diesters are stereoselectively hydrolyzed to optically active half-esters.

d. Prochiral diacyl diols of the general form R_2R_1CH[$(CH_2)_n$ OCOR]$_2$, where R_1, R_2 and R are usually alkyl groups and n

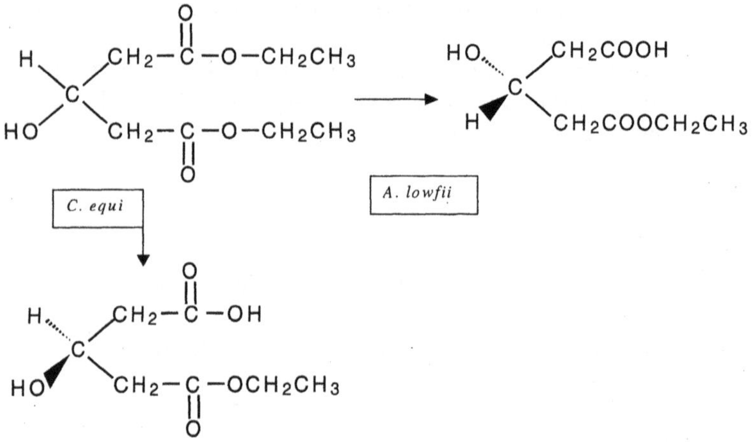

FIGURE 4.55 Stereoselective hydrolysis: variation by changing the microorganisms.

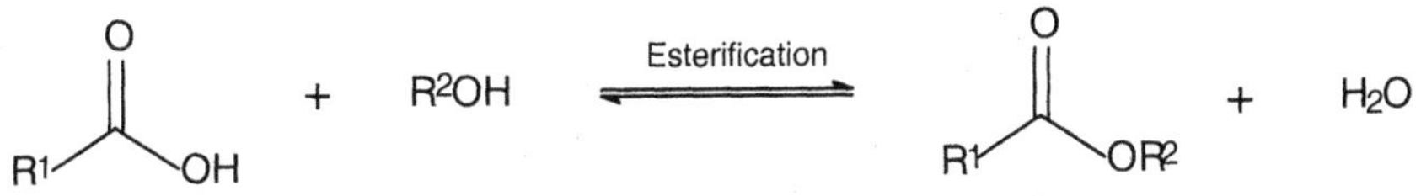

SCHEME 4.6 Ester formation from carboxylic acid and alcohol.

should not be more than 2. Both the acyclic and the cyclic diesters are stereoselectively hydrolyzed to optically active half-esters.

e. Meso diesters of the general form $R_1 CH (COOR) CH (COOR) R_2$, where R_1, R_2 and R are usually alkyl groups. Both the acyclic and the cyclic diesters are stereoselectively hydrolyzed to their corresponding half-esters.

f. Meso diacyl diols of the general form $R_1 CH (OCOR) CH (OCOR) R_2$, where R_1, R_2, and R are usually alkyl groups. Both the acyclic and the cyclic diesters are stereoselectively hydrolyzed to their corresponding half-esters.

4.5.4 Hydrolases in Synthetic Mode

Apart from the usefulness of estrolytic and lipolytic enzymes in the hydrolytic mode, these enzymes can also be used in synthetic mode, i.e., in the synthesis of esters or lactones from carboxylic acids/esters and alcohol.

If the ester formation is from a carboxylic acid and alcohol as shown in Scheme 4.6, water is released, and this leads to loss of homogeneity in the reaction phase.

If the ester is formed from an ester and alcohol as shown in Scheme 4.7, no water is formed, and hence acyl transfer or transesterification is preformed.

Klibanov (11) had shown that the conformation of PPL and other hydrolases is virtually same in low-water media, i.e., organic solvents, as in aqueous media. Because of this, in order to maintain the necessary residual amount of water, which is associated with the surface of the enzyme—the

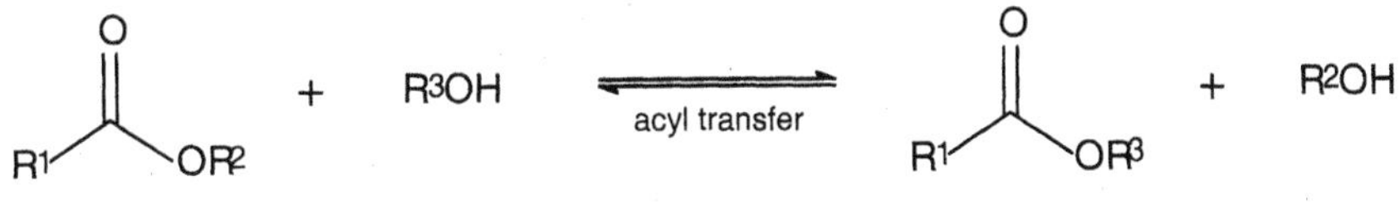

SCHEME 4.7 Ester formation from ester and alcohol.

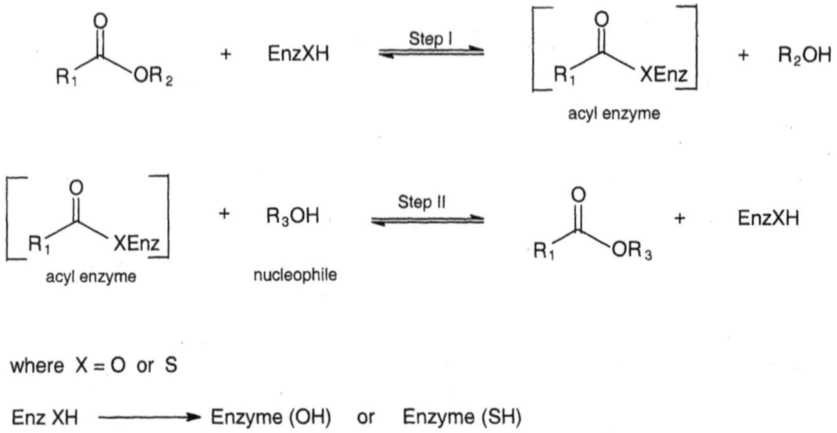

where X = O or S

Enz XH ──────▶ Enzyme (OH) or Enzyme (SH)

FIGURE 4.56 Summary of the mechanism of hydrolases.

bound water—organic solvent of choice for performing enzyme-catalyzed reactions in organic media should preferably be lipophillic. Hence, solvents having a log P value (indicator to hydrophillic/lipophillic character) greater than 3 or 4 (for example, benzene, toluene, cyclohexane, petroleum ether, dichloromethane, and chloroform) support a high degree of activity retention. Whereas more hydrophillic solvents with log P values below 2 (for example, diethyl ether, DMF, DMSO, methanol, and ethanol) are generally unsuitable for biocatalysis, as they remove the bound water necessary to maintain the three-dimensional shape (the native form) of the enzyme.

Since the uniqueness of the enzymes is that they provide a distinct chiral environment, esters, lactones, amides, and lactams can be synthesized in a chemo-, regio-, and enantioselective manner using hydrolases.

Recall the discussion on the mechanism of hydrolases: the same can be summarized as shown in Fig. 4.56. If R_3OH nucleophile happens to be water, we have a hydrolyzing reaction: on the other hand, if the R_3OH is any other nucleophile, we will get a transesterification product. Since, as we have seen earlier, hydrolases can function in organic solvent media where water is removed, we can either add a relevant alcohol to obtain the necessary acyl transfer or the solvent itself can act as the alcohol group/nucleophile.

The transfer of the acyl group is made irreversible by normally taking the enol acetates. The liberated enol tautomerizes to an aldehyde or ketone, which cannot compete for the reverse reaction. (See Scheme 4.8.)

Alternate method of ensuring irreversibility in acyl transfer is the use of anhydride as the acyl donor. In all these cases the main feature is to ensure

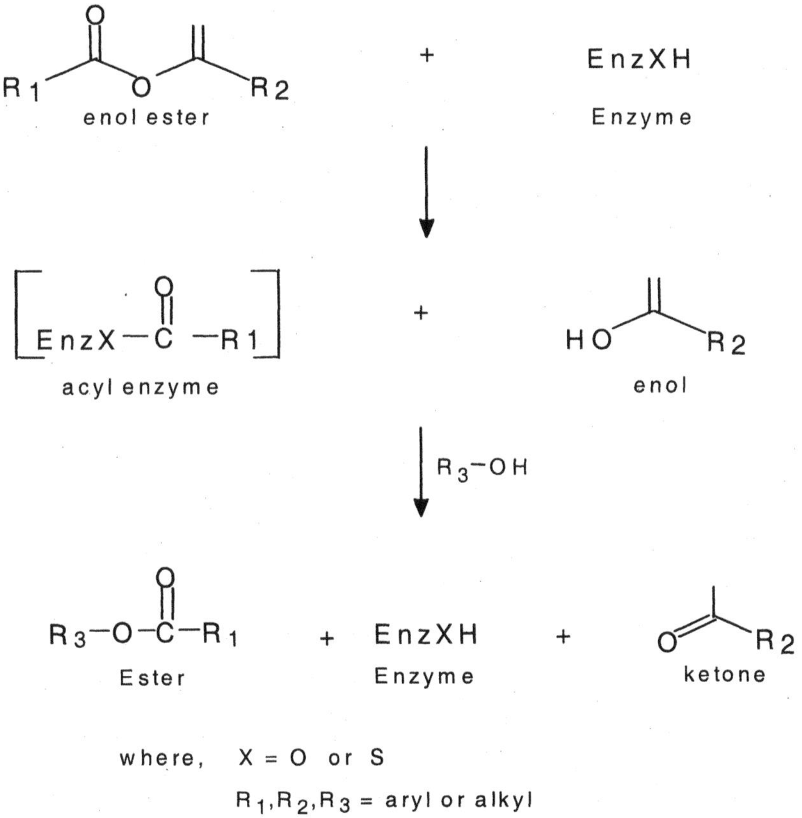

where, X = O or S

R$_1$,R$_2$,R$_3$ = aryl or alkyl

SCHEME 4.8 Irreversible transesterification.

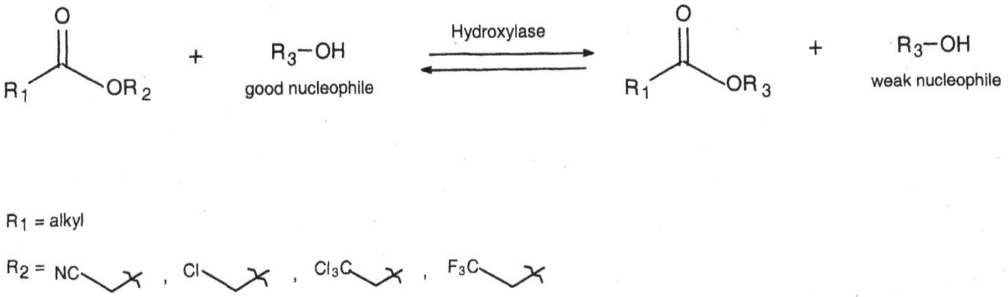

R$_1$ = alkyl

R$_2$ = NC⌒⤬ , Cl⌒⤬ , Cl$_3$C⌒⤬ , F$_3$C⌒⤬

SCHEME 4.9 Irreversible acyl transfer by the use of anhydrides.

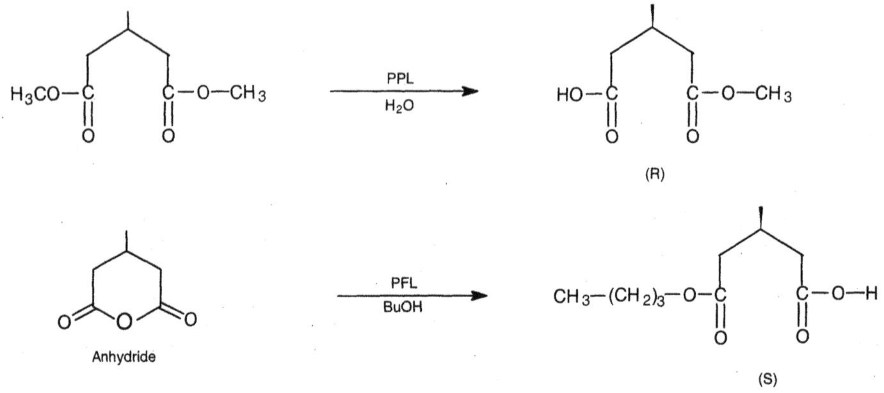

FIGURE 4.57   Synthesis of glutarate half-esters.

that the leaving group from the acyl enzyme intermediate is not as good a
nucleophile as the incoming alcohol group. This is depicted in Scheme 4.9.

Both the enantiomers of a target molecule can be obtained by simply
changing the reaction medium in conjunction with the use of appropriate
substrates (esters or acids). For example, several 3-substituted glutarates were
transformed into either (R) or (S) half-esters as shown in Fig. 4.57. Similarly,
prochiral 2-substituted 1,3-propanediols (PSL) were converted to (R) or (S)
monoesters in excellent yields. The substituent in position 2 and the lipase
used determined the enantioselectivity (of the general reaction shown in
Fig. 4.58), as shown in Table 4.3.

PSL also brought an efficient asymmetric acylation of cyclic meso *cis*-
diols to give the respective chiral monoesters (Fig. 4.59). Vinyl acetate was
found to enhance the rate of the reaction when used as the acyl donor. The
percentage enantiomeric excess (%ee) for various rings is given in Table 4.4.

Numerous primary and secondary alcohols were conveniently resolved
with moderate to excellent selectivities by *Psuedomonas* sp. lipases, using vinyl

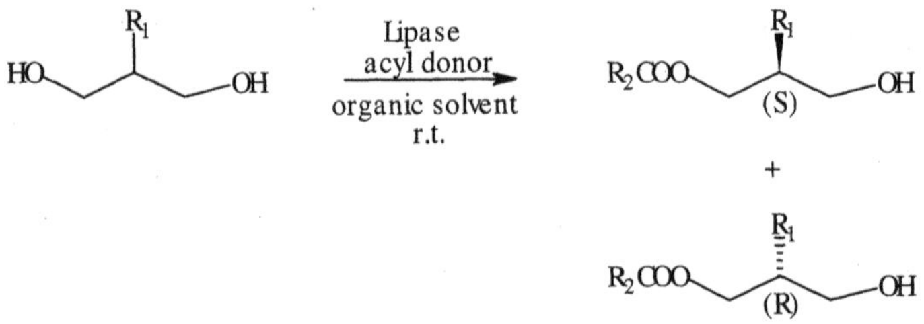

FIGURE 4.58   Synthesis of (R) and (S) mono-esters.

TABLE 4.3 Hydrolysis of Prochiral 2-Substituted 1,3-Propanediols

R_1	Lipase	Acyl donor	R_2	Solvent	Configuration
Me	PSL	Vinyl acetate	Me	CHCl$_3$	S (>98)
Bn	PSL	Vinyl acetate	Me	None	R (>94)
OBn	PSL	Vinyl acetate	n-c$_{17}$H$_{35}$	I-Pr$_2$O	S (92)
OBn	PSL	isopropenyl acetate	Me	CHCl$_3$	S (96)
Bn	PSL	Vinyl acetate	Bu	THF	R (97)

acetate or acetic anhydride as the acyl donor. Excellent selectivities were obtained when the two substituents on the substrate were different in size. For example, the primary alcohol and secondary alcohol gave high ee in the lipase catalyzed acyl-transfer reaction (Fig. 4.60).

Most of the microbial lipases prefer substrates with greater hydrophobicity; hence, this simple modification or derivatization of the substrates in case of enantioselective hydrolysis enhances the ee of the products. For example, the acetate derivative of the hydroxy acid gave a poor ee as compared to the corresponding stearate derivative (Fig. 4.61).

If the substrate is a hydroxy acid or a hydroxy ester, the remote OH-group can act as an internal nucleophile onto the primarily formed acyl-enzyme. This reaction becomes feasible when an organic solvent in the reaction medium replaces water. Using PPL, at low conversion rates (36%), hydroxy methyl esters were converted to (S)-(−)-γ-methyl butyrolactone, as shown in (Fig. 4.62).

5.5 Hydrolases for Protection–Deprotection of Hydroxy Groups in Synthesis

Regioselectivity (modification of one over several groups with similar chemical reactivity) and chemoselectivity (selection of one out of several functional groups with similar chemical reactivity but different in nature) are the two useful properties of hydrolytic enzymes. These properties have been exten-

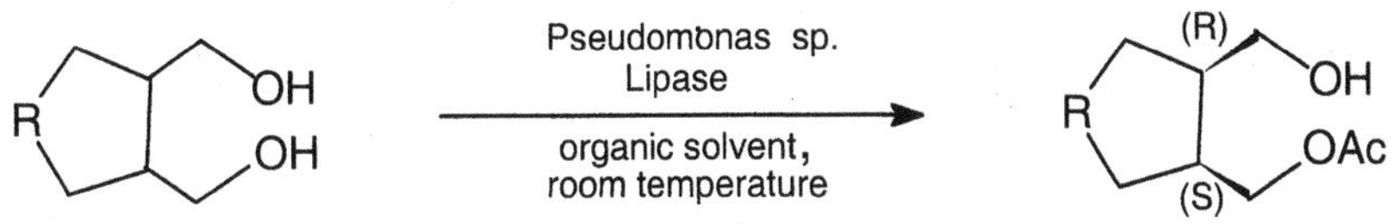

FIGURE 4.59 Asymmetric acylation of meso *cis*-diols.

TABLE 4.4 PSL-Mediated Acylation of Cyclic Meso *cis*-Diols

R	Acyl donor	Solvent	Time	ee (%)
CH_2	EtoAc	None	140	90
CH_2	Vinyl acetate	None	6	>95
CH_2	Vinyl acetate	*t*-BuOMe	7	>95
$(CH_2)_2$	Vinyl acetate	None	5	88
$(CH_2)_3$	Vinyl acetate	*t*-BuOMe	48	>95
$CH{=\!=}CH$	Vinyl acetate	*t*-BuOMe	96	88

sively exploited as a protective step in the modification of poly-hydroxylated and poly-functionalized compound such as carbohydrates, steriods, alkaloids, terperiods, and others. Also, the regioselectivity of these enzymes can be exploited in poly-functionalized compounds because the modification of only one of the identical functional groups in a molecule makes it possible to manipulate a specific position without the need to use protective groups. The regioselective acetylation of (a) varodiol and (b) 12-epi-varodiol with lipases yielded only the 3-acetyl derivative (Fig. 4.63). The results of the action of various lipases are given in Table 4.5.

The selective protection and deprotection of carbohydrates can be achieved by various chemical techniques, but this strategy often involves cumbersome multistep sequences. The possibility of using enzymatic methods to effect regioselective acylation and deacylation is an attractive alternative proposition. Thus, while it is difficult to discriminate by chemical methods between primary and secondary hydroxyl groups in acylation reactions involving carbohydrates, selective acylation of either the primary or the secondary hydroxyl groups is possible using enzymes.

However, the intrinsic polarity of most of these molecules greatly narrows the choice of the solvent, quite often limited to pyridine, DMF, or

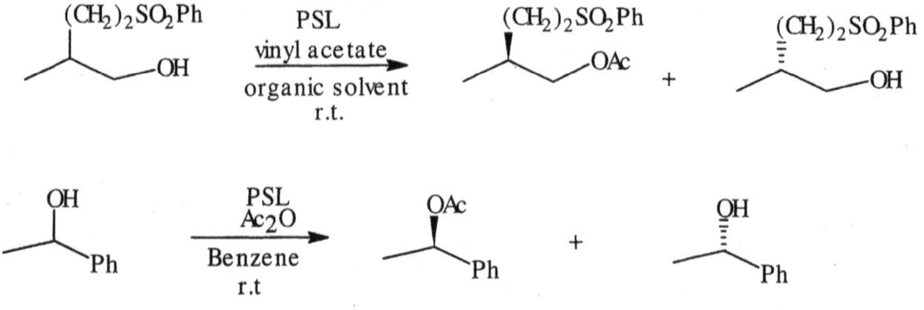

FIGURE 4.60 Resolution of alcohols with lipases.

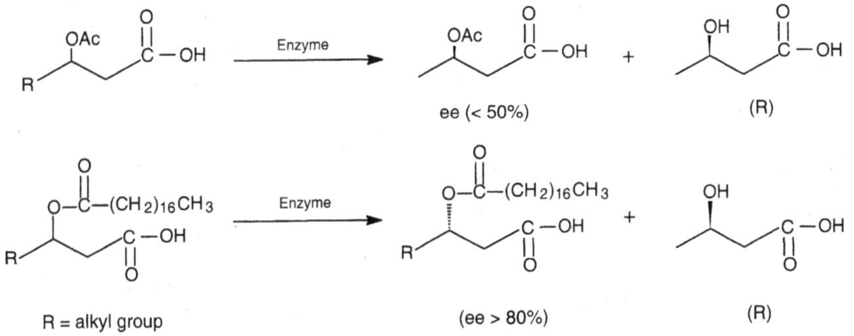

FIGURE 4.61 Hydrophobic substituent enhancing the ee of hydrolysis.

dimethyl acetamide. As a consequence, a minor number of enzymes remain active in such a highly polar environment, and some limitations arise from the nature of the acylating agent to be used. 2-Haloethyl esters are the best acylating agents under these conditions.

PPL catalyzed the regioselective esterification of the primary hydroxyl group of monosaccharides in pyridine with 2,2,2-trichloroethyl ester of aliphatic acid as acylating agent (Fig. 4.64). The results are summarized in Table 4.6.

Most of the enzymes are unstable in polar solvents such as pyridine, DMF, etc, which are needed for dissolving the polar sugars. To circumvent this problem, solvent mixtures were used. For instance, a 2:1 mixture or benzene and pyridine allowed lipase from *Candida cylindracea* to selectively acylate the primary hydroxyl of D-mannose, N-acetyl mannosamine, and methyl-β-D-glucopyranoside, with >90% regioselectivity.

Subtilisin was found to retain its activity in numerous organic solvents, including DMF. Hence, it was found to bring about regiospecific acylation of a number of mono- di-, and tri-saccharides.

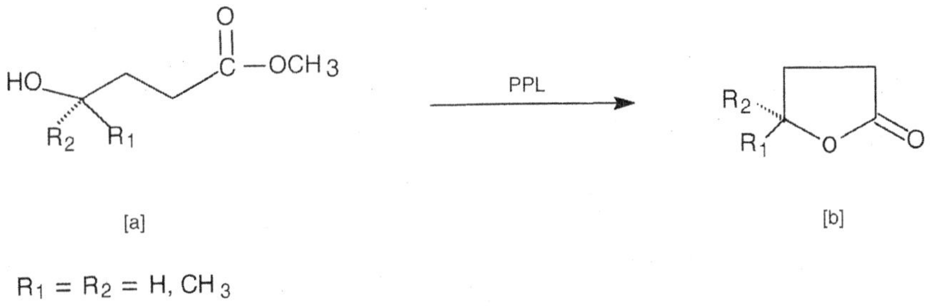

$R_1 = R_2 = H, CH_3$

FIGURE 4.62 Intramolecular esterification.

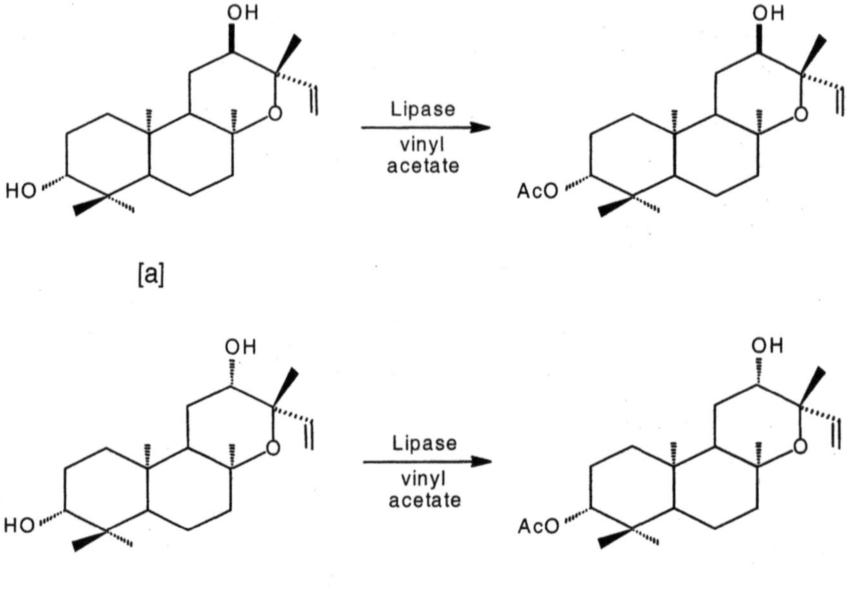

[a]

[b]

FIGURE 4.63 Regioselective acetylation.

By careful choice of lipase and solvent, it is possible to acylate selectively at the C-2 and C-3 positions of C-6 acylated (protected) sugar. Lipases from *A. niger* and *C. viscosum* exclusively acylated the C-3 hydroxy group, while the porcine pancreatic lipase (PPL) displayed a strong preference for C-2 position.

To understand the synthetic potential of secondary hydroxyl acylation, consider the following example. Klibanov (11) modified the C-6 hydroxyl group of α-D-glucopyranose with triphenyl methyl chloride. The resultant 6-0-trityl glucopyranose was quantitatively acylated in the C-3 position by *C*.

TABLE 4.5 Enzyme-Catalyzed Acetylation of [a] and [b] with Vinyl Acetate as Acyl Donor

Substrate	Lipase	Time (days)	Conversion (%)
a	CCL	5	71
a	PPL	5	0
a	Lipozyme IV	2	65
b	CCL	2	85
b	PPL	5	0

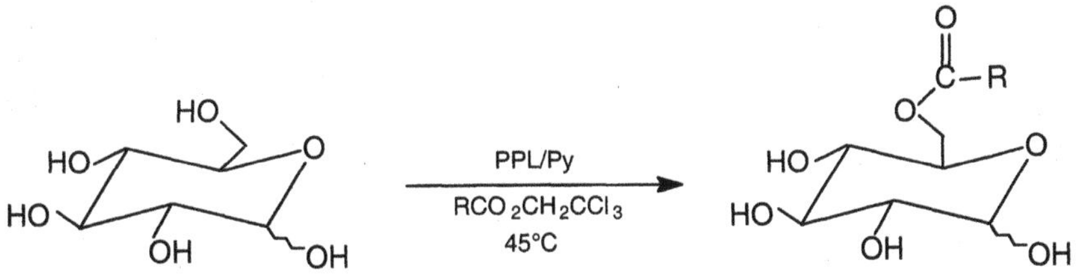

FIGURE 4.64 Acylation of sugars.

viscosum lipase. The resultant product was detritylated to yield 3-O-butyryl glucopyranose in 88% yield (Fig. 4.65).

Since all the hydroxy groups of carbohydrates can be readily acylated by chemical methods, the regiospecific, regioselective enzymatic deacylation provides access to sugars with varying degrees of acylation. For this, the pentanoyl (C-5) derivatives of sugars were found to be the best substrates. CCL was found to react quickly with a high degree of regioselectivity, deacylating the primary hydroxyl group, while lipase from *A. niger* hydrolyzed selectively the acylated secondary hydroxyl groups.

Thus, summarizing the acylation of sugars by enzymes, the primary hydroxyl group is more reactive than the secondary hydroxyl groups. In enzymatic deacylation, the acyl group at the anomeric carbon is more reactive than acyl group at the primary hydroxy center. The acyl group at the primary hydroxy center is more reactive than the acyl groups at the secondary hydroxy centers.

The most common lipases which have been used to catalyze the acylation and deacylation of carbohydrates are *P. cepacia* lipase (PCL), porcine pancreatic lipase (PPL), *C. cylindracea* lipase (CCL), *C. antartica* lipase (CAL), *A. niger* lipase, and *C. viscosum* lipase, while the only protease to have been used is *Bacillus subtilis* protease (subtilisin).

The most common solvents used in the acylation reaction include pyridine, dioxane, THF, DMF, and benzene. For the deacylation reactions, 0.1 mol dm^{-3} phosphate buffer has been used as a cosolvent to improve solubility. The acylating agents of choice have included trichloroethyl butyrate,

TABLE 4.6 PPL-Catalyzed Transesterification of Various Sugars in Pyridine

Carbohydrate	Trichloroethyl ester	Conversion (%)	Product
D-Glucose	Acetate	50	6-0-acetyl
D-Glucose	Butanoate	62	6-0-butanoate
D-Galactose	Acetate	100	6-0-acetyl
D-Mannose	Acetate	38	6-0-acetyl

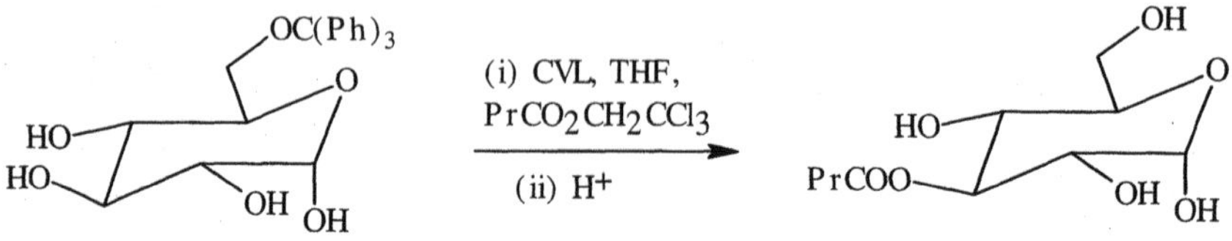

FIGURE 4.65 Regioselective acylation of sugars.

trifluoroethyl butyrate, enol esters, vinyl acetate, and isopropenyl acetate. Most of the acylation and deacylation reactions were carried out at 25–30°C with reaction times averaging 2–12 days.

4.5.6 Nitrile Hydrolysis by Hydrolases

The amidase of *Rhodococus rhodochrous* J1 was found to catalyze the hydrolytic cleavage of a cyano group (CN) to form carboxylic acid and ammonia stoichiometrically. A mutant amidase containing alanine (Ala) instead of ser 195 (Ser) residue, which is essential for amidase catalytic activity, showed no nitrilase activity, demonstrating that this residue plays a crucial role in the hydrolysis of nitriles as well as amides. The mechanism can be schematically shown as given in Scheme 4.10.

Enantioselective biotransformation of racemic *trans*-2-aryl cyclopropane carbonitriles catalyzed by *Rhodococcus* sp. AJ270 cells proceeded efficiently to give good optical yields of (−)-(1R, 2R)-2-aryl cyclopropane carboxamides and (+)-(1S, 2S)-2-aryl cyclopropane carboxylic acids (Fig. 4.66).

5.7 Miscellaneous Reactions Catalyzed by Lipases

Apart from the routine hydrolytic reactions, lipases are also known to catalyze the formation of peracids from acids, epoxides from alkenes, and

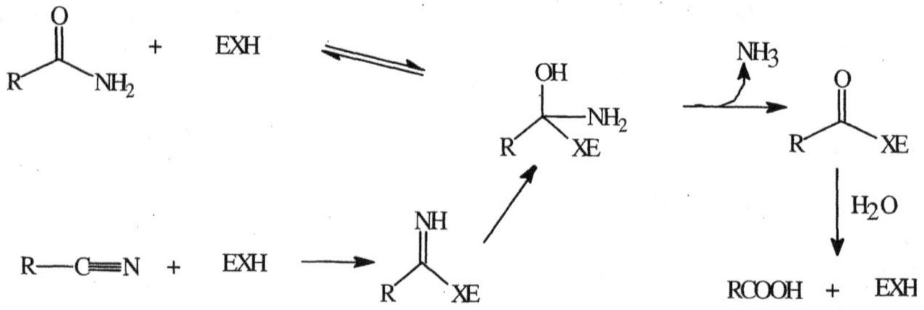

SCHEME 4.10 Nitrile hydrolysis by hydrolases.

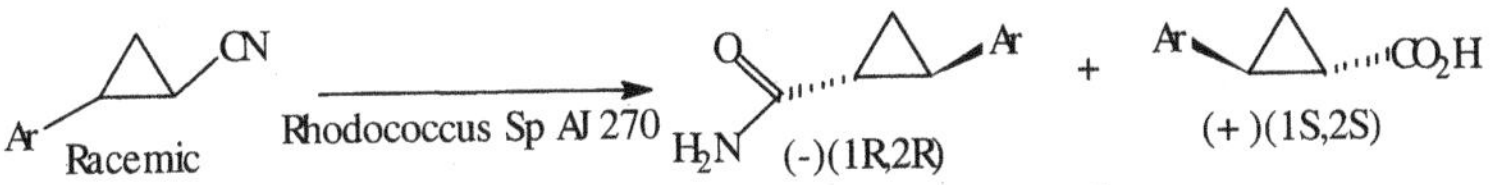

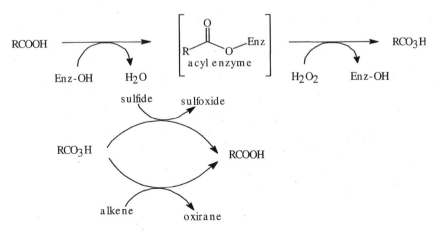

FIGURE 4.66 Nitrile hydrolysis.

sulfoxides from sulfides. These are possible, when a peroxide attacks the acylated enzyme, an acyl enzyme intermediate (Scheme 4.11).

In an unusual report, even a Baeyer–Villiger oxidation was found to be catalyzed by lipase. The *C. antartica* lipase catalyzed the oxidation of some lipohillic cyclic ketones to lactones using myristic acid and peroxide. The results are given in Table 4.7.

4.5.8 Epoxide Hydrolases

Epoxides are usually biotransformed into the corresponding diols by a trans addition of water catalyzed by epoxide hydrolases. These are important enzymes in the xenobiotic detoxifying systems. As required for an enzyme in the metabolism of a wide range of exogenous compounds, microsomal epoxide hydrolases (MEH) have low substrate specificity but are capable of chiral recognition.

In most reported cases epoxide hydrolases catalyze ring opening, with a high degree of enantioselectivity. For example, the racemic (6) epoxides of geraniol N-phenylcarbamate was hydrolyzed to optically pure (6S) diol. Also, the unreacted (6S) epoxides were obtained in ee as high as 96% (Fig. 4.67).

SCHEME 4.11 Unusual reactions catalyzed by lipases.

TABLE 4.7 Baeyer–Villiger Oxidation with Lipase

Substrate ketone	Product	% Yield
	  R=C$_6$H$_{13}$	73
		46 (1:1) ratio
	  R=CH$_3$	57
		20

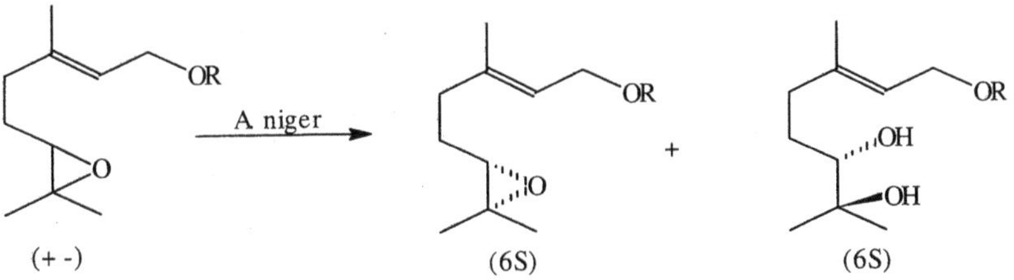

FIGURE 4.67   Epoxide hydrolase-catalyzed enantioselective ring opening.

Bulky substituents on the oxirane rings may not always give the same enantioselectivity. For example,10,11-dihydro-10,11-epoxy-SH-dibenzo (a,d)-cycloheptane was transformed to 9S, 10S diol (Fig. 4.68).

Epoxide hydrolases from different sources have varying enantioselectivity. For example, the epoxide hydrolases of *A. niger* and *Bacillus sulfurescens* showed exact opposite enantioselectivity towards styrene epoxide. Frustoss et al. (12) used this effectively in developing a preparative scale to both the enantiomers of styrene oxide (Fig. 4.69).

4.6 THIAMINE-DEPENDENT ENZYMES

Just as the oxido reductases require a coenzyme/cofactor to bring about the oxidations or reductions, there are a number of other enzymes that also require a nonprotein cofactor or coenzyme to aid in the reaction. Among the sulphur-containing coenzymes *thiamine* (often referred to as *vitamin B₁*) is the most important. The structure of thiamine pyrophosphate (TPP) or thiaminediphosphate (ThDP) is shown in Fig. 4.70. It consists of two subunits, a pyrimidine ring and a thiazolium ring.

Thiamine diphosphate [thiamine pyrophosphate (TPP)]-dependent enzymes vary in their substrate tolerance from rather strict substrate specificity (phosphoketolases, glyoxalate carboligase) to more permissive enzymes (*trans*-ketolase, dihydroxyacetone synthase, pyruvate decarboxylase) and therefore differ in their potential to be used as biocatalysts. TPP participates

FIGURE 4.68   Effect of bulky substituents on the enantioselectivity of EH.

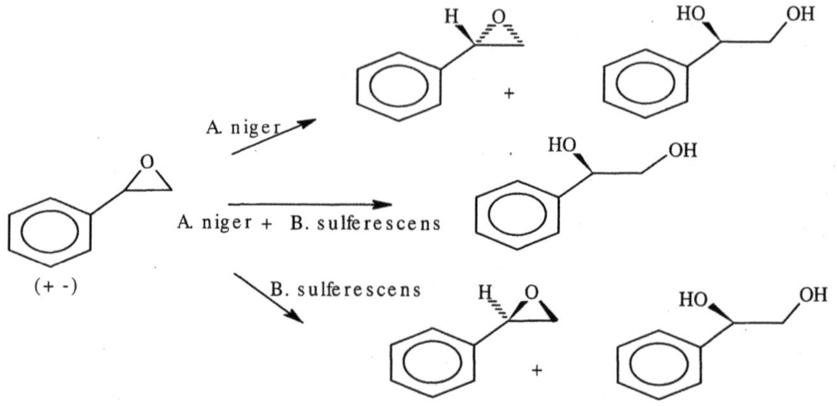

FIGURE **4.69**   Different organisms have varying selectivities.

in reactions involving formation and breaking of carbon–carbon bonds immediately adjacent to a carbonyl group, nonoxidative and oxidative decarboxylations, aldol condensations, etc.

4.6.1   Mechanism of Coenzyme

In attempting to understand the mechanism of these condensations the simplest analogy to be made by the chemist is that thiamine behaves like a cyanide ion in the catalysis of a benzoin type condensation (Fig. 4.71).

When we observe the structure of thiamine pyrophosphate closely, we find that the hydrogen at C-2 position is acidic and hence a easy ylide forms. This is the key intermediate in all the reactions (Fig. 4.72).

Metzler (13) proposed that the charge-relay system from the amino pyrimidine ring helps in the removal of the hydrogen at C-2 position to generate a nucleophillic carbanion that can condense with α- keto acids.

The thiazolium ring then behaves as an electron sink, or electrophile, and decarboxylation follows. Protonation of the enolic intermediate followed

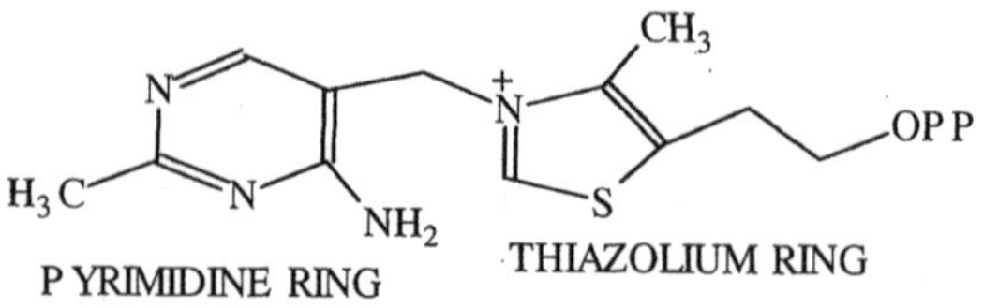

FIGURE **4.70**   Thiamine pyrophosphate (TPP).

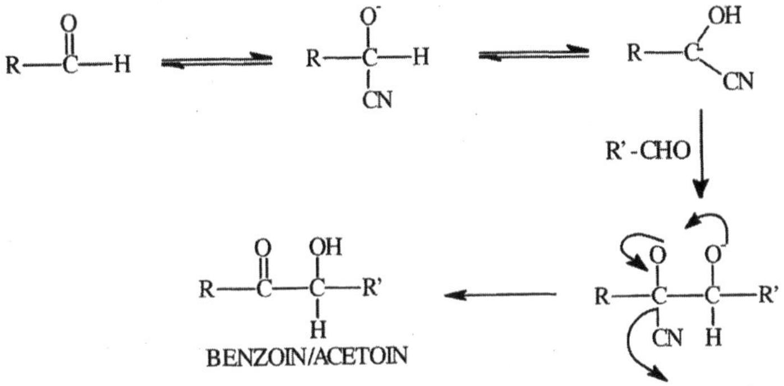

FIGURE 4.71 Benzoin formation.

by the proton uptake from the hydroxy group of the substrate, regenerates the coenzyme in the ylide form and the transformed substrate is produced. The sequence is shown in Fig. 4.73.

On the other hand, if the enolic intermediate (being nucleophillic) adds to a carboxyl compound, a carbon–carbon bond formation occurs to yield an acetoin-type product (Fig. 4.74).

Thus, we see that thiamine pyrophosphate can either (1) bring about decarboxylation of α-keto acids or (2) bring about carbon–carbon bond formation immediately adjacent to carboxyl group.

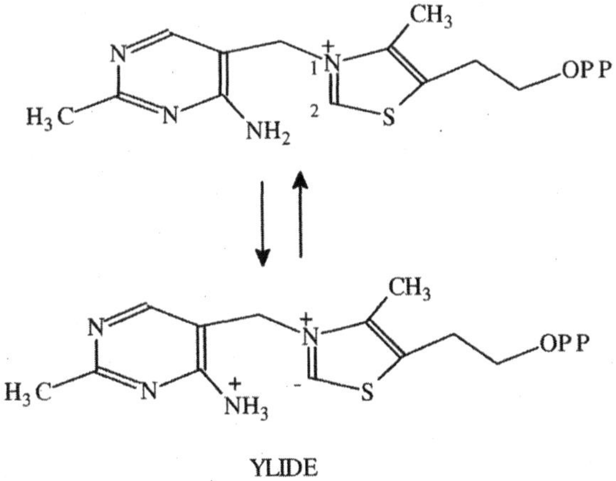

YLIDE

FIGURE 4.72 Ylide formation.

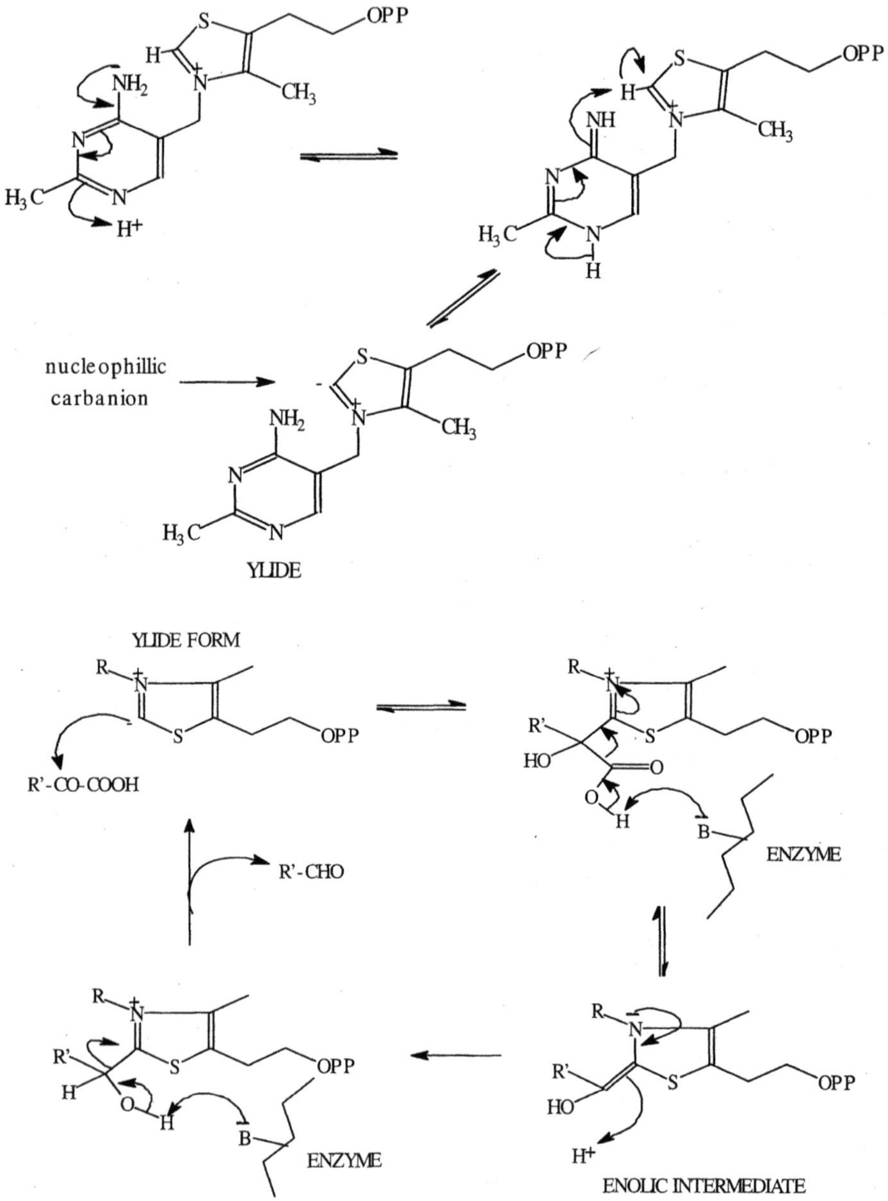

FIGURE 4.73 Catalytic cycle.

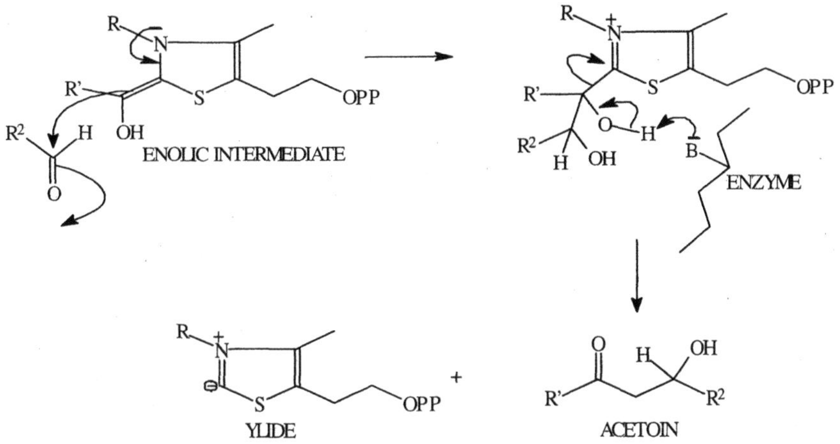

FIGURE 4.74 Acetoin formation.

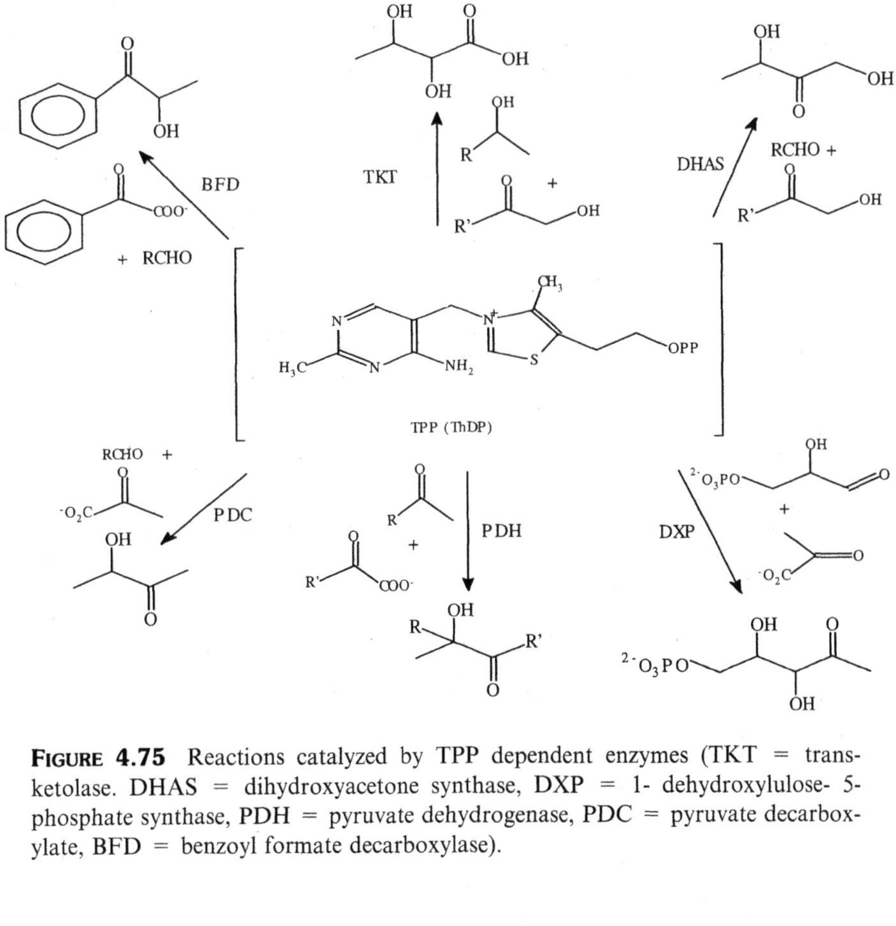

FIGURE 4.75 Reactions catalyzed by TPP dependent enzymes (TKT = transketolase. DHAS = dihydroxyacetone synthase, DXP = 1- dehydroxylulose- 5-phosphate synthase, PDH = pyruvate dehydrogenase, PDC = pyruvate decarboxylate, BFD = benzoyl formate decarboxylase).

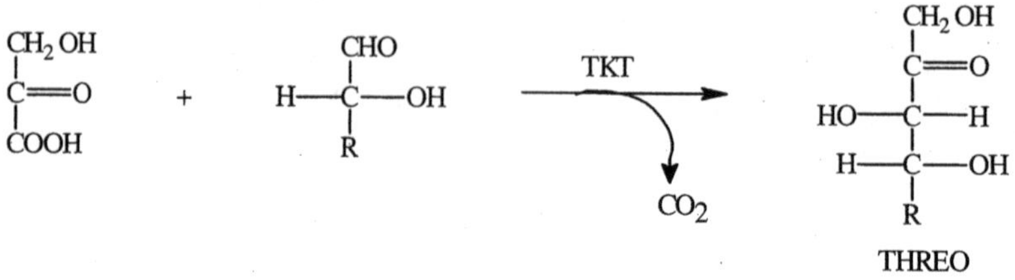

FIGURE 4.76 Transketolase in synthesis.

4.6.2 TPP-Dependent Enzymes in Synthesis

The enzymes, which use TPP and the transformations that they bring about, can be summarized as shown in Fig. 4.75.

Trans-ketolase. It can be used to extend the chain of a variety of aldoses *stereospecifically* by two carbon units. Normally the diol formed during the reaction is the *threo*-diol. This method is used in the synthesis of a number of monosaccharides (Fig. 4.76).

Pyruvate Decarboxylase. The pyruvate decarboxylase in yeast, catalyze the acyloin condensation of aldehydic substrate with an α- keto acid donor. A number of α-keto acids (C_2, C_3, C_4, C_5 units) are accepted by the yeast PDC and the carbanion generated by the decarboxylation of these is enantiotropically added only on the *Si* face of the aldehydic substrate (Fig. 4.77).

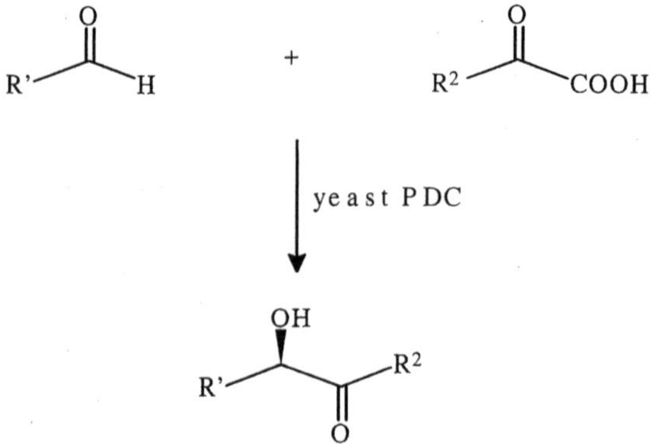

$R' = Ph, p-OCH_3-C_6H_4-, O-Cl-C_6H_4-, 2 or -furyl, 2 or 3-thienyl, etc...$
$R2 = CH_3, C_2H_5, n-C_3H_7$

FIGURE 4.77 Pyruvate decarboxylase in synthesis.

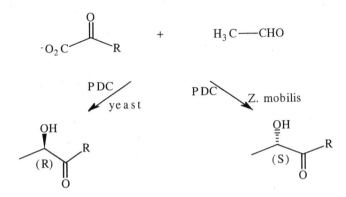

FIGURE 4.78 Stereospecificity of pyruvate decarboxylase.

Interestingly, acetoin and lactaldehyde synthesized from (a) acetaldehyde and (b) pyruvate and glyoxylate as alternative donors by PDC from yeast and *Z. mobilis* showed opposite configurations. The R-configured acyloin was generated with yeast enzyme, while the catalysis with the PDC from *Z. mobilis* led to S configuration. PDC from *Z. mobilis* showed carboligase activity with propanol, heterocyclic aldehydes (as furaldehyder or thiophene-3-aldehyde), and halogenated aromatic aldehydes (Fig. 4.78).

REFERENCES

1. Langmuir, I. J Am Chem Soc 1919, *41*, 1543.
2. Grimm, H.G. Naturwissenschaften 1929, *17*, 557.
3. Erlenmeyer, H.; Leo, M. Helv Chim Acta 1932, *15*, 1171.
4. Friedman, H.L. In *Symposium on Chemical–Biological Correlation*; National Research Council Publication: Washington D. C., 1951.
5. Thornber, C.W. Chem Soc Rev 1979, *8*, 563.
6. Boyd, Derek R.; Sharma, Narain D.; Kerley, Nuala A.; Austin, R.; Mcmordie, S.; Sheldrake, Gary N.; Paul, William; Howard, Dalton. J Chem Soc Perkin Trans I 1995, *1*, 67.
7. Prelog, V. Pure Appl Chem 1964, *9*, 119.
8. Brook, D.W.; Grothaus, P.G.; Irwin, W.L. J Org Chem 1982, *47*, 2820.
9. Shin, J.S.; Kim, B.G. Biotechnol Bioeng 1999, *65*, 206.
10. Then, J.; Bratsch, K.; Deger, M.H.; Grabley, S.; Marquardt, R. U.S. Patent, 5,919,669, 1999.
11. Klibanov, A.M.; Berman, Z.; Alberti, B.N. J Am Chem Soc 1981, *103*, 6263.
12. Pedragosa Moreau, S.; Archelas, A.; Furtoss, R. J Org Chem 1993, *58*, 5533.
13. Metzler, A.E. Biochemistry. *The Chemical Reactions of Livinig Cells*; Academic Press: New York, 1977; Chap. 8.

FURTHER READING

1. Bryan Jones, J. Tetrahedron 1986, *42* (13), 3351.
2. Davies, H.G.; Green, R.H.; Kelly, D.R.; Roberts, Stanley M. *Biotransformations in Preparative Organic Chemistry*; Academic Press Limited: London, 1989.
3. Faber, Kurt. *Biotransformations in Organic Chemistry*; Springer-Verlag: Berlin, 1992.
4. Rossazza, J.P., Ed.; Microbial Transformations of Bioactive Compounds. CRC Press: Boca Raton, Florida, 1982; Vol. 1.
5. Stewart, Jon D. Current Opinion in Chemical Biology 2001, *5,* 120.
6. Schorken, Ulrich; Sprenger, Gearg A. Biochimica et Biophysica Acta 1998, *1385*, 229.
7. Sih, Charles J.; Chen, Ching-Shih. Angew Chem Int Ed Engl 1984, *23*, 570.
8. Herbert, Waldmann; Dagmar, Sebastian. Chem Rev 1994, *94*, 911.
9. Fritz, Theil. Chem. Rev. 1995, *95*, 2203.

5

Experimental Techniques

5.1 EXPERIMENTAL PROCEDURE

Typical laboratory experimental procedure for enzyme- and microbial-catalyzed reactions are described in this section.

EXPERIMENT 1: Biotransformation with and without cosolvent
                conditions

The reduction of 1-phenyl-4-nitro-1-butanone (1) using *Pichia etchellsii* is given as an example.

WITHOUT COSOLVENT.   A solution of nitroketone (75 mg) (1) in ethanol (1.5 ml) was added to a stirred suspension (50 ml) of fresh cells of *P. etchellsii* (2.5g) resuspended in 0.1 M phosphate buffer (pH 6.0) at 28°C. After 48 h the conversion was about 80%. Sodium chloride was added and the suspension extracted with diethyl ether. The organic layer was dried over sodium sulphate, filtered, and concentrated to give crude (S) alcohol (2). The purification was carried out to afford pure(S)-(-)-1-phenyl-4-nitro-1-butanol (2) (57 mg) in 75% yield and 70% ee.

WITH ORGANIC COSOLVENT.   A solution of nitroketone (75 mg) [1] in ethanol (1.5 ml) was added to a suspension of fresh cells of *P. etchellsii* (2.5 g) resuspended in a mixture of 1.0 M phosphate buffer (15 ml, pH 6.0) and organic solvent (35 ml). The mixture was stirred at 28°C for 48 h. Usual workup and purification gave pure (S)-(-)-1-phenyl-4-nitro-1-butanol (2) (57 mg) in 75% yield and 97% ee $[\alpha]_D^{25}$ −50.6° (c = 0.52, CHCl$_3$) (Fig. 5.1).

EXPERIMENT 2: Enantioselective hydrolysis of epoxides

ENZYMATIC HYDROLYSIS OF 10,11-DIHYDRO-10, 11-EPOXY-5H-DIBENZO[A,D]CYCLO-HEPTENE [1]. A solution of epoxide (1) (50 mg. 0.24 mmol) in acetonitrile (1 ml) was added to 10 ml of microsomal preparation containing 40 mg protein/ml, obtained from male New Zealand white rabbits, preheated at 37°C, and the mixture was incubated with shaking. After 12 h, a fresh microsomal preparation (10 ml) was added, and the incubation continued for 12 more hours. The reaction was then stopped by addition of NaCl and the mixture was extracted with EtOAc (3 × 20 ml). The combined extracts were reduced to an exactly known volume by evaporation in vacuo, a proper amount of a stock solution of 9-formylanthrtacene in EtOAc was added to a sample as an internal standard, and the amount of *trans*-10, 11-dihydro-10, 11-dihydroxy-5H-dibenzo[a,d]cycloheptene (2) was determined by HPLC in order to evaluate the extent of hydrolysis of the substrate. This was typically around 25%. The remaining part of the extract was evaporated in vacuo and chromatographed on a column of silica gel (40g) with 7:3 hexane/EtOAc as the eluant. Eluted fractions (5 ml) were analyzed by HPLC. The fractions containing the *trans*-diol (2) were combined and evaporated to give 10 mg of pure (HPLC) 2: mp 147–150°C; $[\alpha]_D^{20}$-69° (c = 0.55, MeOH) (Fig. 5.2).

EXPERIMENT 3: Biohydrolysis of styrene oxide with Asperigillus niger

Hydrolysis of racemic styrene oxide were carried out in Erlenmeyer flasks (0.5 l) containing phosphate buffer (0.1 l, 0.1 M, pH 8) and 10% by weight of a fungal cake obtained from a 2-l fermentor. A solution of styrene oxide (1) (range of 0.1–1 g) in EtOH (1 ml) was poured into the medium, and the flasks were stirred at 27°C. The course of the bioconversion was followed by withdrawing samples (2 ml) at time intervals. After saturation with NaCl,

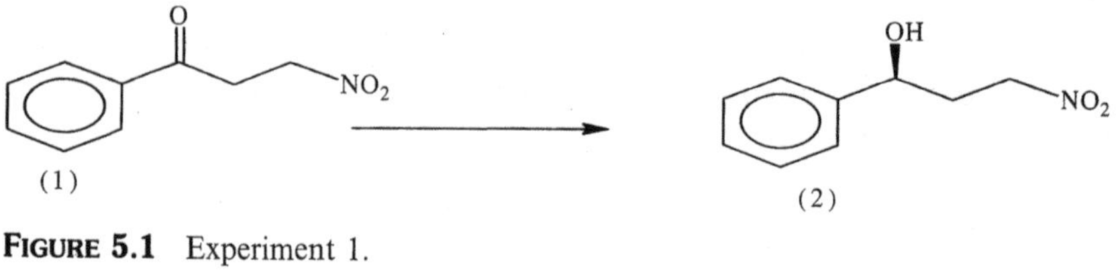

FIGURE 5.1 Experiment 1.

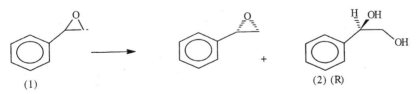

FIGURE 5.2 Experiment 2.

the samples were extracted each with a solution of ethyl acetate (2 ml) containing tetradecane as an internal standard for subsequent direct GC analysis of epoxide and diol. At the same time, samples (2 ml) were extracted with pentane aliquots, and were purified through silica gel chromatography, to yield (R)-phenylethanediol (2) in 54% yield (Fig. 5.3).

EXPERIMENT 4: General procedure for enzymatic acylation using enol esters

Substrate alcohol was dissolved in a lipophillic organic solvent (preferably hexane, cyclohexane or tolune) to give a solution of about 5–10%. Then, acyl donor (2–10 mol equiv, vinyl acetate, butanoate, or the corresponding trifluoroethyl ester) was added in one portion. The crude solid enzyme preparation (equal to the weight of substrate, usually a lipase or a protease) was suspended, and the mixture was stirred (or better shaken) at room temperature. At various intervals samples were withdrawn and the course of the reaction was analytically monitored by Gas-Liquid Chromatography (GLC) or High-Pressure Liquid Chromatography (HPLC). When the desired point of conversion was reached, the enzyme was filtered and washed with a small amount of lipophillic solvent. Evaporation of the solution and conventional column chromatography (if necessary) gave the products in overall yields of 80–100%.

GENERAL PROCEDURE FOR ENZYMATIC ACYLATION USING ACID ANHYDRIDES. In the above described procedure the vinyl ester was replaced by an aliphatic acid anhydride (1–2 mol equiv, preferably acetic or butanoic anhydride). In case the reaction proceeded with a low selectivity, addition of a weak organic base (e.g., 2,6-lutidine) at the same molar amounts as the

FIGURE 5.3 Experiment 3: *A. niger*-catalyzed epoxide hydrolysis.

acid anhydride may help. After filtration of the enzyme, the solution was washed (5% $NaHCO_3$) to remove the by-product acid before evaporation of the solvent. Comparable yields of 80–100% were obtained.

EXPERIMENT 5: *Candida antarctica-catalyzed Baeyer–Villiger oxidation of cyclic ketones*

To a solution of the ketone (1.11 mmol) (1) in toulene (10 cm^3) was added myristic acid (tetradecanoic acid) (0.32 g, 1.43 mmol) and immobilized *C. antarctica* lipase (0.25 g). To this hydrogen peroxide (30%; 1 cm^3, 4.4 mmol) was added over 10 h via syringe pump addition. The reaction was monitored by Gas Chromatography (GC). After 6 days, the reaction mixture was filtered through a plug of Celite and washed with dichloromethane (100 cm^3). The filtrate was dried over $MgSO_4$, and the solvent removed. Column chromatography over silica afforded the pure lactone. Alternatively, sodium hydroxide (2 mol dm^{-3}, 3 cm^3) can be added at the end of the reaction. The resulting mixture was filtered through Celite, the organic phase removed, dried, and evaporated to afford the crude product, which was purified as above (Fig. 5.4).

EXPERIMENT 6: *Biotransformations of organic sulfides*

Two slopes of *Helminthosporium* species NRRL 4671 were used to inoculate 15 1-l Erlelmeyer flasks each containing 200 ml of an autoclaved medium composed of V-8 vegetable juice (200 ml) and calcium carbonate (3 g) per liter of distilled water, adjusted to pH 7.2 prior to sterilization by the addition of 1 M sodium hydroxide. The flasks were allowed to stand overnight at 27°C, then placed on a rotary shaker at 180 rpm, and growth continued for a further 72 h at 27°C. The fungus was then harvested by vacuum filtration (Buchner funnel), and resuspended in 15 1-l Erlemmeyer flasks each containing 200 ml of distilled water, resulting in 90 g (wet weight) of mycelial growth per flask. Substrate (1 g in 30 ml of 95% ethanol) was then distributed among the flasks, which were replaced on the rotary shaker at 180 rpm, 27°C for a further 48 h. The fungus and aqueous medium were then separated by filtration as before,

FIGURE 5.4 Experiment 4: microbial-mediated Baeyer–Villiger reaction.

the aqueous medium extracted with dichloromethane (continuous extraction, 72 h), and the fungus discarded. Concentration of the medium extract gave the crude product, which was treated as described below.

BIOTRANSFORMATIONS WITH MORTIERELLA ISABELLINA. These were performed as described above, with the following modifications. The fungal growth was composed of glucose (40 g), soybean flour (5 g), yeast extract (Sigma 5 g), sodium chloride (5 g), and dibasic potassium phosphate (5 g) per liter. The fungus was separated from the medium by centrifugation (IEC chemical centrifuge), resuspended as above resulting in 85 g (wet weight) of biomass per flask, and biotransformation carried out for a period of 24 h to 40 h.

ISOLATION AND CHARACTERIZATION OF PRODUCTS. The crude biotransformation products obtained as described above were examined on Thin Layer Chromatography (TLC), using ether or 10% methanol/ether as the solvent, and then submitted to flash chromatography using a benzene–ether 10% stepwise gradient, followed by an ether–methanol 5% stepwise gradient. The ee values of the product (2) was > 95% (Fig. 5.5).

5.2 ENZYME IMMOBILIZATION

Two main types of immobilization include immobilization by physical means (namely, entrapment or encapsulation) or through chemical forces (achieving binding through adsorption or covalent forces).

5.2.1 Physical Means

Physical entrapment is a technique most widely used and consists of using porous polymers such as agar, alginate chitosan, and gelatin; silica gel; and polystyrene. The beads must be prepared in the presence of cells so that they get entrapped during the preparation process. Preparation techniques include

1. Gelation of polymers: Here liquid polymer is mixed with a cell suspension and templates are used to create beads

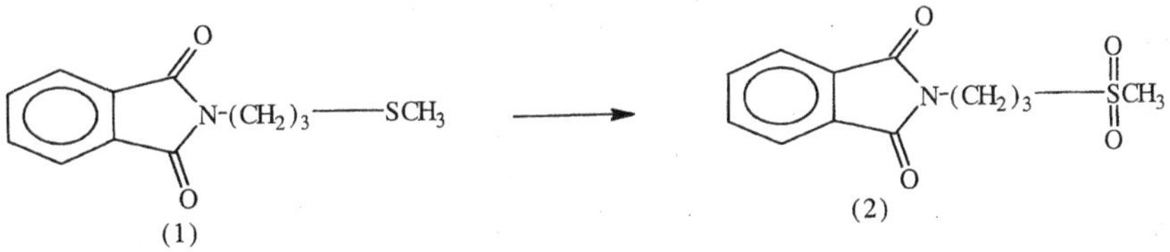

FIGURE 5.5 Experiment 5: biotransformation of organic sulfides.

2. Precipitation of polymers: As the polymer is mixed with the cell suspension, the pH or the solvent used is changed causing the polymer to precipitate out
3. Ion exchange gelation: A water soluble polyelectrolyte is mixed with a salt solution which causes solidification into a gel
4. Polycondensation: Epoxy based resins
5. Polymerization: Cross-linking of polymers using monomers like acrylamide and methacrylate, to form soft solid block.
6. Encapsulation of the cells into microcapsules. The microcapsule membrane could be of nylon or ethyl cellulose.
7. Entrap cells in hollow fibre tubes with the cells held outside the tubes and nutrients being pumped through the tubes. The nutrients would diffuse through the membrane and products diffuse back out.

5.2.2 Chemical Forces

Adsorption is a technique widely used here achieving high cell loadings. Internal mass transfer becomes a problem with porous support particles. Adsorption capacity and strength of binding are important parameters in designing such systems.

Covalent binding is stronger than adsorption and also avoids cell loss through shear. However, when cells grow exponentially, they will be lost into the suspension. Further strong binding could affect cellular metabolism and also enhance diffusional limitations.

Biocatalytics of Burbank, California has licensed a salt-immobilization technique in which, the enzyme is freeze-dried in concentrated aqueous potassium chloride, retaining the confirmation of the enzyme similar to solution.

Altus Biologics (Mass.) purifies enzymes in single crystal form, then treats with glutaraldehyde to form cross-linked protein chains. These crystals are stable to extreme conditions of temperature and pH.

Another approach is to form cross-linked enzyme aggregates, which involves precipitating the enzyme solution using ammonium sulphate, polyethylene glycol, or butanol followed by addition of glutaraldehyde to cross-link the protein chains. Cross-linking can be carried out between enzyme via cross-linked enzyme crystals, thereby increasing stability compared to non-immobilized enzymes and improving separation efficiencies for chiral molecules. Biocatalytics has commercialized the technology for salt enzyme complexes which not only serves as an immobilization technique but also increases their activity by a factor of 100 to 1000 than the native enzymes. Nippon Shokubai has immobilized enzymes via polymers, where the polymer does not interfere with the catalytic activity and helps recovery without the need for cofactors. Enzymes can be trapped in microemulsion aqueous droplets using surfactants. The droplets are suspended in the organic media that is undergoing reaction.

Another technique to immobilize enzyme in hollow polymeric microcapsules. These capsules are compatible with organic solvents as well as water. Other immobilization techniques include adsorption to a support (through physical or ionic bonding, chelating, or covalent binding) entrapment within membranes films, gel, or fibers. Gels or fibers are also used for entrapment of enzyme.

5.3 LABORATORY REACTOR SYSTEMS

The simplest laboratory apparatus for kinetic studies is the shake flask where the reactants and the biocatalyst are mixed together and shaken for a long period of time. The flasks are heated from the bottom. Samples can be taken from time to time and analyzed for conversion. Shake flasks suffer from many problems, and they do not simulate the actual industrial environment. Nevertheless they are used in the labs for screening new enzymes or biocells because of their simple construction and very high mixing patterns inside the apparatus. Most of the microorganisms change the pH of their medium during the course of the experiment, and pH control is very inefficient or not possible here. Oxygen transfer is very poor compared to aerated stirred tank bioreactors and thus cells will often be oxygen limited during the fermentation. Sampling during shaking is not possible, and if shaking is stopped for sampling, the lack of oxygen can lead to cell death or a change in cell metabolism. Shake flask is also suitable for collecting batch kinetic data.

Another laboratory apparatus that can be used for reactions involving supported solid enzyme is a tubular reactor (Fig. 5.6). The supported enzyme is packed into the tube and held in place either by using sintered disks on either sides or by using glass wool. The reaction mixture is fed from one end using a peristaltic pump and the product is collected at the other end. The required residence time can be achieved by varying the flow rate of the solution or by altering the volume of the tube. This reactor is generally operated in the continuous mode. Part of the product collected can be recycled back into the reactor to increase the conversion. If the entire product stream is recycled back to the reactor entrance the unit operates in the batch mode. This apparatus cannot be used for studying reaction kinetics but is an alternate to agitated vessel. It may be difficult to carry out aerobic reactions in this set up. The pressure drop will increase with decrease in the size of the packing material: in such a situation the supported catalyst is mixed with large diameter inert material.

Microfluidic devices help to conduct bench-top biological and chemical assays directly in a miniaturized and automated way. These devices detect and

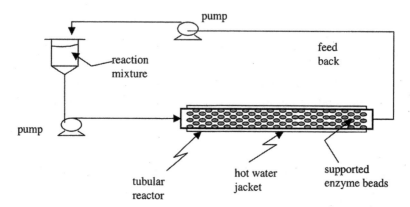

FIGURE 5.6 Schematic of a lab-scale tubular reactor assembly.

analyze the products using biosensors, which convert observed chemical changes into measurable signals. These signals could be electrical, light, or sound. This conversion process is achieved through an enzyme, which acts as a biological catalyst for certain reactions and can bind itself to specific substrates to form a product.

Several biological processes in the liquid-phase like cell activation, enzyme reactions, protein folding require analysis of the intermediates in the time frame of the order of milliseconds. Microreactors, which are now finding applications in the areas of chemical process development, are ideally suited for carrying out such fast reactions. The typical dimension of such a reactor will be 1 cm × 1 cm × 1 mm micromachined silicon mixer for reactions of 100 µs duration and Reynolds number of the order of 2000–6000. The mixer is fabricated by etching the silicon wafers with potassium hydroxide followed by anodic glass bonding.

Microfermentors on chip are being developed by Bio Process Corp. (Woburn, MA) to optimize production and secretion of engineered protein that are developed for therapeutic purposes. Traditionally such studies will require 10,000–50,000 experiments using lab bottles or shaker flasks, which are labor intensive and require large quantity of the reagents and enzymes. Whereas, thousands of experiments can be run in a microfluidic device in a fully automated way, cutting down discovery time and cost. The system consists of a silicon wafer on which about 16 microfermentors are mounted. The bioreaction chamber is made of polymer composites like polycarbonate or polysiloxanes that can hold about 100 µl of liquid. Fluidic circuits built inside the chip enable nutrients and other reagents in and out of the reaction chamber. Facilities are available to control the temperature and pH. The cost of such unit is of the order of few dollars.

Stopped-flow experiments involve mixing two (or more) solutions together as rapidly as possible and positioning the mixture under an optical spectroscopy (absorbance or fluorescence) quickly to observe the extent of reaction (Fig. 5.7). The apparatus consists of two or more holding chambers filled with reactant/nutrient and/or enzyme solutions. They are rapidly injected into a mixing device using syringe pumps. The mixture flows into the observation chamber displacing the previous sample. A stop syringe limits the volume of the solution injected. The mixture in the cell is illuminated by a light source and the change in optical property is measured as a function of time. The property could be absorbance, fluorescence, light scattering, turbidity, fluorescence anisotropy, etc. Variable-ratio mixing can be achieved if syringes of different sizes are used. This apparatus operates in the continuous mode.

In the quench flow system the reagents are mixed together very precisely, allowed to age for a predefined period of time, and the reaction is halted (or quenched) with a third solution (Fig. 5.8). This quenched mixture is collected for ex situ analysis. Conversion is determined quantitatively by standard chromatographic methods. This unit also contains an aging chamber prior to the quenching zone. Aging or delay times of 1 or 2 ms to many hundreds of milliseconds can be provided by varying the speed of solution flow through the aging chamber or by changing the size of the aging chamber or both. The reaction profile as a function of time can be generated by producing a series of solutions, quenched at different times after initial mixing.

Temperature jump is another kinetic technique ideally suited to follow fast reactions with half-lives of just a few microseconds. Reagents are mixed and allowed to equilibrate, and the mixture is rapidly heated by passing short pulse of current at high voltage, to induce temperature rise of approximately $6\,^{\circ}C$ within a few microseconds. The system reaches a new equilibrium at the higher temperature, which is monitored using spectroscopic techniques, and the rate constant of the reaction is determined.

Another type of relaxation method is giving a concentration jump to the reaction mixture that is at equilibrium (at steady state). The dynamic changes

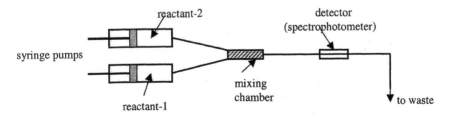

FIGURE 5.7 Stopped-flow apparatus.

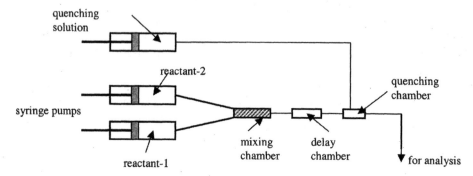

FIGURE 5.8 Quenched-flow apparatus.

in the output due to the step change in concentration are measured as a function of time, which is used to determine the reaction mechanism and also the rate constants.

A flow microcalorimeter is a versatile and practical tool for investigating the catalytic properties of immobilized biocatalysts and enzymes. The technique is useful for both native as well as supported enzymes and the experiments do not require large quantity of the reaction medium. An isothermal reaction calorimeter consists of two cells with sensitivity of the order of 1 μcal. The reaction mixture under study is injected into the reaction cell of volume of the order of 150 to 200 μl. Change in heat is measured via electrical compensation of the thermal effect. A Gill titration calorimeter is useful for measuring the equilibrium constant and enthalpy change of biochemical reactions. Batch type microcalorimeters can study very small volume of liquids (of the order of 200 μl) with high heat sensitivity (measure 1 μcal). The challenge in the design of such system is the design of a low-volume thermally equilibrated injection system.

In many fast reactions measurable changes in absorbance or fluorescence may not be observed, whereas there could be changes in conductivity as the reaction progresses. Examples of such reactions where changes in conductivity are observed include organic and inorganic oxidation and reductions, proton exchange, metal–ligand complex formation, surfactant aggregation to form micelles, and ion exchange. A stopped-flow system equipped with conductivity measurement facilities can further enhance the scope of this technique. Conductivity measurements, in addition offers greater signal-to-noise ratio, which means it can be operated in the presence of large buffer concentrations and during the use of nonaqueous ionizing solvents such as ethanol and acetone. Also, it has a much better dynamic range than spectrophotometric measurements.

Anisotropy measurements as well as total fluorescence measurements can be made with stopped-flow instruments equipped with a fluorescence polarization technique. This is achieved by simultaneously detecting the fluorescence contributions of a sample from light polarized perpendicular and parallel to the parallel plane of the excitation beam.

5.4 SPECTROSCOPY

A biological process is extremely complex consisting of sequence of several elementary reactions. Also at the molecular level the dynamics of the single-bond-breaking and bond-formation reactions, energy and electron transfer reactions are very fast, generally in the order of femtoseconds to picoseconds. Several techniques that can follow such fast dynamics are available. Ultrafast spectroscopy is one such tool for studying these events, as well as the transition from reactants to products. Laser pulses with femtosecond (~10–100 fs) duration are necessary to monitor the evolution from reactants to products and to capture initiation of a chemical reaction.

These short pulses offer high temporal resolution, which can be used to obtain information about the dynamics of the molecular system. The ultrashort pulses that are now generated with solid-state laser technologies have several other very useful properties like very high peak power, broad spectral bandwidth (due to the short time duration), very widely tunable wavelength, and tunable temporal and phase characteristics.

Several areas in chemistry and biology that can be studied by various forms of ultrafast spectroscopy include reaction dynamics of isolated molecules as well as of molecules in clusters and in the condensed phase, energy and electron transfer in photosynthesis, photoreceptor dynamics, single-molecule spectroscopy, biological tissue imaging, time-resolved molecular structures, coherent control of reactions, and design and synthesis of new materials.

The effect of intramolecular properties on the reaction path may be studied in isolated molecules using molecular beams and ultrafast (femtosecond) spectroscopy. One of the successes of this technique is the detection of transition states and reaction intermediates along a reaction path.

Understanding a chemical or biochemical reaction at the atomic level requires depiction of the reactants and products, and also the description of the solvent medium in which the reaction is taking place. There is an interaction between the solvent medium and the reactants, and it is termed *solvent–solute* interactions. The solvent plays several roles during a chemical reaction namely, reactive molecules are stabilized (solvation), reaction products are caged (caging), and vibrational excitation is exchanged (vibrational relaxation). Diffusion of the reactants and products also take place in the medium, which is at higher time scales, much longer than femtoseconds, generally of the order of milliseconds. With femtosecond lasers it is possible to make direct observation of these phenomena in the liquid phase. Many papers have been published pertaining to solvation, caging, and vibrational relaxation of solute molecules.

One of the approaches to study how solute–solvent interactions modify a chemical reaction in the condensed phase is to attach surrounding molecules successively to the reactant, thereby gradually increasing the intermolecular interactions. This is done by molecular-beam techniques by incorporating the reactant in a molecular cluster. The isolated reactant molecule and the condensed phase are the two extremes in density.

Microscopic control of chemical reactions can be achieved through selective cleavage or formation of chemical bonds on a molecular level using monochromatic laser light. By exactly tuning the light according to the frequency of a specific chemical bond enough energy could be pumped in to cause selective bond breakage. Selectivity is poor because of rapid intramolecular energy redistribution. Several control schemes have been proposed that make use of the coherent nature of laser radiation.

Weak interactions (van der Waals forces) between molecules control a wide range of phenomena and these forces include dispersion forces, electrostatic interactions, and hydrogen bonding. For example, the folding of a protein and the positioning of a substrate near the enzyme active site in a biochemical reaction take place because of these forces. A Fourier transform microwave (MW) spectrometer is a technique available to study such interactions. Microwave radiation is pulsed at molecules and the reemitted signal is captutured and analyzed for rotational energy levels, from which detailed structural information can be deduced. This technique can also be used to study dipole moments, internal dynamics and interaction forces for a variety of dimers, trimers, and small clusters held together by very weak forces. Weak binding of species could be achieved by expanding the gas through a pinhole in vacuum. As the gas expands the intermolecular forces between them decrease.

The kinetics of intramolecular cyclization, for example, of aspratame can be studied using a Fourier transform IR microspectrophotometer equipped with differential scanning calorimeter.

If redox processes and complex biochemical reactions generate paramagnetic intermediates, then such processes could be understood using insitu spectroscopic techniques like insitu ESR (electron spin resonance spectroscopy) or in situ ultraviolet (UV)-visible NIR. ESR is a sensitive technique and also gives information on radical intermediates.

Many kinetic reactions are initiated using a high intensity, short duration pulse of light, as in traditional flash photolysis experiments. These techniques have recently found applications in the areas of physiology and pharmacology to study caged compounds. The pulse of light is used to release the compound of interest from a molecular "cage" within a short period of time in order to generate a concentration jump. Xenon flash lamps can focus a high-energy pulse into the solution mixture to initiate this jump.

5.5 CHEMINFORMATICS

Cheminformatics is a key technology for new discoveries. Experimental techniques such as high throughput screening and combinatorial chemistry have led to generation of vast amount of data in the laboratory (in the form of structural, physical, and chemical), and the management of this information is vital for the success of the project and hence a challenging business. Analysis of this vast data also gives clues for further studies and path forward. An effective cheminformatics system ensures that scientists can maximize the value of the data and information available to them. There are many commercial softwares available to manage the data, filter it and present it in a easily interpretable way. The scientist can query the database by providing structural information or key words and get leads.

5.6 EXPERIMENTAL DESIGN PLAN

The lab chemist or the pilot plant engineer will be interested in studying the reaction system by varying several operating conditions such as temperature, biocatalyst amount, and solvent polarity. Experimental data collection may be necessary to understand the reaction mechanism and determine the constants in the kinetic equation or to develop a mathematical model. These experiments should be in an organized or planned way so that more information could be extracted with less number of experiments. Carrying out a large number of experiments is time consuming, costly, and may generate large quantity of waste, which needs disposal. A planned experimental design will help in identifying the key or critical variables from not so critical ones and also reduce the maximum number of experiments that need to be done. This methodology is known as *design of experiments*. The effect of noise variable in the system performance can also be clearly identified by this approach. Varying one factor at a time may not lead to correct understanding of the reaction mechanism and system behavior. Also varying one factor at a time cannot reveal interaction between parameters. Whereas a more accurate picture could be obtained if the experiments are planned in such a way that all the variables are altered simultaneously and the resulting performance measured. The technique of design of experiments also helps the experimenter in coming up with the best way of changing these independent variables. Design of experiments is dealt in detail by many books on industrial statistics, and it is outside the scope of this book.

6

Frontiers in Biotransformations

6.1 CROSS-LINKED ENZYMES AND NOVEL CATALYTIC MATERIALS

Enzymes by and large are not readily accepted in chemical industry due to

High cost
Limited substrate specificity and
Low enantioselectivity for unnatural synthetic substrates.

Apart, from these, the main reason often being the lack of enzyme stability at elevated temperatures and rigorous process conditions. One approach by which this limitation is overcome is by cross-linking the enzymes and crystallizing these cross-linked enzyme systems. These are termed *cross-linked enzyme systems*, or *CLEC*s. Cross-linking "locks" the protein in a particular conformation. Historically, glutaraldehyde (1) has been by far the most popular cross-linking agent. However, a number of other reagents are now known (Fig. 6.1).

Chemically, CLECs are significantly more stable against denaturation by heat, organic solvents, and proteolysis than the corresponding soluble proteins. In certain cases the enzyme crystal may even be more active than the same enzyme in solution. For example, CLECs of lipases and subtilisin formulated with surfactants exhibited specific activity in organic solvents higher than that of native enzymes. Crystals of *Candida Rugosa* Lipase (CRL)

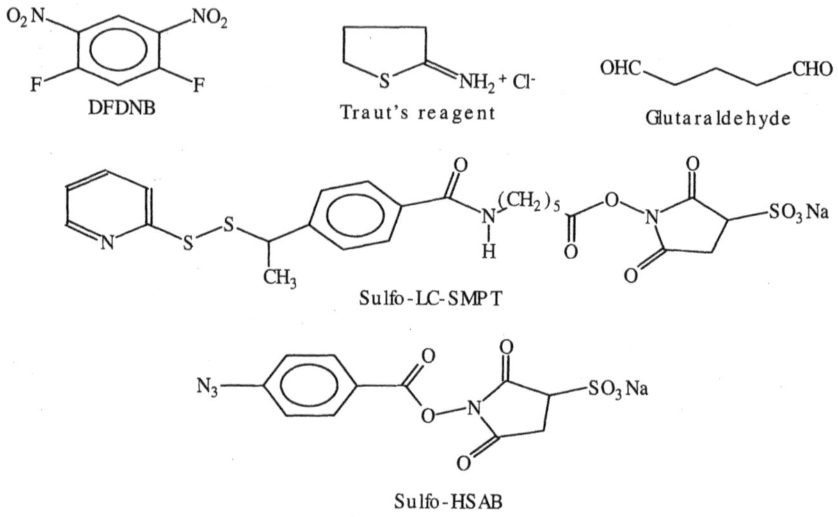

FIGURE 6.1 Reagents for cross-linking proteins.

cross-linked with sulpho-LC-SMPT were five to seven times more active than pure native CRL in the olive oil assay.

Apart from this, the CLECs remain insoluble throughout the process and can be recycled many times, thereby, increasing the productivity of the catalyst. A good example of the high productivity of CLEC catalysts in organic solvents is the resolution of 1-phenyl ethanol with vinyl acetate in toulene catalyzed by CLECs of *Pseudomonas cepacia* lipase (LPS-CLECs). In this reaction a substrate to catalyst ratio of 4600:1 was achieved.

Various enzymes, such as lipases, esterases, and dehydogenases, have been cross-linked and all of them showed higher activity and greater stability. In the case of dehydrogenases, the cocrystallization with the cofactor was adopted. Hence, this nascent science of "crystalomics" provides us with a general strategy for putting proteins to work rapidly and efficiently in the harsh conditions typical of most practical industrial processes.

6.2 RATIONAL DESIGN OF ENZYMES

One of the requirements of enzymatic organic synthesis is to develop protein catalysts with tailored activities and selectivities. The various approaches toward the attainment of this goal are as follows:

1. Construct (synthesize) a protein with a designed catalytic activity and selectivity from a designed sequence of amino acids.

2. Synthesize nonprotein enzyme models (biomimics) with altered/required activity.
3. Catalytic antibodies.
4. Site-selective modification of existing proteins.

Our understanding of the dynamics involved in the folding of protein molecules is not complete, despite a significant amount of work directed toward the correlation of primary structure (sequence of amino acids) with the three-dimensional structure. Therefore, it is not yet possible to construct a protein with the desired amino acid side chains in its active site from a designed sequence of amino acids.

With regards to the biomimetic approach, a lot of work is being done and a great deal of understanding about the actual mechanics of enzymatic action has been obtained. But still, the design of a nonprotein organic molecule having the same efficiency and selectivity as the enzyme is far from being realized.

6.2.1 Catalytic Antibodies: Abzymes

Antibodies and enzymes share the ability to bind to compounds with great specificity and affinity. Antibody–antigen and enzyme–substrate interactions are similar. Since one of the mechanism by which enzymes act as structure-selective catalysis is to provide steric and electronic complementarity to a rate-determining transition state of a given reaction, an antibody elicited against a stable transition-state analog of a reaction should catalyze that reaction. In a pioneering work done by R. A. Cemer, mice were immunized with a hapten–carrier complex in which the hapten was structurally made to resemble the transition state of an ester hydrolysis reaction. Thus, with appropriate design of the haptens (in the hapten–carrier antigen), specific functional groups can be induced in the binding site of an antibody (active site) to perform catalysis.

The most successful reactions using abzymes are selective ester hydrolysis, transesterification and pericyclic reactions. For example, the prochiral diester (1) was enantiospecifically hydrolyzed to (2) in >98% ee, by an abzyme developed for the hapten (3) (Fig. 6.2).

This approach has an unlimited scope in the design of enzyme (protein catalyst) for the reaction. However, the bottleneck in this rapidly growing field is the inefficiency in the production of monoclonal antibodies in large quantities. The advances made in the study of FABs are providing a solution to this problem. In conclusion, it can be said that this technique has great potential and may provide the solution to the rational design of active site geometry.

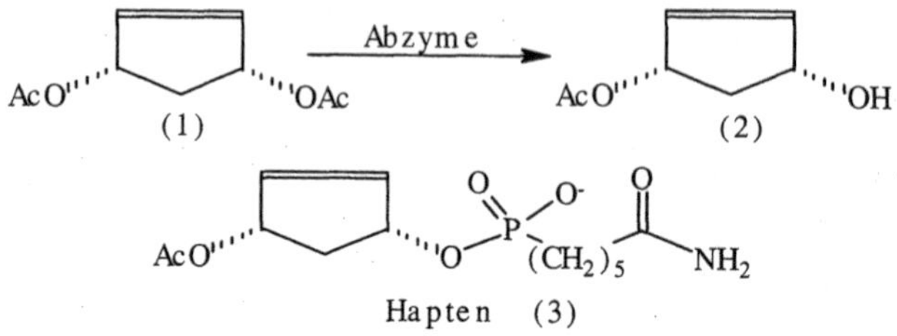

FIGURE **6.2** Ester hydrolysis-catalyzed by abzymes.

6.2.2 Site-Selective Modification of Enzymes

The active site of an enzyme has a number of side chains of the amino acids residues, arranged in a definite manner, so as to have *specific binding* with the substrate and bring about the *catalytic activity*. Based on this understanding of the mechanism of protein binding and catalysis, it is possible to alter the active site rationally to accommodate new substrates or to catalyze new reactions. Site-directed mutagenesis has been used in such alterations. These are observed in nature. Nature itself brings about these minor differences among the proteins of various organisms. The same can now be done by gene manipulation. This approach has advantages, but it has a limitation of limited variation of the side chains, because only 20 amino acids can be coded by the genetic code.

Alternative approaches to this site-directed variations are by chemical means. Attachment of oligonucleotides, metal ions, cofactors (flavin) or small alkyl chains to either the active site or to the surface of the enzyme molecule, have been done (4).

A working together of these two principles—abzymes and site-directed modifications—will lead to *rational design* of *new enzymatic catalysts*. This obviously involves the joining of the expert hands in immunology, molecular biology, synthetic organic chemistry, and enzymology. Coming together of theses various experts and working for a common cause of "ecofriendly reactions," for the safety and prosperity of the world at large is the need of the hour.

REFERENCES

1. Margolin, Alexey L.; Nausia, Manuel A. Ang Chem Int Ed 2001, *40*, 2204.
2. Ikeda, S.; Weinhouse, M.I.; Janda, K.D.; Lerner, R.A.; Danishefky, S.J. J Am Chem Soc 1991, *113*, 7763.
3. Noren, C.J.; Anthony-Cahill, S.J. Griffith, M.C.; Schultz, P.G. Science 1989, *224*, 182.
4. Wilson, M.E.; Whitesides, M.G. J Am Chem Soc 1978, *100*, 306.

7

Enzyme and Biocell Kinetics

7.1 KINETICS OF REACTION

For example, in the following reaction,

$$A + 3B \rightarrow 2P$$

The numbers -1, -3, and 2 are the stoichiometric coefficients of A, B, and P, respectively, in this reaction; by convention they are positive for products and negative for reactants. For a reaction of this kind the rate of reaction is the rate of consumption of reactants or formation of any product divided by the appropriate stoichiometric coefficient, i.e. (in this example),

$$-\frac{dC_A}{dt} = -\frac{1}{3}\frac{dC_B}{dt} = +\frac{1}{2}\frac{dC_p}{dt}$$

An elementary reaction is one in which no reaction intermediates have been detected or need to be considered to describe the chemical reaction on a molecular scale. Such reactions are said to occur in a single step. The term *molecularity*, which applies only to an elementary reaction, refers to the number of molecular particles involved in the microscopic chemical process. With reactions in solution, solvent molecules are counted in the molecularity only if they enter into the overall process, but not if they merely provide an environment. For example,

$$A + B \rightarrow P$$

has a molecularity of 2 and is said to be bimolecular. The reverse process in solution,

$$P \rightarrow A + B$$

has a molecularity of unity and is said to be unimolecular.

For a simple reaction of the form

$$A \rightarrow P$$

$$-\frac{dC_A}{dt} = +\frac{dC_p}{dt} = -r_A = kC_A^n \tag{7.1}$$

where C_A and C_p are concentration of reactant and product respectively. r_A is the order of the reaction, and k is the rate constant and is a function of temperature. For Arrehenius type of reaction it is given as

$$k = k_0 e^{-E/RT} \tag{7.2}$$

The initial rate or initial velocity v_0 of the reaction can be determined from

$$v_0 = \frac{([C_A]_0 - [C_A]_1)}{t_1} \tag{7.3}$$

$[C_A]_0$ is the initial concentration of the substrate and $[C_A]_1$ at time $= t_1$

The relationship between initial rate and initial substrate concentration for 0, 1, and 2 order reactions are shown in the Fig. 7.1. Hence this technique

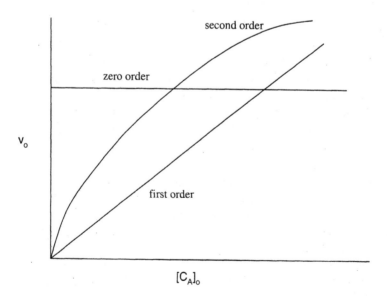

FIGURE 7.1 Effect of initial concentration on initial rate as a function of reaction order.

could be followed in the laboratory to differentiate between various reaction orders.

If the reaction is of the form A + B → P, the general form of the rate equation may be $-dC_A/dt = kC_A^n C_B^m$. Once again the orders n and m can be estimated by determining the initial rates with respect to A and B independently, by keeping the initial concentration of the other substrate constant.

A reaction of the form A → P → Q is called a *series reaction*, where P is the intermediate, whereas a reaction of the form A → P and A → Q is known as a *parallel reaction*.

Cell growth process is an autocatalytic reaction, where the reaction exhibits an induction period with very low rate at initial time, followed by a sudden increase in rate until all the substrate is consumed. The rate curve exhibits a sigmoidal-shaped curve with an induction period. Autocatalytic behavior is exhibited during the exponential growth phase of the biocell, and it is described in more detail later.

7.2 ENZYME-CATALYZED REACTION

Enzyme behaves like any other catalyst, by forming enzyme-substrate complex, but this is not similar to the transition state, which organic molecules pass through. The formation of enzyme-substrate complex is based on two mechanisms, lock and key and hand and glove. In the lock-and-key mechanism it is assumed that the fit happens because the size and shape of the active site in the enzyme matches exactly with that of the substrate, analogous to a lock and key. In the hand-and-glove mechanism the enzyme active site adjusts itself both in size and/or shape to suit that of the substrate. Factors like pH, temperature, chemical agents (such as alcohol and urea), irradiation, mechanical shear stress, and hydrostatic pressure alter the active site and affect the performance of enzymes.

7.2.1 Michaelis-Menten Kinetics

For a single-substrate reaction catalyzed by an enzyme, there are several steps involved:

1. Substrate binds to the enzyme at specific site (known as *active site*) to form an enzyme substrate complex.
2. Formation of a transition state.
3. Enzyme product complex.
4. Separation of products from the enzyme and freeing of the active enzyme site. The active enzymes site is once again available for the reaction.

The free-energy diagram for such a multistep reaction will be as shown in Fig. 7.2. These steps can be mathematically represented as

$$E + A \rightleftharpoons EA \qquad EA \rightleftharpoons EP \qquad EP \rightleftharpoons E + P$$

Generally the second equation is ignored, and the last equation is assumed to be proceeding only in the forward direction to arrive at simplified form as shown below. The formation of enzyme-product complex from a free enzyme and a product molecule can take place only at very high product concentration (generally at the end of the reaction).

$$E + A \underset{k_{-1}}{\overset{k_1}{\rightleftharpoons}} EA \qquad EA \overset{k_2}{\longrightarrow} E + P$$

k_1, and k_{-1} are the rate constants for the forward and backward reactions for the formation of the enzyme-substrate complex. k_2 is the step that gives the product as well as the liberation of the free enzyme.

Material balance for the accumulation of the enzyme-substrate complex can be written as

$$\frac{d[EA]}{dt} = k_1[E][C_A] - k_{-1}[EA] - k_2[EA] \qquad (7.4)$$

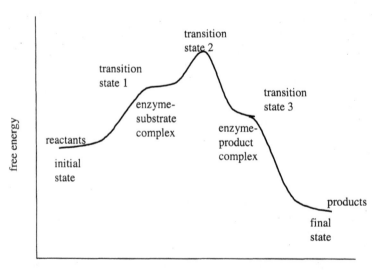

FIGURE 7.2 Free-energy diagram for an enzyme-catalyzed reaction.

At steady state the equation is equal to zero. Also, $[E] = [E]_0 - [EA]$, where $[E]_0$ is the initial enzyme concentration. Substituting this in the previous equation and rearranging gives

$$[EA] = \frac{[E]_0[C_A]}{K_m + [C_A]} \tag{7.5}$$

and rate is given by

$$r_A = k_2[EA] = \frac{V_{max}[C_A]}{K_m + [C_A]} \tag{7.6}$$

This equation is called the *Michaelis-Menten equation*.
  Here,

$$V_{max} = k_2[E]_0 \text{ and } K_m = (k_{-1} + k_2)/k_1 \tag{7.7}$$

V_{max} is the limiting or maximum rate possible, and K_m equilibrium constant (Michaelis constant for A; the alternative name Michaelis concentration) is a function of enzyme type and substrate. Units of V_{max} will be same as that of rate and K_m of that of concentration and are functions of temperature. K_m is also equal to the substrate concentration at which the rate $r_A = 0.5\ V_{max}$. Reducing enzyme concentration will alter V_{max}, since it is dependent on the initial enzyme concentration, $[E]_0$, but K_m remains unchanged.
  Integration of the Eq. (7.6) leads to a form

$$V_{max}t = [C_A]_0 - [C_A] + K_m \ln \frac{[C_A]_0}{[C_A]} \tag{7.8}$$

The Michaelis-Menten equation [Eq. (7.6)] could be extended to relate initial rate and initial substrate concentration as

$$r_{A_0} = \frac{V_{max}[C_A]_0}{K_m + [C_A]_0} \tag{7.9}$$

At very low substrate concentrations denominator of Eq. (7.9) can be approximated to K_m, then

$$r_{A_0} = V_{max} \frac{[C_A]_0}{K_m} \tag{7.10}$$

which means, that it is first order with respect to initial concentration. At very high initial concentrations denominator of Eq. (7.9) $\approx [C_A]_0$, then

$$r_{A_0} = V_{max} \tag{7.11}$$

which means it is a constant (zero order). This is pictorially represented in Fig. 7.3.

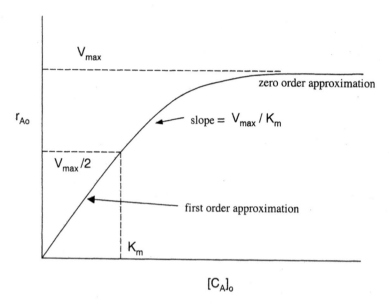

FIGURE 7.3 Effect of initial substrate concentration on initial rate for reactions obeying Michaelis-Menten kinetics.

The pre-steady-state kinetic equation [Eq. (7.4)] will be of the form (i.e., before $d[EA]/dt$ becomes zero)

$$\frac{d[EA]}{dt} = k_1([E]_0 - [EA])[C_A]_0 - k_{-1}[EA] - k_2[EA] \qquad (7.12)$$

7.2.2 Estimation of Michaelis-Menten Constant

Several different approaches are available to determine the Michaelis-Menten constants V_{max} and K_m from the laboratory experiments.

Nonlinear Approach

This consists of an optimization technique, where the constants are adjusted so that the sum of square of the errors between the predicted rate and the observed rate is minimum. Rate is predicted from the Michaelis-Menten model. The approach can be represented mathematically as minimize

$$\sum_{N}^{i=1} \left\{ r_A(\text{experimental}) - \frac{V_{max}[C_A]}{K_m + [C_A]} \right\}_i^2 \qquad (7.13)$$

N is the number of data points collected during the experiments.

Nonparametric Approach

Two algebric equations could be generated by considering two sets of $[C_A]$s (namely $[C_A]_1$ and $[C_A]_2$) and corresponding r_A s (r_{A_1} and r_{A_2}), and the Michaelis-Menten constants V_{max} and K_m are estimated by solving the two simultaneous equations given below:

$$r_{A_1} = \frac{V_{max}[C_A]_1}{K_m + [C_A]_1} \tag{7.14}$$

$$r_{A_2} = \frac{V_{max}[C_A]_2}{K_m + [C_A]_2} \tag{7.15}$$

This procedure could be repeated several times, namely, considering two sets of $[C_A]$s and corresponding r_A s to get several estimates of the *Michaelis-Menten* constants and then the average of these set of values.

Graphical Approaches

Four different approaches are described below for determining the Michaelis-Menten constant from experimental data using graphical techniques.

Figure 7.3 describes the first graphical approach. Here the constants are determined from a graph of initial rate (r_{A_0}) and initial substrate concentration ($[C_A]_0$).

Rearranging the Michaelis-Menten equation [Eq. (7.9)] gives

$$r_{A_0} = V_{max} - K_m \frac{r_{A_0}}{[C_A]_0} \tag{7.16}$$

The second approach consists of drawing a graph of r_{A_0} vs. $r_{A_0}/[C_A]_0$, which will give a straight line with an intercept of V_{max} and slope K_m. This is called the *Eadie-Hofstee transformation*.

If the Michaelis-Menten equation [Eq. (7.9)] is inverted, it takes up the form

$$\frac{1}{r_{A_0}} = \frac{1}{V_{max}} + \frac{K_m/V_{max}}{[C_A]_0} \tag{7.17}$$

In the third graphical approach a graph is drawn between $1/r_{A_0}$ vs. $1/[C_A]_0$. This will once again give a straight line with a slope of K_m/V_{max} and intercept $1/V_{max}$ (see Fig. 7.4). This plot is known as *Lineweaver-Burke plot*.

Equation (7.17) can be further rearranged to give

$$\frac{[C_A]_0}{r_{A_0}} = \frac{[C_A]_0}{V_{max}} + \frac{K_m}{V_{max}} \tag{7.18}$$

A graph of $[C_A]_0/r_{A_0}$ vs. $[C_A]_0$ will once again give a straight line with an intercept of K_m/V_{max} and slope $1/V_{max}$. This is the fourth graphical approach and is known as *Hanes plot*.

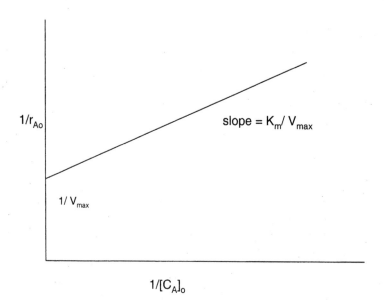

FIGURE 7.4 Lineweaver-Burke plot for single-substrate enzyme-catalyzed reaction.

7.2.3 Non-Michaelis-Menten Kinetics

The assumptions in deriving the Michaelis-Menten kinetics are that

1. It reaches a steady state fast (hence one can ignore the unsteady dynamics).
2. The formation of the enzyme substrate complex is an equilibrium reaction.
3. The product formation step is not reversible (i.e., product does not bind to an enzyme to generate a product enzyme complex).

Deviations from these assumptions could lead to kinetic expressions that are different from the Michaelis-Menten form.

Allosteric enzymes often show sigmoid kinetics (i.e., non-Michaelis-Menten), which means that such enzymes are very sensitive to small changes in substrate concentration. Sigmoid kinetics is a consequence of interaction between sites or due to the presence of sites that bind substrate other than the active site. For example, studies with cytochrome P450 3A4 on diazepam, temazepam, and nordiazepam indicated that the enzyme contains two substrate-binding sites in a single active site that are both distinct and cooperative.

Presence of inhibitor can also lead to deviations from Michaelis-Menten type of behavior.

7.2.4 Inhibition

Two types of inhibitions are observed namely, reversible and irreversible. An inhibition is termed reversible if it is possible to cancel out the inhibition action by decreasing its concentration (e.g., by dilution or dialysis). The bonding betwee the enzyme and the substrate is generally weak. In the latter there is no reversal of the inhibition action, due to the formation of covalent bonds.

Several types of reversible inhibitions are observed and they are discussed below.

Competitive Inhibition

Two types of inhibition are possible here:

Active Site Binding. Here the inhibitor is a compound, which has a close structural and chemical similarity to the substrate of the enzyme. Because of this similarity the inhibitor binds to the active site in place of the substrate. The inhibitor simply blocks the active site, and it is impossible for both of them to bind to the active site at the same time. Malonate is a competitive inhibitor of the reaction succinate to fumarate catalyzed by succinate dehydrogenase. Sulpha-nilamide, once used as a drug, is a competitive inhibitor for bacterial enzymes involved in the biosynthesis of coenzyme tetrahydrofolate from p-aminobenzoic acid. Both sulphanilide and p-aminobenzoic acid have similar structures. Gleevac, a drug, targets a specific ab-normal tyrosine kinase, which is made from a cancer-causing gene. It is highly active in chronic myelogenous leukemia and an uncommon type of gastrointestinal cancer.

Conformational change. In the second type, the inhibitor binds not to the active site but to another site that causes a conformation change in the active site of the enzyme such that the substrate can no longer bind to it.

The inhibition can be represented in the equation form as

$$E + I \rightleftharpoons EI$$

The inhibition constant is $K_i = [E][I]/[EI]$. The binding takes place only with the free enzyme producing an inactive dead-end complex.

The Michaelis-Menten kinetic equation can be written as

$$r_A = \frac{V_{\max}[C_A]}{[C_A] + K_m(1 + [I]/K_i)} \tag{7.19}$$

Inverting the equation and substituting initial conditions gives

$$\frac{1}{r_{A_0}} = \frac{1}{V_{max}} + \frac{K_m}{V_{max}} \frac{1 + [I]_0/K_i}{[C_A]_0} \tag{7.20}$$

A Lineweaver-Burke plot of $1/r_{A_0}$ vs. $1/[C_A]_0$ give straight lines with varying slopes and constant intercept. The slope increases with increasing inhibitor concentration. Once again a plot of slopes vs. $[I]_0$ will give a straight line with intercept K_m/V_{max} and slope $K_m/V_{max} K_i$. All three constants can be estimated graphically by this technique (Fig. 7.5).

Another graphical approach is the Dixon plot (Fig. 7.6), which consists of plotting $1/r_{A_0}$ vs. $[I]_0$ for different starting substrate concentrations $[C_A]_0$. The slope of these straight lines will be $K_m/(V_{max} [C_A]_0 K_i)$ and the intercept $(K_m/[C_A]_0 + 1)/V_{max}$.

The Hanes equation for competitive inhibition will be

$$\frac{[C_A]_0}{r_{A_0}} = \frac{[C_A]_0}{V_{max}} + \frac{K_m}{V_{max}}\left(1 + \frac{[I]_0}{K_i}\right) \tag{7.21}$$

A plot of $[C_A]_0/r_{A_0}$ vs. $[C_A]_0$ will give a straight line, with the intercept varying as a function of inhibitor concentration.

Uncompetitive Inhibition

Here the inhibitor binds only to the enzyme-substrate complex producing a dead-end complex, thereby preventing product formation. This type of inhibition is not generally observed, and one example is the inhibition of arylsulphatase by hydrazine. The inhibition can be represented as

$$EA + I \rightleftharpoons EAI$$

The Michaelis-Menten kinetic equation will be of the form

$$r_A = \frac{V_{max}[C_A]}{[C_A](1 + [I]/K_i) + K_m} \tag{7.22}$$

The Lineweaver-Burk equation at initial conditions will be

$$\frac{1}{r_{A_0}} = \frac{1 + [I]_0/K_i}{V_{max}} + \frac{K_m/V_{max}}{[C_A]_0} \tag{7.23}$$

The intercept will vary while the slope will remain constant for different inhibitor amounts and, and the three constants can be estimated graphically as before (Fig. 7.7).

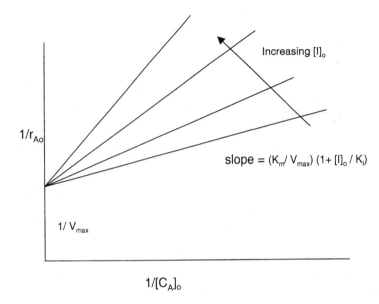

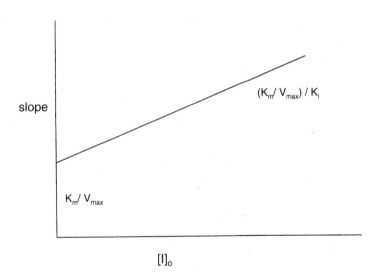

FIGURE 7.5   Lineweaver-Burke plot for single-substrate enzyme-catalyzed reaction with competitive inhibition.

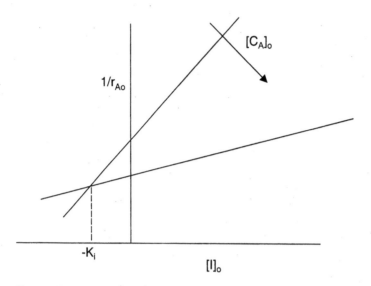

FIGURE 7.6 Dixon plot for single-substrate enzyme-catalyzed reaction with competitive inhibition.

Noncompetitive Inhibition

Here the inhibitor is assumed to bind with equal affinity to both the free enzyme as well as the enzyme-substrate complex. Inhibition constant is assumed to be the same for both the equations.

$$E + I \rightleftharpoons EI \qquad EA + I \rightleftharpoons EAI$$

The enzyme *chymotrypsin*, which has an active site that can accept a proton, can be inhibited by increasing hydrogen ion concentration. Other examples of this inhibition are heavy metral ions and organic molecules that bind to the $-SH$ group of cysteine residue in the enzyme and cyanide that binds to the metal ions of metalloenzymes.

The Michaelis-Menten kinetic equation will be

$$r_A = \frac{V_{max}[C_A]}{([C_A] + K_m)(1 + [I]/K_i)} \qquad (7.24)$$

Lineweaver-Burke equation at initial conditions will be

$$1/r_{A_0} = \frac{(1 + [I]_0/K_i)}{V_{max}} + \frac{K_m(1 + [I]_0/K_i)}{V_{max}[C_A]_0} \qquad (7.25)$$

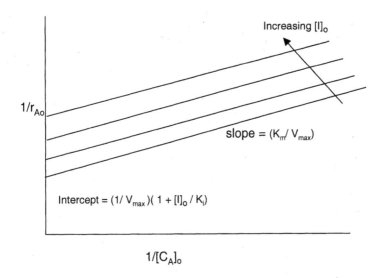

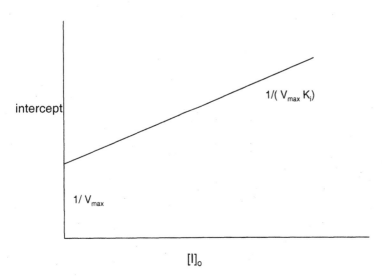

FIGURE 7.7 Lineweaver-Burke plot for single-substrate enzyme-catalyzed reaction with uncompetitive inhibition.

Both slope and the intercept will change for varying concentrations of inhibitor, and the three constants can be estimated graphically as before (Fig. 7.8) from the slope vs. $[I]_0$ and intercept vs. $[I]_0$ graphs. In this case all the lines will meet on the negative axis of $1/[C_A]_0$.

Mixed Inhibition

The action is termed as *mixed inhibition* if the inhibitor constant for both the steps is different.

$$E + I \rightleftharpoons EI \quad \text{(inhibitor constant} = K_i)$$

$$EA + I \rightleftharpoons EAI \quad \text{(inhibitor constant} = K_I)$$

In the Lineweaver-Burke plot the lines will not intersect at the $1/r_{A_0}$ or the $1/[C_A]_0$ axes. If $K_I > K_i$, then the straight lines will cross to the left of the $1/r_{A_0}$ axis, but above the $1/[C_A]_0$ axis as shown in Fig. 7.9a. If $K_I < K_i$, then the straight lines will cross below the $1/[C_A]_0$ axis (Fig. 7.9b).

Another mechanism that is observed in enzyme catalysis is called *allosteric control*. Such enzymes have two sites, an active and a control site. Regulators can bind to the control site and change the structure of the enzyme or actually improve the geometry of the active site and make the enzyme more effective. Such regulators are called *positive regulators*. Other regulators reduce the effectivenss of the enzyme and are called *negative regulators* and are thus like noncompetitive inhibitors.

Substrate and Product Inhibition

In the reaction succinate to fumarate, by the enzyme succinate dehydrogenase, at very high concentrations of the substrate, two carboxyl groups of two separate substrate molecules bind to the active site leading to deactivation. Substrate inhibition is observed during the hydrolysis of ethyl butyrate by sheep liver carboxylesterase and during cholineester splitting process. The inhibition could be uncompetitive type or competitive type. The kinetic model for substrate inhibition is

$$r_A = \frac{V_{\max}[C_A]}{K_m + [C_A] + [C_A]^2/K_i} \tag{7.26}$$

Product inhibition is a phenomena observed at the end of the reaction after accumulation of sufficient quantity of the product. Ethanol and several other alcohols act as inhibitor in processes where it is produced (such as in fermentation). The model will be of the form

$$r_A = \frac{V_{\max}[C_A]}{K_m + [C_A] + [C_A][C_p]/K_i} \tag{7.27}$$

where $[C_p]$ is the product.

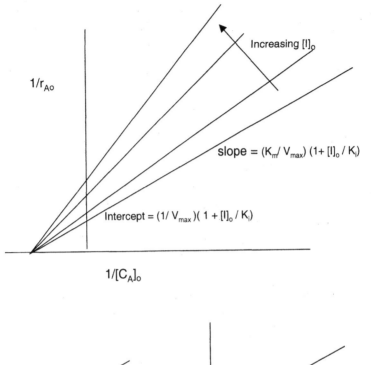

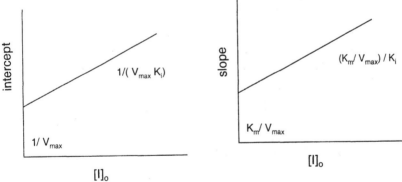

FIGURE 7.8 Lineweaver-Burke plot for single-substrate enzyme-catalyzed reaction with noncompetitive inhibition.

7.2.5 Enzyme Deactivation

Many weak secondary intramolecular forces, giving rise to flexibility, stabilize the enzyme structure but this also leads to facile denaturation and deactivation. Factors like temperature, solvents, chemicals, etc., can effect denaturation. Many proteins start denaturing irreversibly around 45°C.

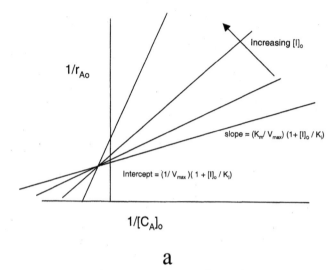

a

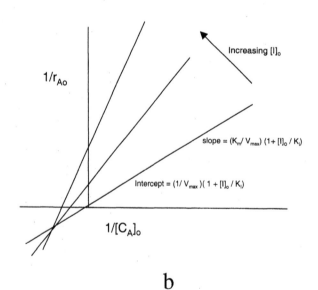

b

FIGURE 7.9 Lineweaver-Burke plot for single-substrate enzyme-catalyzed reaction with mixed inhibition.

Nonaqueous solvents affect the enzyme in a variety of ways including shifting the equilibrium from active conformation to inactive conformation, competing with substrate binding, dissociating the multiple units of the enzyme, and altering the amount of helix. Solvents may affect the water balance, which keeps lipase active thereby causing deactivation. Penicillin (GAG 14.17) covalently reacts with an essential serine residue in the active site of glycoprotein peptidase. This enzyme cross-links peptidoglycan chains during synthesis of bacterial cell walls.

Simplest deactivation model could be of the exponential decay form

$$\frac{de}{dt} = -k_d e \tag{7.28}$$

The decay constant depends on temperature and may follow an Arrehenius type of behavior, with activation energy in the range of 40 to 80 kcal/mol.

Certain enzymes exhibit rapid inactivation initially as a result of minor conformational changes followed by a decelerated decay, approaching a plateua. These are two most common modeling approaches for the deactivation mechanism: The first is A parallel deactivation scheme in which the native enzyme is thought to exist in multiple forms (in equal amounts) with different activity and stability, so that the overall activity (e) consists of summation of decaying exponentials of different time constants as shown below:

$$\frac{de_i}{dt} = -k_{di} e_i \tag{7.29}$$

$$e = \sum e_i \tag{7.30}$$

(at time = 0, all e_i = same). In the alternative mechanism, the native enzyme is assumed to deactivate through a series of intermediates, which are considered to be in equilibrium with one another.

$$\frac{de_i}{dt} = k_{di-1} e_{i-1} - k_{di} e_i \tag{7.31}$$

$$e = \sum e_i \tag{7.32}$$

(at time = 0, all e_i = 0, except e_1). The third postulated reason could be due to the formation of aggregates. As the enzyme particles start aggregating, their activity could decrease.

7.2.6 Bisubstrate Reactions

If one of the substrates is in large excess, then pseudo-first-order conditions can be assumed, and the models discussed in the previous sections could be used. Two types of mechanisms are possible in a bisubstrate reaction, namely, ternary complex and the ping-pong.

Ternary Complex Model

In this model it is assumed that a complex is formed between the enzyme and both the substrates to form a ternary complex. This proximity leads to the formation of the products and their subsequent release.

$$E + A \rightleftharpoons EA$$

$$EA + B \rightleftharpoons EAB$$

$$EAB \rightarrow P + Q + E$$

In the random order ternary complex model the reactants have no preferred order for absorption, which means either substrate A or B could be absorbed first, followed by the second substrate. In the compulsory order ternary complex model, one of the substrates is assumed to be absorbed first, followed by the second, and the order of absorption does not get reversed. Glycidol butyrate synthesis in chloroform using porcine pancreatic lipase, action of alcohol dehydrogenase during the reversible reaction between NAD+ and ethanol to form acetaldehyde and NADH, reaction between citronellol and lauric acid in heptane as solvent with lipase, and protein kinases used as an anticancer agent are assumed to proceed through this mechanism.

The rate equation for the ternary complex model is

$$r_{A_0} = \frac{V_{\max}[C_A]_0[C_B]_0}{K_m^B[C_A]_0 + K_m^A[C_B]_0 + [C_B]_0[C_A]_0 + K} \tag{7.33}$$

This model has four constants, unlike the single-substrate model, which has only two. At very large $[C_B]_0$, the model simplifies to single-substrate Michaelis-Menten kinetic equation [Eq. (7.9)]. The graphical plot of $1/r_{A_0}$ vs. $1/[C_A]_0$ will give straight lines as shown in Fig. 7.10a with slope and intercept varying with $[C_B]_0$.

Ping-Pong Mechanism

In this mechanism unlike the previous case both the substrates do not absorb at the same time:

$$E + A \rightleftharpoons EA$$

$$EA \rightarrow P + E'$$

$$E' + B \rightleftharpoons E'B$$

$$E'B \rightarrow Q + E$$

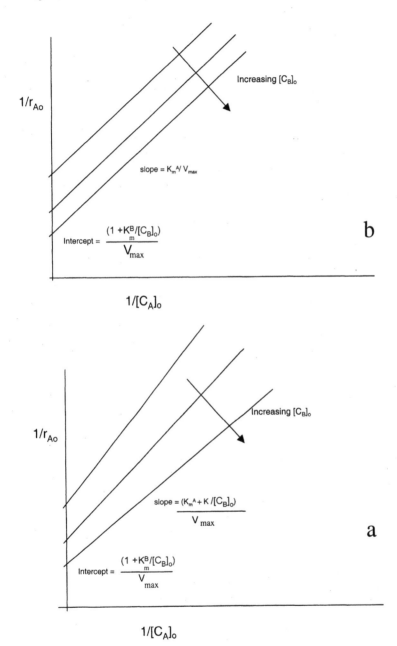

FIGURE 7.10 Lineweaver-Burke plot for bisubstrate enzyme-catalyzed reaction: (a) ternary complex model and (b) ping-pong mechanism.

E' is the modified form of the enzyme. The rate equation for this mechanism is given as

$$r_{A_0} = \frac{V_{max}[C_A]_0[C_B]_0}{K_m^B[C_A]_0 + K_m^A[C_B]_0 + [C_B]_0[C_A]_0} \qquad (7.34)$$

A plot of $1/r_{A_0}$ vs. $1/[C_A]_0$ will give straight lines with intercept varying with $[C_B]_0$ as shown in Fig. 7.10b. This mechanism is used to describe the kinetics of triglyceride synthesis in cyclohexane using *C. rugosa* lipase, transfer of an amino group from an amino acid to an α-keto acid by amino transferases and isoamyl synthesis using lipozyme in *n*-hexane. Yeast hexokinease has a ternary complex mechanism, but brain *hexokinease* proceeds by ping-pong mechanism involving an E-P intermediate.

Production of ethyl oleate using *M. miehei* lipase immobilized on resin beads can be explained by a ping-pong mechanism with an inhibition term to account for the formation of dead-end complex by ethanol. The kinetic equation will then be

$$r_{A_0} = \frac{V_{max}[C_A]_0[C_B]_0}{K_m^B[C_A]_0 + K_m^A[C_B]_0(1 + [C_B]_0/K_i) + [C_B]_0[C_A]_0} \qquad (7.35)$$

Classical kinetic approach is adopted for the synthesis of oleic acid ester of isopropylidene glycerol with *M. miehei* lipase.

$$A + B \rightleftharpoons C + D$$

$$\frac{dC_A}{dt} = k_1 C_A^2 + k_{-1} C_c^2 \qquad (7.36)$$

Both the forward reaction and backward are assumed to be second order. A similar modeling approach is adopted for the transesterification of substituted ethanols with ethyl acetate using *C. rugosa* and PPL lipases.

Inhibitors, which compete with one of the substrates for a site on the enzyme, can give useful mechanistic information. For example, a product of the reaction if present in excess at the start of the reaction itself can compete with one of the substrates and slow down the rate of the forward reaction.

7.2.7 Batch Cell Growth: Structured and Unstructured Models

A process model helps in the process improvement and design stages. In some cases simplified models for process design can be sufficient. However, in other cases it may be advantageous to use structured, fuzzy-rule-based, or hybrid models. An unstructured model could describe the basic behavior of a biochemical process. First principle models are derived from the physics and the chemistry of the process and are useful when the experimental data set is limited. Black box models relating the system inputs with the system outputs through mathematical transfer functions can also be developed, where the parameters in the equations may not have any physical significance. Accurate data collection is the key to the success of this modeling exercise.

The rate of substrate consumption (C), cell growth (x), and product formation (P) are related by the equations

$$-\frac{dC}{dt} = \frac{1}{Y_{xc}} \frac{dx}{dt} \tag{7.37}$$

and

$$-\frac{dP}{dt} = \frac{1}{Y_{pc}} \frac{dC}{dt} \tag{7.38}$$

Y_{xC} is the yield factor relating substrate utilization to the cell number produced and Y_{PC} the yield factor relating product formation and the substrate utilization.

One of the simplest models for rate of increase in cell number (x) in a batch culture is

$$\frac{dx}{dt} = \mu x \tag{7.39}$$

Incorporation of an inhibition factor could lead to

$$\frac{dx}{dt} = \mu x(1 - \alpha x) \tag{7.40}$$

where μ is the specific growth constant and is given as

$$\mu = \mu_{max} \frac{C_A}{C_A + K_s} \tag{7.41}$$

This substrate limited growth equation is called *Monod equation* (described more in detail in Chapter 9).

L-Sorbose production from sorbitol is inhibited both by the substrate and product and the equation for the specific growth is given below:

$$\left[\frac{\mu_m C_s}{K_s + C_s}\right]\left[1 - \left(\frac{C_s}{C_{sm}}\right)^a\right]\left[\frac{K_p}{K_p + C_p}\right] \tag{7.42}$$

A delayed response model of the form given below describes continuous culture of *E. coli* 23716 buildup, under conditions of limiting glucose concentration. The rate equation considers the substrate concentration from the previous time step and not at the present time.

$$\frac{dx}{dt} = fn(x(t - \tau), C(t - \tau)) \tag{7.43}$$

The function could contain terms like $[1 - e^{-1/\tau}]$.

Oxygen plays a major role in aerobic processes and appears as a limiting term in the kinetic equation

$$\frac{dx}{dt} = \left[\frac{\mu_{max}C_A}{C_A + K_s}\right]\left[\frac{\mu_{max,O_2}C_L}{C_L + K_{O_2}}\right] x \tag{7.44}$$

The rate of oxygen transfer from the gas phase is given by

$$K_L a(C_L^* - C_L) \tag{7.45}$$

In many aerobic and anaerobic processes, pH changes during the course of the reaction and, if not controlled, could influence cell growth and act as an inhibitor. Then the equation for specific growth rate could be

$$\mu = \frac{\mu_{max}C_A[H^+]}{K_s + C_A[H^+] + K_{HS}C_A^2[H^+]^2} \tag{7.46}$$

The maximum growth rate is generally dependent on temperature in the form of an Arrehenius-type relation.

7.3 SUPPORTED ENZYMES

Due to immobilization, external and internal mass transfer resistances hinder the accessibility of the substrates to the enzyme surface. The former occurs when the rate of diffusion of the substrates from the bulk medium to the surface is less than the rate of reaction. This generally depends on degree of mixing of the reaction mixture and thickness of the stagnant film surrounding the enzyme particle. The internal mass transfer resistance arise when the rate of substrate diffusion from the support surface to the enzyme surface becomes rate limiting or controlling. This happens since the substrate has to diffuse through the pores of the support to reach the enzyme surface.

The rate of mass transfer from the bulk of the liquid to the surface of the solid particle is given by the equation $K_s(C_A - C_A^s)$. The concentration of the substrate in the bulk (C_A), and on the particle surface (C_A^s) will be different, the latter less than the former (as shown in Fig. 7.11). K_s is the mass transfer coefficient, which is a function of the diffusion coefficient (D_s) and thickness of

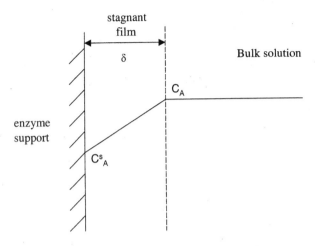

FIGURE 7.11 Substrate mass transfer from bulk to enzyme surface.

the stagnant layer surrounding the enzyme (δ). From Fick's law of diffusion $K_s = D_L/\delta$. If the reaction follows Michaelis-Menten kinetics, then rate of consumption of the substrate will be

$$r_A = \frac{V_{\max}[C_A^s]}{K_m + [C_A^s]} \tag{7.47}$$

The reaction is dependent on the surface concentration, and not on bulk concentration. At steady state both these rate terms have to be equal which leads to

$$\xi = \frac{C_A^s}{C_A} = \frac{\beta}{2}\left(\pm\sqrt{1 + \frac{4\kappa}{\beta^2}} - 1\right) \tag{7.48}$$

where

$$\beta = \mathrm{Da} + \kappa - 1 \tag{7.49}$$

$$\mathrm{Da} = \frac{V_{\max}}{K_s C_A} \tag{7.50}$$

$$\kappa = \frac{K_m}{C_A} \tag{7.51}$$

Da, the Damköhler number, is the ratio of maximum reaction to maximum mass transfer rates. If it is less than unity, the maximum mass transfer rate is larger than the maximum reaction rate, indicating low mass transfer resistance and the system is in the reaction-limited regime. If Da is greater than unity, maximum reaction rate is greater than the maximum mass transfer rate, indicating low reaction resistance and the system is in the diffusion- or mass-transfer-limited regime. Increasing agitation or decreasing the particle size can speed up the mass transfer process.

In the presence of mass transfer resistance the effectiveness factor of the process is defined as

$$\eta = \frac{\xi/(\kappa + \xi)}{1/(\kappa + 1)} \tag{7.52}$$

which lies between 0 and 1. Diffusion resistance becomes controlling at high enzyme loading on the surface, low bulk substrate concentration, low substrate diffusivity, low K_m, high enzymic specificity, low rate of stirring or agitation, and large average particle diameter.

Most of the supports are porous in nature so the diffusion of the substrate through the pores of the support also may play some role in the reaction rate process. The internal diffusion process depends on the shape of the particles, pore size distribution, shape of the pores, porosity, tortuosity, the effective diffusivity of the reactants and products within the pores, and the degree of uniformity of the enzyme's distribution within the particles. The concentration gradient of the substrate inside a spherical particle can be determined by solving the following equation:

$$\rho^2 \frac{\delta^2 C}{\delta \rho^2} + 2\rho \frac{\delta C}{\delta \rho} = \frac{R^2 V_{max}}{[C_A^R] D_{rL}} \frac{\rho^2 \beta C}{1 + \beta C} \tag{7.53}$$

where $\rho = r/R$, $\beta = [C_A^R]/K_m$, and $C = [C_A^r]/[C_A^R]$. $[C_A^R]$ is the concentration of the substrate at the surface of the pellet, which is assumed to be spherical in shape with radius $= R$ (the surface concentration is also equal to the bulk concentration if the external mass transfer resistance is small), and $[C_A^r]$ is the substrate concentration inside the pellet at $r = r$.

Thiele modulus φ is defined as

$$\phi = R\sqrt{\frac{V_{max}}{K_m D_{rL}}} \tag{7.54}$$

The square of the Thiele modulus is the ratio of first-order reaction rate to diffusion rate. Another observable modulus (ϕ') is defined as

$$\phi' = \frac{R^2 r_A}{9 D_{rL}[C_A^R]} \tag{7.55}$$

where r_A is the observed rate. All the parameters defined here are measurable, and Fig. 7.12 shows graphically the relationship between ϕ' and effectiveness factor η. At very low ϕ' (i.e., small particle diameters, low reaction rates or high diffusion coefficient) effectiveness factor is equal to 1 and it decreases sharply as ϕ' increases beyond 1.

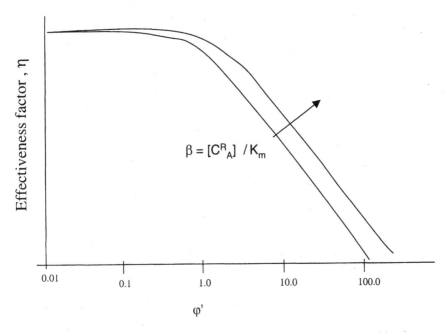

FIGURE 7.12 Effectiveness factor for substrate utilization with enzyme kinetics.

PROBLEMS

1. For the binding of a certain ligand to a protein at 25°C, $K_d = 2 \times 10^{-7}$ M. What is the free-energy change for association of the ligand with the protein under standard conditions? (For the dissociation reaction, $\Delta G^0 = -RT \ln K_d$.)

2. If $K_d = 0.5 \times 10^{-7}$ M for the binding of a certain ligand to a protein at 25°C, what is the free-energy change for dissociation of the ligand with the protein under standard conditions?

3. If the free-energy change and enthalpy change of denaturation of a protein are $+13$ and 40 kcal/mol at 30°C, determine the corresponding entropy change.

4. If at 25°C, the equilibrium constant of a reversible reaction is $= 3.5 \times 10^{-7}$, determine the free-energy change. Explain whether the reaction will proceed forward or in reverse.

5. Calculate the ratio of rate constants for an enzyme-catalyzed reaction and uncatalyzed reaction when difference in activation energy between the two is 35.0 kJ/mol (temperature $= 30°C$).

6. For a system obeying Michaelis-Menten kinetics, if rate $= 35$ mmol/min and $[C_A] = K_m$, what is the value of V_{max}?

7. Derive the kinetic equation for a competitive inhibition.

8. Derive the kinetic equation for a noncompetitive inhibition.
9. Derive the kinetic equation for the uncompetitive inhibition.
10. Derive the kinetic equation for the mixed inhibition.
11. Pictorially describe what the Dixon plot will look like for these inhibitions.
12. How can one differentiate compulsory-order ternary complex bisubstrate reaction and random-order ternary complex bisubstrate reactions? Which is more complex described kinetically? Why?
13. Two different enzymes having same V_{max} are able to catalyze reaction, A → B. For enzyme 1, the K_m is 1.0 mM, for enzyme 2 the K_m is 10 mM. When Enzyme 1 was used with 0.1 mM A, it was observed that B was produced at a rate of 0.0020 mmol/min. What is the value of V_{max}? What will be the rate of production of B when Enzyme 2 is used with 0.1 mM A?
14. A competitive inhibitor ($K_I = 2 \times 10^{-5}$ M) binds to an enzyme that has a true $K_m = 1.8 \times 10^{-6}$ M for its substrate and a V_{max} of 2.1×10^{-4} mol/min. Calculate the apparent K_m value in the presence of 1.5×10^{-3} M inhibitor. What is the maximum velocity that could be observed in the presence of the competitive inhibitor?
15. If $V_{max} = 145$ mmol/min and $v_0 = 80$ mmol/min at 78 mM substrate for an enzyme that obeys Michaelis-Menten kinetics, what is its K_m?
16. Calculate the ratio $[C_A]/K_m$ when the velocity of an enzyme-catalyzed (no inhibitor) reaction is 15% of V_{max}?
17. Estimate V_{max} and K_m for the following enzyme catalyzed reaction data.

Initial substrate concentration (mmol/l)	1	4	6	8	12	15	20
Initial rate (mmol/l/min)	0.001875	0.002609	0.002727	0.002791	0.002857	0.002885	0.002913

18. What will be the value of V_{max} for doubling the reaction rate given below?

Initial substrate concentration (mmol/l)	1	4	6	8	12	15	20
Initial rate (mmol/l/min)	0.001389	0.001802	0.001863	0.001896	0.001929	0.001943	0.001957

19. The initial rate without and with an inhibitor is given below. Determine the nature of inhibition and estimate V_{max}, K_m, and K_i.

Initial substrate concentration (mmol/l)	1	4	6	8	12	15	20
Initial rate (mmol/l/min)	0.000571	0.000727	0.00075	0.000762	0.000774	0.000779	0.000784
Initial rate in the presence of 0.2 mmol/l of inhibitor	0.000125	0.000131	0.000132	0.000132	0.000133	0.000133	0.000133

20. The initial rate without and with an inhibitor is given below. Determine the nature of inhibition and estimate V_{max}, K_m, and K_i.

Initial substrate concentration (mmol/l)	1	4	6	8	12	15	20
Initial rate (mmol/l/min)	0.000571	0.000727	0.00075	0.000762	0.000774	0.000779	0.000784
Initial rate in the presence of 1 mmol/l of inhibitor	1.93E-05	7.21E-05	0.000103	0.000132	0.000183	0.000217	0.000265

21. The initial rate without and with an inhibitor is given below. Determine the nature of inhibition.

Initial substrate concentration (mmol/l)	1	4	6	8	12	15	20
Initial rate (mmol/l/min)	0.000429	0.000545	0.000563	0.000571	0.000581	0.000584	0.000588
Initial rate in the presence of 0.1 mmol/l of inhibitor	0.000143	0.000182	0.000188	0.00019	0.000194	0.000195	0.000196

22. The initial rate without and with an inhibitor is given below. Determine the nature of inhibition.

Initial substrate concentration (mmol/l)	1	4	6	8	12	15	20
Initial rate (mmol/l/min)	0.000393	0.0005	0.000516	0.000524	0.000532	0.000536	0.000539
Initial rate in the presence of 0.2 mmol/l of inhibitor	8.59E-05	0.000103	0.000105	0.000106	0.000107	0.000108	0.000108

23. The initial rate data for a bisubstrate reaction is given below. Determine the type of mechanism.

Initial C_{A_0} concentration (mmol/l)	1	4	6	8	12	15	20
Initial rate with $C_{B_0} = 0.2$ mmol/l	0.000163	0.000384	0.000452	0.000496	0.000549	0.000574	0.000601
Initial rate with $C_{B_0} = 0.1$ mmol/l	9.59E-05	0.000272	0.000341	0.000392	0.000459	0.000493	0.000532

24. The initial rate data for a bisubstrate reaction is given below. Determine the type of mechanism.

Initial C_{A_0} concentration (mmol/l)	1	4	6	8	12	15	20
Initial rate with $C_{B_0} = 0.3$ mmol/l	0.000124	0.00015	0.000154	0.000156	0.000158	0.000158	0.000159
Initial rate with $C_{B_0} = 0.2$ mmol/l	7.02E-05	8.49E-05	8.7E-05	8.8E-05	8.91E-05	8.95E-05	9E-05
Initial rate with $C_{B_0} = 0.1$ mmol/l	4.9E-05	5.92E-05	6.06E-05	6.13E-05	6.21E-05	6.24E-05	6.27E-05

25. The V_{max} and K_m for the unsupported enzyme catalyzed reaction are 0.002 mmol/l/min and 0.44 mmol/l. For the supported enzyme the bulk mass transfer coefficient is 0.05/min. What is the substrate concentration if the bulk concentration is 1.0 mmol/l?

26. In the presence of a isomerase enzyme V_{max} is 3.6 and 7 μmol/min at 25 and 35 °C. In the absence of the biocatalyst, the rate is 5.5 and 16 nmol/min. Estimate the activation energies of the catalyzed and the uncatalyzed reactions.

NOMENCLATURE

C_L^*	Equilibrium oxygen solubility in the solvent
C_L	Oxygen concentration in the solvent
$[C_A]_0$	Initial concentration of the substrate
$[C_A]$	Substrate concentration
$[C_A^s]$	Substrate concentration on the support surface
$[C_A^r]$	Substrate concentration inside the pellet at r = r.
$[C_A^R]$	Substrate concentration inside the pellet at r = R.
Da	Damköhler number
E	Activation energy
$[E]_0$	Initial enzyme concentration
$[E]$	Free enzyme active site
$[EA]$	Enzyme substrate complex
K_s	Substrate mass transfer coefficient to the particle surface
R	Gas constant
R	Radius of the pellet
T	Temperature
V_{max}, K_m	Michaelis-Menten constants
a	Gas-liquid interfacial area per volume
r_A	Rate of reaction
k_L	Oxygen gas-liquid mass transfer coefficient
k_0	Frequency factor
k_1, k_{-1}, k_2	Rate constants
D_L	Substrate diffusion coefficient
D_{rL}	The internal (through the pellet) diffusion coefficient
δ	Film thickness
η	Effectiveness factor
φ	Thiele modulus

8

Biochemical Reactor

A wide range of antibiotics, vitamins, amino acids, fine chemicals, and food-stuffs are manufactured biochemically. Detoxification of industrial and domestic waste water is also carried by biochemical means. The heart of a biochemical process is the reactor. In a bioreactor, the transformation of raw materials into desired products is carried out by the enzyme systems, living microorganisms or by isolated enzymes. The reaction products are formed by three basic processes, namely

1. The product is produced by the cells either as extracelluar or intracellular. Examples of the former include alcohols or citric acid
production, and examples of the latter are metabolite or enzyme
production.
2. Production of cell mass, like baker's yeast or single-cell proteins for
food industries.
3. Biotransformations, where the cell catalyzes the conversion via
dehydrogenation, oxidation, hydrogenation, amination, or isomerization. Steroids, antibiotics, or prostaglandins are produced by this
approach.

The type of reactor depends on the nature of the process (including reaction kinetics), operating conditions (namely, mode of operation and gas liquid flow patterns), and physical and chemical properties of the substrates and the

169

microbe. A plethora of biochemical reactors are available commercially, from which the most suited one could be selected based on certain selection criteria. Chapters 11 and 12 deals in more detail agitated and tower reactors, respectively.

In general the special features required by a biochemical reactor are reliable sterilization and ease of maintenance of sterility, rapid mixing (homogenization) of reactor contents, sufficient and economical oxygen supply to the bioculture (aeration), good removal of heat of reaction, efficient retention of the bioculture in the reactor, permitting a high cell concentration, and sophisticated level of instrumentation to maintain constant operating conditions. Good mixing is achieved by proper design of stirrer, revolutions per minute (rpm), and stirrer motor. Oxygen flow rate and correct design of gas distributor in the reactor can ensure high gas-to-liquid mass transfer rate. Reactors are provided with jacket and cooling coils to rapidly remove the exothermic heat generated during the reaction, so that enzymes or the organisms are not exposed to excessive temperatures. As microorganisms are very sensitive to medium pH, reliable process control is very essential in bioreactors.

Biochemical rectors are made of stainless steel to maintain sterile conditions. Simple geometric shape, minimum number of flanges, welds and measuring and sampling nozzles, elimination of dead zones, and minimum surface roughness are also very essential for sterile operation. Valves are flush bottom diaphragm valves to avoid dead pockets and facilitate ease of sterilization.

8.1 REACTOR SELECTION CRITERIA

Reactor selection depends on the reaction kinetics, mode of operation of the reactor, nature of organism, and properties of the medium. This section deals with these aspects.

8.1.1 Mode of Operation

Operationally reactors can be classified as batch, continuous, fed batch, extended fed batch, or repeated fed batch. A batch reactor has neither feed nor product streams. The substrates and biocatalyst are introduced at the start of the reaction, the reaction temperature and/or pressure raised in a preprogrammed manner, maintained at those conditions for a specified amount of time (or until the completion of the reaction), brought back to ambient conditions again in a preprogrammed manner, and finally the contents are discharged. The reactor is cleaned and sterilized using steam or hot water before the next batch is charged. In the continuous reactor, nutrients and organisms are fed and the liquid and gaseous products or effluents are removed continuously, thereby maintaining the reactor volume constant. In the fed batch reactor, feed is introduced either continuously or intermittently while there is no continuous product removal. This leads to an increase in reactor volume with time. The products are discharged at the end of the cycle. These reactors are also called *semibatch reactors*. In the extended fed batch, the feed to the reactor is maintained constant so that there is no change in substrate concentration inside the reactor. In the repeated fed batch a small amount of fermentation broth is left behind as inoculum for the next batch.

Batch reactors have simple construction and are suitable for small production, but may lead to increased batch cycle time on account of downtime due to charging and discharging. Continuous reactors provide high production rates and better product quality due to constant reaction conditions. Continuous reactor can be tubular or a stirred tank. In the tubular reactor the feed enters through one end of a tube and the products exit from the other end. In the continuous stirred tank reactor (CSTR), the products leave the reactor as overflow. Also, the reactor contents are agitated to achieve good mixing. Fermentors are operated in batch, continuous or fed batch mode, where air is sparged continuously into the reactor, and exhaust gases are removed continuously.

Fed batch mode is desirable in fermentors when low glucose concentrations (10 to 50 mg/l) are needed to maintain the productivity of microorganisms. Continuous fermentors are not generally used due to the difficulty of maintaining monosepsis conditions (i.e., preventing contamination by foreign organisms) and challenges posed in controlling the different phases of the growth curve of the microorganism. In fed batch the feed rates and volume-vs-time pattern could be varying, leading to varying substrate concentration inside the reactor, while in the extended fed batch the feed is continuous, thereby maintaining constant substrate concentration.

Continuous bioreactors provide high degree of control and uniform product quality than batch reactors. Whereas the disadvantage are controlling or minimizing the production of non-growth-related products and loss of original product strain over time, if it is overtaken by some faster growing one. In many cases a fraction of the product stream is recycled back to the reactor inlet. Recycle leads to higher conversions.

Kinetic considerations dictate the reactor choice. Mode of operation is immaterial for a zero-order reaction, whereas batch or tubular reactor is desirable for first-order or Michaelis-Menten-type reaction kinetics. Continuous stirred tank reactor is ideally suited for reaction with substrate inhibition, while the other two reactors can be used for product inhibition.

A cascade of continuous stirred tank reactor followed by a tubular reactor can be used if the reaction exhibits a combination of autocatalytic (an initial low yield followed by a sudden increase in the rate) and ordinary kinetic behavior. While autocatalytic behavior predominates during the exponential growth phase, ordinary kinetics will prevail during the stationary and death phases. In type II and type III fermentations (see Chapter 9) large rates of product synthesis and substrate uptake are exhibited after the decline of the growth rate. Although a CSTR may be better from the viewpoint of biomass production, a combination of CSTR-tubular design will maximize product yield.

8.1.2 Nature of Organism

The living cells used in bioreactors can be bacteria, yeasts, hypha, fungi, plant or animal cells. They differ in the demands they make on the environment, nutrient supply and product formation. This results in a wide range of requirements for the bioreactors and the process layout. Stability of the strain determines the mode of operation of the reactor. Mutations of the micro-organisms can occur under suboptimal conditions. Only strains that are sufficiently stable can be used in continuous reactor.

Type of reactor depends on whether the organism is aerobic or anaerobic. In the breeding of aerobic organism, adequate amount of dissolved oxygen must be ensured in the medium. Since the solubility of oxygen in the medium is very low, it must be supplied continuously and gas-to-liquid mass transfer should be maintained high. A minimal critical concentration of dissolved oxygen has to be maintained in the substrate to keep the microorganism active. The critical oxygen concentration values are in the range of 0.003 to 0.05 mmol/l, which is 0.1 to 10% of oxygen solubility values in water. But presence of salts like NaCl decrease oxygen solubility by a factor of 2. The oxygen utilization by the microbe includes cell maintenance, respiratory oxidation for further growth (biosynthesis), and oxidation of substrates into related metabolic end products. Oxygen uptake rate for bacteria and yeast are the highest (0.2–2×10^{-3} kg/m^3/s), followed by fungi (0.1 to 1) and rest of the biocultures (0.01 to 0.001). Metabolic heat generation rates for fungi, bacteria, and yeast are in the range of 2–20 kW/m^3, and for other biocultures, in the range of 0.02–0.15.

The size and shape of the cells also have an influence on the type of reactor. Spherical cells are smaller and less sensitive to shear than filamentous organisms. Small dimension also means a high surface-to-volume ratio, which will lead to high rate of uptake of substrate and oxygen leading to rapid growth. Hence the reactor should be able to operate at very high transfer rates; whereas filamentous organisms generally agglomerate leading to low surface-to-volume ratio, low substrate utilization, and low oxygen uptake. The low rate of oxygen uptake permits the use of reactors, which disperse air only to a slight or moderate degree.

Cells that form agglomerates are easier to separate from the media and returned to the reactor; hence, they permit the continuous operation. With low mechanical stress in the reactor, many filamentous organisms (like mold or fungi), form agglomerates interlocking with one another leading to high apparent viscosity of the medium. When the mechanical stress is increased the same organism may form pellets. This leads to reduction in the apparent viscosity of the medium.

Many organisms tend to grow on surfaces as in the case of metabolite producing organisms or those that are used in sewage treatment. Surface reactors have to be used for systems that exhibit such behavior. Film formation is generally not desirable, and if cells show such a tendency, then stagnant regions at the surface must be avoided by suitable reactor design.

Agitator tips can tear filamentous organisms. Also, shear stresses arising in agitated vessels can cause undesired effects on mycelial morphology and hence the product yields. Tower-type reactors like air lift bioreactors without agitation are suitable for such bioorganisms.

8.1.3 Properties of Medium

The physical properties of the substrate and products produced also have an influence on the selection of reactor. For example, if methane is produced excessively during the reaction, then reactors with large interconnected gas volumes should not be used, since methane air form an explosive mixture. In the case of volatile substrates, cocurrent flow with little axial mixing or multistage units should be used in order to minimize losses in the waste gases. Higher paraffin has a reducing effect on the oxygen transfer rate; hence surface reactors are unsuitable for the fermentation of oil emulsions. Stirred tanks with helical stirrer can be used for highly viscous media. Fine long threads in dilute suspension lowers the flow resistance, and large particles or fibrous materials may block openings. In this case inserts with constrictions must be avoided.

In the case of substrates showing inhibition of growth, the process must be carried out either in a semi continuous reactor with sustained feed of the substrate (fed batch culture), or in a continuous culture. If products induce inhibition, it is desirable to use a multistage arrangement.

When coalescence-suppressing agents like surfactants are used, it is appropriate to apply high rates of energy dissipation in a gas-dispersing device (two-phase nozzle with single stirrer). If the media has coalescence-enhancing properties like antifoaming agents, the rate of energy dissipation must be uniform throughout the reactor.

Viscosity of the medium for fungi is in the range of 0.1 to 1.5 Pa s, and less than 0.1 for bacteria, yeast, and mixed cultures. Low-viscosity media presents no problems with respect to mixing times and oxygen transfer rates. When substrate or product is responsible for high viscosity, then the medium generally has a Newtonian behavior. Whereas, if the high viscosity is due to the organism, then such a system will exhibit a non-Newtonian type of behavior. For low-viscosity fermentations at 50 to 500-m^3 scale, a bubble column reactor is chosen because it is the cheapest, and at 220 to 10,000-m^3 scale, an air lift reactor is chosen because it permits controlled substrate addition. Large-scale stirred fermentors will require very high agitator motors, which could be uneconomical. If the viscosities can rise above 0.1 Pa s (mycelial biopolymer fermentation), a stirred vessel is desired up to a volume of 500 m^3.

Excessive foam formation may lead to loss of material from a fermentor. Foams could be reduced by adding antifoaming agents or by mechanical means.

8.2 TYPES OF REACTORS

The biochemical reactors can be broadly classified into submerged and surface reactors [2,6]. In surface reactors the culture adheres to a solid surface and oxygen is supplied from the gas phase to the continuously wetted solid surface. Wastewater treatment employs such designs. In submerged reactors gas-to-liquid mass transfer is achieved by dispersing the gas in the liquid through continuous input of energy. Submerged reactors can be divided into three groups depending upon the nature of the energy input, viz., (1) mechanically stirred systems with agitators (Chapter 11 deals with this type in more detail), (2) forced convection of the liquid using recirculating pumps, and (3) pneumatic operation by pumping compressed air (bubble columns or airlift reactors). Chapter 12 deals in more detail on tower reactors.

The simplest and widely used reactor is a stirred tank with baffles (Fig. 8.1). Marine propeller, hydrofoil, or pitched blade turbine stirrer is used to achieve axial motion, and flat blade turbine, back swept, or Rushton turbine stirrers are used to achieve radial motion. Rushton turbine is good for gas dispersion. Helical or anchor impellers is used for viscous substrates, while low rpm ribbon agitator is good for ultra-high-viscosity and anchor stirrer is good for high viscosity, blending, and heat transfer applications.

Propeller is used for low to medium viscosities, pitched blade turbine is generally used for solid suspension and blending. Air is introduced from the bottom through a sparging arrangement. The sparger could be single hole, multiple hole, perforated plate, or sintered plate. Gas-aspiring or self-sucking agitator has a hollow shaft through which air is sucked from the top space (or head space) and distributed from the liquid bottom. These agitators are common in food technology, especially in the manufacture of vinegar, yeast,

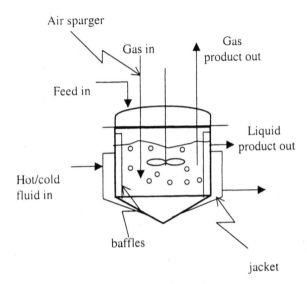

Air sparger

Gas in

Gas product out

Feed in

Liquid product out

Hot/cold fluid in

baffles

jacket

FIGURE 8.1 Continuous stirred tank reactor.

vitamins, and amino acids. The height to diameter ratio of such reactors is less than 3. The reactor is provided with jacket or coils for heating and cooling purposes.

In baffled stirred reactors, complex flow patterns are observed, which can be overcome by providing coaxially arranged cylindrical tube (draught tube). The circulating flow is well defined (Fig. 8.2), and these reactors consume less power than the conventional stirred tank reactors. Also, the power uptake is lower in overflow operation than in the completely filled state.

Stirred cascade reactors are column reactors in which several sections arranged above one another are formed by intermediate plates (Fig. 8.3) Mixing and dispersion of gas takes place in each chamber.

In plunging jet reactors, the gas is dispersed by the free jet from the nozzle impinging perpendicularly on the surface of the liquid. In jet loop reactors the liquid phase is returned from the outlet of the reactor to the inlet. Recycling can take place through a draught tube placed inside the vessel (Fig. 8.4). A circulation pump produces the driving jet, with both the phases in cocurrent. Submerged reactors of these types are used for wastewater treatment.

Perforated plate or sieve plate cascade reactors have countercurrent gas–liquid contact (Fig. 8.5), in which the liquid flows down from one stage to another by overflow pipes. The liquid collected at the bottom is externally recycled back with the help of pump.

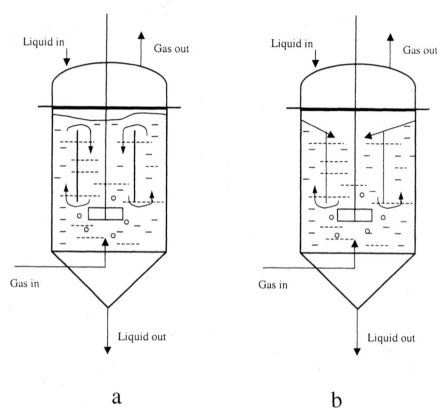

FIGURE 8.2   Draught tube reactor: (a) fully immersed tube, (b) overflow.

In air lift loop reactors the circulation of the liquid is due to the density difference between the mixed phases in the aerated tower and the liquid in the down comer (Fig. 8.6). These reactors have internal draught tube, internal partition or external loop as shown in the figure. The draft tube gives the airlift reactor a number of advantages like preventing bubble coalescence by causing bubbles to move in one direction, distributing the shear stresses uniformly throughout the reactor and thus providing a healthier environment for cell growth. Also circular movement of fluid through the reactor increases the heat transfer rates. As the bubbles in the draft tube rise they also carry the liquid up with them, and when they disengage at the top, the liquid travel down in the down comer section. The heating/cooling jacket is located on the walls of the airlift reactor. A reactor with an external riser will have the advantage of having greater turbulence near the jacket and thus better heat

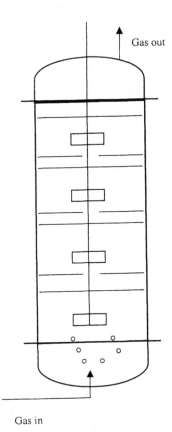

FIGURE 8.3 Multicompartment stirred column reactor.

transfer efficiency. Also the amount of foam produced with an external riser is less than the one with an internal riser. Deep shaft reactors are 50–150 m long and are made of concrete. They are buried underground and are used for sewage treatment. The air in this reactor is not introduced at the bottom, but in the middle. These reactors are also suitable for shear sensitive, foaming, and flocculating organisms (Fig. 8.7).

Bubble columns are slender columns with gas distributor at the bottom (Fig. 8.8). Construction of bubble columns is very simple, and higher mass transfer coefficient than loop reactors can be achieved with them. These reactors can be as large as 5000 m³. Since they have broad residence time distribution and good dispersion property, they can be used for aerobic wastewater treatment and production of yeast.

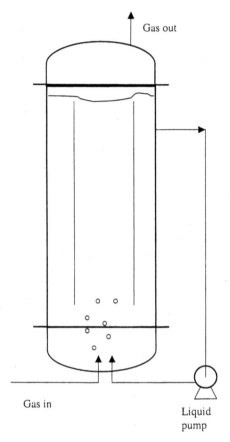

FIGURE 8.4 Jet loop reactor with external liquid recycle.

Gas mixed reactors are not provided with impellers to decrease bubble diameter and increase k_L, whereas they have a larger height to diameter ratio so as to improve oxygen transfer efficiency. This design helps in increasing the pressure at the base of the reactor which increases the saturation concentration of oxygen at the base, increasing the bubble residence time and gas holdup. A very tall reactor will have oxygen starved conditions at the top of the reactor and also large bubble circulation times. These reactors are provided with a disengagement zone at the top with a larger cross section, which leads to the reduction in the fluid velocity. So bubbles thus rise less rapidly and disengage slowly leading to less damage to the cells. Aerosol formation and evaporation are reduced. Bubbles disengage from the liquid and proceed to the reactor exit instead of returning down.

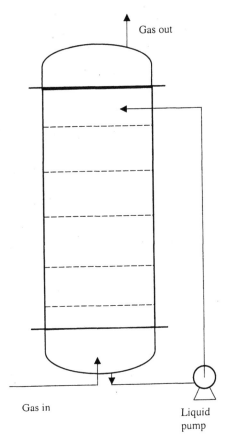

Gas out

Gas in

Liquid pump

FIGURE 8.5 Sieve plate column reactor with liquid recycle.

Packed bed column reactors (Fig. 8.9) are used in enzyme-catalyzed re-actions, sewage treatment or vinegar production. The nutrient or substrate is evenly distributed over the packing through distributor. Air is introduced countercurrent to the liquid flow. In enzyme-catalyzed reactions, supported enzyme is packed in the reactor tube. The pressure drop in this reactor is generally high, but because of the low voidage, the effective enzyme concen-tration is high, and hence the reactor volume is generally smaller than the stirred tank reactor. The lower degree of back mixing found in this reactor when compared to stirred reactor also results in a lower enzyme requirement. Industrially pressurized packed column reactors are used for the production of L-alanine and L-aspartic from ammonium fumarate and magnesium chloride. The reactors are packed with immobilized *E. coli* and *P. dacunnae.*

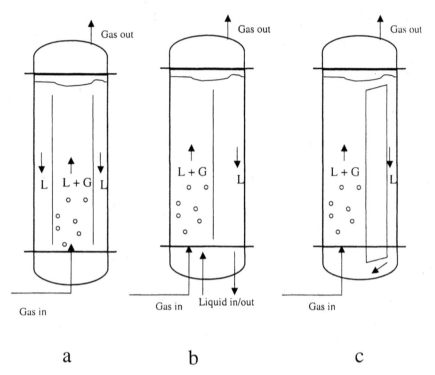

FIGURE 8.6 Air lift loop reactor: (a) internal draught tube, (b) internal partition, (c) external loop.

Conitinuous production of L-maleic acid from fumaric acid using immobilized *B. ammoniagenes* is achieved in a packed bed reactor. 6-Aminopenicillanic acid is prepared from penicillin G industrially in immobilized enzyme columns.

In fluidized bed reactor the fluid flow is such that the solids are in a suspended or fluidized state leading to good gas–liquid contact and mixing. Pressure drop here is low when compared to packed bed reactor. Particles carry over is an issue in this design. These reactors, unlike packed bed reactor, show less tendency to blockage and problems associated with heat and mass transport.

Trickle bed reactor contains packing, which act as a support for the growing of the biocell. The substrate, nutrient, and air flow slowly over the packing in a trickling flow. These reactors are used in the manufacture of vinegar and in waste water treatment.

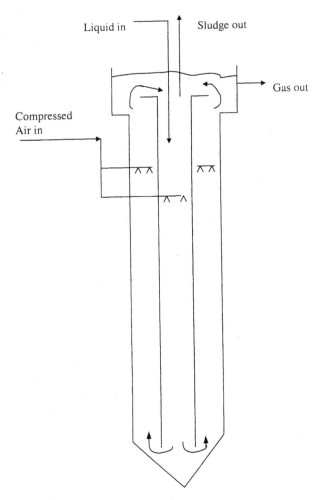

FIGURE 8.7 Deep shaft reactor.

Tubular membrane or hollow fiber reactors consists of tube made of membranes that permit radial diffusion of reactants and products in and out of the membrane while retaining the soluble enzyme inside.

Film reactors consist of vertically arranged tubes or channels, upon which nutrient solution flows from above, while the gas is introduced from the bottom. Unlike packed bed reactors, the gas velocity in these reactors can be varied within wide range. Fermentation is also carried out in tray reactors in which the substrate overflows from one tray to another and the cultures

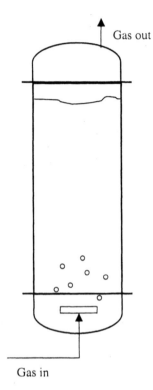

Gas out

Gas in

FIGURE 8.8 Bubble column reactor.

generally can float on the liquid surface. These reactors are ideal for aerobic wastewater treatment and production of acetic acid. The oxygen transfer occurs through the film and reaches the biomass. The rate of oxygen mass transfer here depends on the diffusion coefficient in the medium and the rate at which the fluid near the surface is renewed by the liquid circulation pattern. The oxygen mass transfer coefficient is given by

$$k_L = \sqrt{\frac{D_g}{\pi \tau}} \tag{8.1}$$

where recirculation time τ is equal to the ratio of stream depth to average velocity,

Stirred tank and draft tube reactors have very low oxygen transfer rates (of the order of 4 g/l h), while jet loop, airlift loop, bubble column, and sieve plate column reactors have very high oxygen rates (around 10–12 g/l h). Oxygen transfer rates of packed tubular reactors are of the order of 0.5 g/l h. Power requirements of these reactors are the lowest (0.5 kW/m^3), followed by

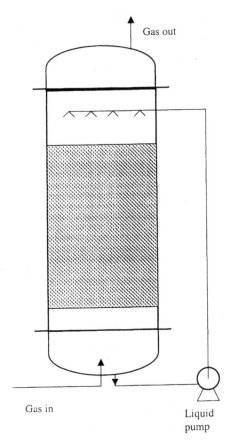

Gas out

Gas in

Liquid
pump

FIGURE 8.9   Packed bed column reactor with liquid recycle.

loop and column reactors (3–5 kW/m^3), and highest for stirred tank and draft tube reactors (8 kW/m^3). Overall heat transfer coefficient for reactors operating with most of the organisms is in the range of 200–1500 W/m^2/k.

8.3 DESIGN EQUATIONS

8.3.1 Batch Reactor

Since there is no in and out flow, whatever is produced is accumulated inside the reactor at constant volume. The batch reaction time is given by

$$t = \int_{C_A(0)}^{C_A(t_b)} \frac{dC_A}{-r_A} \tag{8.2}$$

where $C_A(0)$ and $C_A(t)$ are the initial and final concentrations of the substrate, respectively, inside the rector and r_A is the reaction rate. The rate equation could be a simple one of the form kC_A^n or of the Michaelis-Menten type. The total batch cycle time not only depends on the batch reaction time but also on the time required for charging the reactor, discharging the contents, reactor cleaning, and heating and cooling the contents. The reactor volume depends on the annual production, and the estimation procedure is shown in the Example 8.1.

8.3.2 Fed Batch Reactor

In the fed batch reactor there is only an inflow, but no outflow, leading to a change in reaction volume (V) with time. The feed flow is intermittent leading to fluctuating concentration inside the vessel. The mass balance equation is given as

$$V \frac{dC_A}{dt} + C_A \frac{dV}{dt} = F_i C_{Ai} - r_A V \tag{8.3}$$

F_i and C_{Ai} are the inlet flow rate and input concentration.

8.3.3 Extended Fed Batch

Here the reactor feed is maintained constant leading to constant substrate concentration (i.e., $dC_A/dt = 0$). Then the mass balance equation simplifies to

$$C_A \frac{dV}{dt} = F_i C_{Ai} - r_A V \tag{8.4}$$

8.3.4 Continuous Stirred Tank Reactor

Here the inlet and outlet flow rates are assumed equal (F), and hence the reactor volume is constant. The concentration of the out going fluid (C_{A_0}) is assumed to be that of the concentration of the fluid inside the vessel. ($C_{A_0} - C_{A_i}$) is also called *conversion*. The reactor residence time is given as

$$\tau_{CSTR} = \frac{V}{F} = \frac{(C_{A0} - C_{A_i})}{r_A} \tag{8.5}$$

The flow rate F can be estimated from the annual production, and the reactor volume can be determined from Eq. (8.5) for a proposed conversion.

8.3.5 Plug Flow Reactor

In an ideal reactor the continuous phase flows as a plug: hence, the residence time is given as

$$\tau_{PFR} = \frac{V}{F} = \int_{C_{A_i}}^{C_{A_0}} \frac{dC_A}{-r_A} \tag{8.6}$$

8.3.6 Monod Chemostat

This is an extension of the CSTR model to account for the cell growth. At steady state, if x and x_0 are the cell concentration in the exit stream and in the feed and r_x is the cell formation rate equal to μx, the material balance equation for the cells will be

$$(\mu - D)x + Dx_0 = 0 \tag{8.7}$$

and for the substrate/nutrient will be

$$D(C_{A_i} - C_{A_0}) - \frac{\mu x}{Y} = 0 \tag{8.8}$$

The relationship between the specific growth rate and the substrate concentration is similar to the Michaelis-Menten equation for enzyme-catalyzed reaction {namely, $\mu = \mu_{max}[C_A/(C_A + K_s)]$}.

Often the feed stream to a continuous unit may contain only nutrient, so that $x_0 = 0$. In such a situation $D = \mu$, and the exit cell and substrate concentrations are given as

$$x_{sf} = Y\left(C_{A_i} - \frac{DK_s}{\mu_{max} - D}\right) \tag{8.9}$$

$$C_{A_{sf}} = \frac{DK_s}{\mu_{max} - D} \tag{8.10}$$

As the dilution rate is increased, the exit cell concentration (x) decreases, and at a certain value of D, x reaches zero, which is termed as *wash out*. The value of D at wash out is given as

$$D_{max} = \frac{\mu_{max} C_{A_i}}{C_{A_i} + K_s} \tag{8.11}$$

The rate of cell production per unit reactor volume is equal to Dx_{sf}. The maximum cell production rate is given as

$$Y\mu_{max}\left(C_{A_l} - \frac{DK_s}{\mu_{max} - D}\right)\left(1 - \sqrt{\frac{K_s}{(K_s + C_{A_l})}}\right) \qquad (8.12)$$

8.3.7 Recycle Reactor

When a part of the product stream is recycled back to the reactor, and if the feed does not contain any cells then

$$D(1 - f) = \mu \qquad (8.13)$$

where f is the recycle ratio. The exit cell and substrate concentrations are given as

$$C_{A_{sf}} = \frac{D(1 - f)K_s}{\mu_{max} - D(1 - f)} \qquad (8.14)$$

$$x_{sf} = Y\left[C_{A_l} - \frac{D(1 - f)K_s}{\mu_{max} - D(1 - f)}\right] \qquad (8.15)$$

Higher conversion (or lower $C_{A_{sf}}$) could be achieved by increasing the recycle ratio f. Also higher cell concentration could be achieved through recycle.

EXAMPLE 8.1: ESTIMATION OF REACTOR VOLUME FOR A BATCH PROCESS.

If the annual demand is a tones, then per day production d (in kg) = $a \times 1000/300$ (assuming 300 working days per year) per shift production, s = d/number of batches per day, reactor volume = $s \times$ density of the fluid/0.7 (assumes 70% of the reactor is occupied).

PROBLEMS

1. Compare the performance (residence time to achieve 90% conversion) in a plug flow reactor and in a continuous stirred tank reactor for a first-order reaction.

2. Estimate the wash out flow rate in a continuous-flow bioreactor of volume 100 m^3, feed concentration is 0.15 mol/lt, μ_{max} = 0.15/min and K_s = 0.25 mol/l.

3. Estimate the volume of an extended fed batch reactor for first-order reaction operating to achieve a conversion of 90% within a residence time of 1 h. The initial concentration of the feed is 0.25 mol/l.

4. Estimate the volumes of a CSTR and a plug-flow reactor to achieve 99% conversion, for zero-order, first-order, and second-order reaction kinetics. Comment on the results (initial substrate concentration = 1 M, rate constant = 0.01).

5. The inflow and outflow BOD to a reactor are 280 and 20 mg/l. Determine the reactor volume for a flow of 2 million m^3/day and active solids concentration of 250 mg/l. Y = 0.66, μ_{max} = 3.5/day, K_s = 20 mg/l (the process is aerobic and assume there is no deactivation).

6. Estimate the maximum flow to a reactor of size 10,000 m^3. Decrease in BOD desired is from 1000 to 20 mg/l (Y = 0.45, μ_{max} = 2.4/day, K_s = 90 mg/l). Active solids concentration is 400 mg/l.

7. If the fraction of product recycled back to the reactor is 0.25, how many days are required to achieve the same decrease in substrate amount?

8. Manufacture of an ester at 50,000 l/h is carried out in a supported enzyme packed bed reactor. Assume the liquids to have physical properties similar to that of water. If the liquid–solid mass transfer coefficient is 0.01/min and the reaction follows Monod kinetics with μ_{max} and K_s = 0.2/min and 0.1 mmol/l, respectively. Determine the dimensions of the tubular reactor (assume fraction of packed volume = 0.4).

9. Sometimes tower reactors or tubular reactors are modeled as tanks in series. Estimate the conversion at the exit for a reactor of volume 10 m^3 and flow rate of 3 m^3/h, for two cases where you assume the reactor is divided into three sections and five sections. A first-order reaction is taking place inside the reactor with the rate constant equal to 0.5/h and inlet concentration is 5 mol.

10. Determine the volume a CSTR to produce 1000 kg/h of product. The conversion is 99% and the feed concentration = 1 kg/m^3. μ_{max} = 2.5 × 10^{-4}/m^3 catalyst/s and K_s = 5.5 × 10^{-3} kg/m^3. Reactor contains 0.15 m^3 catalyst per m^3 of reactor volume.

11. Estimate the number of CSTRs needed in series of the same volume to achieve a reduction in concentration form 40 mol/l to 2 mol/l if the organism follows Monod kinetics with μ_{max} and K_s = 0.45/min and 0.2 mol/l, respectively, when the dilution rate is 80% of the wash-out rate.

12. Estimate the exit cell concentration in a CSTR, for a feed substrate concentration of 20 mol/l, if the dilution rate is 70%, 75%, 80%, 85%, and

90% of the washout rates. Comment on the results, The organism follows Monod kinetics with μ_{max} and K_s = 0.66/min and 0.22 mol/l, respectively.

NOMENCLATURE

A_{HT}	Tube heat transfer area = πdL
A_t	Empty tube cross sectional area
C_{A_i}	Substrate/nutrient concentration in the feed
C_{A_0}	Substrate/nutrient concentration in the exit
$C_{A_{sf}}$	Exit substrate concentration when $x_0 = 0$
C_s	Concentration at the surface of the enzyme
D	$F/V = 1/\tau_{CSTR}$
D_g	Diffusion coefficient
E_t	Grams enzyme per unit volume
F	Feed rate
Q	Volumetric flow
V	Reactor volume
V_{max}, K_m	Enzyme rate parameters
Y	The yield factor = mass of cells formed per mass of substrate consumed
d and L	Reactor diameter and length
k_s	Substrate mass transfer coefficient
k_L	Gas–liquid mass transfer coefficient
r_x	Cell formation rate = μx
x	Cell concentration in the tank and in the exit stream
x_0	Cell concentration in the feed
x_{sf}	Exit cell concentration when $x_0 = 0$
μ	Specific growth rate (assumed to follow the Monod equation $\mu = \mu_{max} C_{A_0}/(C_{A_0} + K_s)$)
μ_{max}, K_s	Cell growth rate parameters
τ_{CSTR}, τ_{PFR}	Residence time (τ = recirculation time)
r_A	reaction rate
f	recycle ratio

REFERENCES

1. Prave, P.; Faust, U.; Sittig, W.; Sukatsch, D. *Fundamentals of Biotechnology*; VCH: Federal Republic of Germany, 1987.
2. Prenosil, J.E.; Dunn, I.J.; Heinzle, E. Biocatalyst reaction engineering: Applications of enzymes in biotechnology; Rehm, H.J., Reed G., Kennedy, J.K., Eds.; VCH: Federal Republic of Germany, 1987.

3. Enzymes. In *Ullman's Encyclopedia of Industrial Chemistry*; Gerharts, W., Yamamoto, Y.S., Eds.; VCH, Federal, Republic of Geramany, 1987; Vol. A9.

4. Fermentation in Kirk and Othmer. In *Encyclopedia of Chemical Technology*; Kroschwitz, J.I., Grant, M.H., Eds; John Wiley: New York, 1993; Vol. 10.

5. Bioreactors. In *Ullman's Encyclopedia of Industrial Chemistry*; Elvers, B., Hawkins, S., Schulz, G., Eds.; VCH: Federal Republic of Germany, 1987; Vol. B4.

6. Klaas van't Reit Hans Tramper, Basic Bioreactor Design; Marcel Dekker: New York, 1991.

7. Levenspiel, O.*Chemical Reaction Engineering,* 3rd Ed.; John Wiley & Sons: New York, 1998.

9

Fermentation

Different types of commercial products that are produced by fermentation can be classified as

1. Biomass, animal feed additive, yeast
2. Metabolic products or primary metabolites—antibiotics, amino acids, vitamins
3. Products based on biotransformation
4. Production of microbial enzymes—lipases, amylases

Advantages of fermentation technology are (1) complex compounds like vitamin B12 and penicillin can be produced in single step and economically feasible means, (2) use of cheap raw materials, and (3) compounds of greater specificity and purity can be produced.

Fermentation has an old history. Initially wooden vats, barrels, shallow trays, and clay pots were used for preparation of alcohols, vinegar, and yogurt. Later stirred steel vessels were used for production of yeast and acids, and aerated stirred tank vessels were used for vitamins and antibiotics. The reactor design has currently shifted from stirred vessels to fluidized bed, airlift, and pressure jet fermentors. Pressure cycle fermentors are being used to minimize heat loss and increase oxygen transfer rates.

Most of the fermentation operation can be broken down into

1. Formulation of the culture medium
2. Preparation of the sterile medium
3. Sterilization of the fermentor and other related equipments.
4. Propagation of the pure and active culture to seed the production.
5. Culturing of the microorganism in the fermentor under optimum conditions for the formation of the product
6. Discharging of the product mixture

Steps 3 and 5 are generally carried out in several steps as shown in Fig. 9.1. This step-by-step procedure is followed to ensure uniform growth of the microorganism from the small scale (preserved state) to the large scale.

The microorganisms require water, carbon source (e.g., molasses, cereal grain, starches, glucose, sucrose, and lactose), energy, nitrogen source (corn steep liquor, soya bean meal, cotton seed flour, fish meal, ammonia, ammonium salts), minerals, vitamins, and growth factors. Soya bean oil, lard, and cottonseed oil serve to control foam in aerated fermentors and silicon oils and PEG act as defoamers. Brewer's yeast serves as a source for nitrogen, vitamin, and growth factor.

The batch medium is sterilized by heating it to 120°C at 15 to 18 psig for 15 min. Air used in the aerobic fermentors is also sterilized and filtered

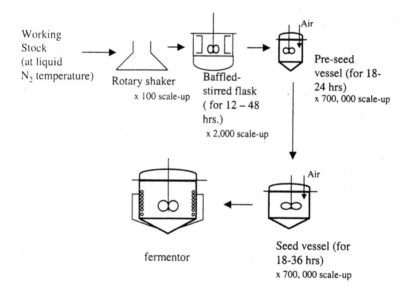

FIGURE 9.1 Propagation of cell from small-scale to industrial-scale fermentor.

through stainless steel mesh or bacteriological filters to prevent contamination. Following sterilization the pH and temperature of the fermentor broth are adjusted to the desired value before introducing the seed material or the innoculum containing the actively growing culture of the desired microorganism. The innoculum volume is of the order of 1 to 10% of the fermentation volume. During the fermentation process the temperature and the pH are controlled within tight limits. While the former is adjusted with the help of steam, hot water, and cold water, the latter is controlled using acids and bases that include sodium hydroxide, sulphuric acid, hydrochloric acid, ammonium hydroxide, and ammonia.

9.1 FERMENTOR DESIGN

A typical agitated fermentor with the process control system is given in Fig. 9.2. The vessel and the accessories are designed so that they can operate for about 30 days at monoseptic conditions. The fermentor height to diameter (aspect ratio) is generally in the range of 1.2 to 3. If the number is too high, mixing is improper, leading to low oxygen concentration at the upper portion of the fermentor and also accumulation of carbon dioxide. Fermentors have both jacket and coils to achieve the desired heat transfer area. Fermentors with aspect ratio greater than 1.5, may require multiple impellers. The lower

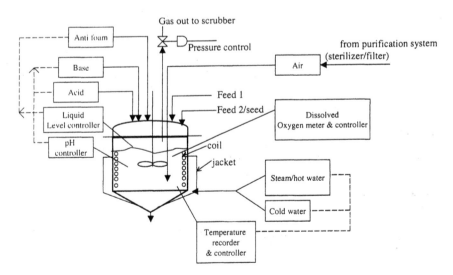

FIGURE 9.2 A typical fermentor with associated control systems.

impeller addresses shear and the top one bulk mixing. The power imparted to the fluid in an ungassed system is

$$P \alpha \rho N^3 D^5 \tag{9.1}$$

An impeller-to-tank diameter ratio of 0.3 to 0.5 is employed for microbial fermentations, and lower values are used for cell cultures, since they are susceptible to shear.

The reactor is generally operated in the batch or fed batch mode. The later mode is useful when there is a limitation on the heat removal rate. By pumping in limited amount of nutrient the growth rate of the organism could be controlled there by the heat generation and the oxygen uptake rates. Continuous operation requires continuous sterilization of the feed and nutrients and continuous downstream processing for the removal and recovery of products. One also has to ensure that no contaminants enter the fermentor and change the characteristics of the strain.

The reactor can be considered as a three phase system where the liquid phase contains dissolved nutrients, substrates, and metabolites; the solid phase consists of individual cells, pellets, insoluble substrates, or precipitated metabolic products, and the gaseous phase contains oxygen and CO_2.

Air is continuously introduced in aerobic fermentors (generally at the rate of about 0.5 to 1 fermentor broth volume per minute), because of the low O_2 solubility in water (oxygen solubility in water is about 10 mg/l). Highest O_2 partial pressure is attained during aeration with pure oxygen compared to the value in air: 43 mg O_2/l dissolves in water when pure oxygen is used as against 9 mg O_2/l with air. As temperature rises, the O_2 solubility decreases (the solubility at 33°C is 7.2 mg O_2/l). An active yeast population may have an oxygen consumption of the order of 5 mg O_2/min per gram of dry cell mass. In addition, presence of dissolved salts decreases oxygen solubility. These issues indicate that it is imperative to use high oxygen transfer rates to maintain the cell population.

Sparger located inside the vessel and mechanical agitation distributes the air uniformly. Oxygen starvation is the limiting factor in culture growth and productivity and hydrofoil mixers in conjunction with rushton turbine overcomes some of the problems of oxygen mass transfer. While agitated vessels are suitable for small fermentors, airlift or bubble column designs are adapted for large-scale fermentors.

9.2 BIOCELL GROWTH

When a small quantity of living bio cells is added to a solution containing essential nutrients at the correct operating conditions, the cells will grow.

Unicellular organisms divide as they grow; hence, the increase in number of cells is also accompanied by increase in the biomass. As the cells take in nutrients from the surroundings they release metabolic end products. The growth of the cell population with time is shown in the Fig. 9.3.

Upon change of environment the cells go though a phase, where there is no appreciable change in their number and this is known as *lag phase*. The length of the lag phase is determined by characteristics of the bacterial species and the media—both the medium from which the organisms are taken and the one to which they are transferred. Some species adapt to the new medium within an hour or 2, while others take several days. This phase is followed by a rapid increase in the cell number, almost exponentially with time. This phase is called *exponential growth* phase or *logarithmic* phase. During this stage the number of cells present at any given time is a multiplicative function of the number of cells present at a previous time. This multiplication factor is constant.

This phase can be mathematically represented as

$$\frac{dn}{dt} = \mu n \tag{9.2}$$

where n is the number of cells and μ the specific or exponential growth factor. This can be integrated to get

$$n = n_0 e^{\mu t} \tag{9.3}$$

The time required for doubling the cells is given as $\ln 2/\mu$, n_0 is the number of cells at the beginning of the exponential growth rate period.

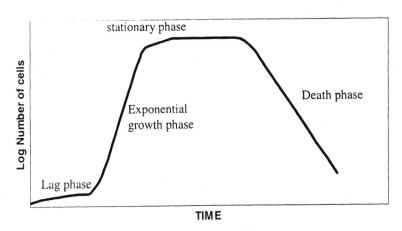

FIGURE 9.3 Growth curve for a microbial culture.

The specific growth rate is a function of the nutrient concentration and is related as

$$\mu = \mu_{max} \frac{c_i}{K_i + c_i} \qquad (9.4)$$

This relationship is called *Monod's equation* and is an ideal representation of the cell growth process. This equation appears similar to the Michaelis-Menten kinetic equation for enzyme-catalyzed reactions. μ_{max} is the maximum growth rate achievable when $c_i \gg K_i$. c_i is the concentration of the nutrient. The nutrient level is generally high in the initial stages, leading to true exponential growth rate at early stages. As the nutrient gets consumed and the quantity decreases the growth rate deviates from Monod equation. If the nutrient consumption is rapid, the internal cell nutrient levels could be much below the medium concentration, which in itself is changing with time. In order to account for these nonidealities, several other models have been proposed, and they are

$$\mu = \mu_{max} \frac{c_i}{K_i c_{i_0} + c_i} \qquad (9.5)$$

$$\mu = \mu_{max} \frac{c_i}{K_{i1} + K_{2i} c_{i_0} + c_i} \qquad (9.6)$$

If the substrate is inhibiting the growth, then

$$\mu = \mu_{max} \frac{c_i}{K_i + c_i + c_i^2/K_p} \qquad (9.7)$$

and if product is inhibiting the growth (as in alcohol fermentation), then

$$\mu = \mu_{max} \frac{c_i}{K_i + c_i} \frac{c_p}{K_p + c_p} \qquad (9.8)$$

The growth of the bacteria or biocell can be manipulated by various means namely, culture conditions, nutritional requirements, pH, temperature, oxygen, and water availability. Using cultures, which are in the exponential growth phase, or continuously supplying fresh nutrients and removing the spent ones may manipulate culture conditions. Depending upon the type of bacteria the carbon source needed as nutrition for the bacteria for its growth could be from CO_2, organic or inorganic compound, or light. The optimum temperature for bacterial growth varies between 15° to 80°C and pH from 5 to 9. Water requirement for bacterial growth varies from very dry conditions, salty, and environments with high sugar. Type of bacteria varies from those that prefer oxygen, those that don't prefer oxygen, killed by oxygen, and those that grow both ways but prefer oxygen.

The third phase is called *stationary* phase, which arises due to the depletion of the nutrient or accumulation of undesired toxin. If the rate of depletion of nutrient is proportional to the cell number, then

$$\frac{dC_i}{dt} = -kn \tag{9.9}$$

k is a constant. When the nutrient gets fully exhausted, the growth enters the stationary phase, and the maximum number of cells at that time is given by

$$n_s = n_0 + \frac{kC_{i_0}}{\mu} \tag{9.10}$$

n_0 is the initial number of cells at time $= 0$ and C_{i_0} the initial nutrient concentration. Equation (9.10) is obtained by substituting Eq. (9.2) in Eq. (9.8) and integrating.

If growth halts due to the exhaustion of the nutrient, dilution with an inert solvent will not affect n_s. On the contrary, if it is due to accumulation of toxin, dilution will increase the number of cells.

The fourth stage is the death phase, where the number of cells declines in an exponential fashion following the relation

$$\frac{dn}{dt} = -k_d n \tag{9.11}$$

k_d is the decay constant.

9.3 GROWTH OF MOLDS AND FILAMENTOUS ORGANISMS

In microbial populations increase in biomass is accompanied by increase in number of cells. Whereas in the case of molds and other filamentous organisms there is no change in the number during the growth process but an increase in their size and the mass. The mass of a spherical pellet as a function of time (t) is given as

$$M = \left(M_0^{1/3} + \frac{\alpha t}{3} \right)^3 \tag{9.12}$$

where M_0 is the initial mass and α is a constant. Generally M_0 is small. Equation (12) indicates that the mass increases rapidly with time (at the rate of cubic power of time).

9.4 TYPES OF BIOPROCESSES

Gaden classified fermentation processes into three types. In type 1 process the production is due to result of primary metabolism. The rate of product formation is directly related to the rate of substrate consumption and also to the rate of cell mass produced. Examples of aerobic systems that follow this classification are acetic acid (using *Gluconobacter suboxidans*), single-cell protein (using *Candida utilis, S. cerevisiae*) and baker's yeast (using *S. cerevisiae*). The first system is operated under batch mode. The second under batch or fed batch mode, while the third system is operated at fed batch mode. Examples of anaerobic processes are ethanol (*S. cerevisiae, Zymomonas mobilis*), acetone/butanol (*C. acetobutylicum*), lactic acid (*Lactobacillus bulgaricus*) and propionic acid (*Propionibacterium shermanii*). The first two products are run in continuous mode of operation and the next two in batch mode.

Type 2 production process is due to primary metabolisms, but the reaction rates could be complex. The product may be an intermediate and not the end product of a metabolic pathway. It has something to do with catabolism/energy metabolism. The production phase can be distinguished from the growth phase. Examples of type 2 classifications are alkaline protease (*Bacillus lycheniformis*), amylase (*Aspergillus oryzae*), pectinase (*Aspergillus niger*), cellulase (*Trichoderma resii*), citric acid (*Aspergillus niger*), amino acids (*Corynebacterium glutamicum*), and riboflavin (*Eremothecium ashbyi*). All these processes are aerobic and are operated in batch mode. The first four systems are operated in fed batch mode as well. One example of an anaerobic system is vitamin B12 (*Pseudomonas nitrificans*), which is operated in batch mode.

Type 3 process is a nongrowth-associated production. Sometimes production only begins when the main carbon source is exhausted and a secondary carbon source is used. The product is not connected to catabolism or energy metabolism, e.g., production of secondary metabolites like antibiotics, vitamins. There are distinct growth and production phases, with negative specific growth rate. Examples of aerobic processes in this category are alkaloids (using *Claviceps paspali*), immunoglobulins (using hybridoma cells), interferons (mammalian cells), penicillin (*Penicillium chrysogenum*), cephalosporin (*Acremonium chrysogenum*), tetracycline (streptomyces strains). The first example is operated in the batch mode, the second two in the continuous mode and the last three in the fed batch mode.

Deindoerfer proposed another type of classification, which relates product formation with nutrient consumption during the fermentation process. This classification is useful for designing batch fermentors.

1. Simple type, where nutrient is converted to products with fixed stoichometry as in the case of glucose getting converted to gluconic acid:

   $A \rightarrow B$

   Here

   $$\frac{-dC_A}{dt} = \frac{dC_B}{dt}. \tag{9.13}$$

   where C_A and C_B are concentrations of the two species. At $t = 0$, $C_A = C_{A_0}$ and $C_B = 0$.

2. Simultaneous: The nutrient is converted to two products in variable stoichometry without the formation of intermediates. Formation of cell mass and yeast from glucose is an example of this type.

   $C \rightleftharpoons A \rightleftharpoons B$

   $$\frac{-dC_A}{dt} = r_{AB} + r_{AC} \tag{9.14}$$

   r_{AB} and r_{AC} represent the rate corresponding to the conversion of A to B and A to C, respectively. The rate constants for the forward and backward reaction between A and B are larger than those relating A and C.

3. Sequential: Also known as *consecutive reaction*, where the formation of final product is superceded by an intermediate. An accumulation in intermediate is observed here.

   $A \rightarrow B \rightarrow C$

   Here $-dC_A/dt = r_A$, $dC_B/dt = r_A - r_B$, and $dC_C/dt = r_B \cdot r_A$ and r_B are the rate of reactions.

   An example of sequential reaction is the conversion of glucose to glucolactone and then to gluconic acid in the presence of the enzyme *Pseudomona ovalis*.

4. Stepwise: Here nutrients get completely converted to intermediate first, before this intermediate gets converted to the final product. As the nutrient is fully consumed, the biocell adapts itself to use the intermediate as the new nutrient. The reaction sequence here is

similar to sequential type, but the formation of C will occur only after complete exhaustion of A. Two examples of stepwise production are (1) diauxic growth of *E. Coli* in glucose-sorbitol medium and (2) growth of *Acetobacter suboxydans* in glucose. In the second example the biocell first grows on glucose and as soon as it is consumed on gluconic acid, which is formed from glucose. During this phase, gluconic acid gets converted to 5-ketogluconic acid.

9.5 DOWNSTREAM PROCESSING

The separation of bioactive compounds from crude fermentation broths or plant extracts requires a combination of traditional chemical engineering unit operations and those that are quite different. It is important to maintain the compound's bioactivity while maximizing its purity and yield. Two major problems to be addressed are that the compounds is usually present in very low concentrations, namely, a few hundred parts per million to 10% of the broth, and the impurities could have physicochemical properties that are similar to the desired compound, making conventional separations difficult.

The primary downstream operation is the removal of insolubles (e.g., by sedimentation, filtration, centrifugation, membranes), followed by isolation of the compound (e.g., precipitation, extraction, adsorption, membranes), final purification (e.g., crystallization, chromatography), and drying (conventional dryers or spray dryer). These unit operations will be discussed in Chapter 13.

Major unit operations involved in the downstream of manufacture of several fermentation processes are listed below:

Process	downstream purification processes
Food stuff	filtration, separation, mixing, distillation, sterilization, drying
Ethanol	filtration, distillation
Amino acids, L-lysine	cell separation, chromatographic purification, extraction, crystallization
Citric acid	separation, precipitation, activated carbon purification
Lactic acid	filtration, decantation, vacuum evaporation, centrifugation
Butanol	filtration, distillation
Vitamin B12	activated carbon treatment, extraction, ion exchange
Biopolymers	pasteurization, precipitation, decantation, drying, grinding

Tetracycline	filtration, precipitation, adsorption, crystallization, centrifugation, vacuum drying
Bacitracin	filtration, extraction, vacuum drying
Steroids	extraction, crystallization
Bioprotein	filtration, centrifugation, drying

9.6 PROCESS ANALYSIS AND CONTROL

The fermentation process is managed using on line computer control strategies for (1) setting the initial process conditions during start-up, (2) close monitoring to ensure whether the process is following the required path, (3) manual adjustments to the process variables, (4) deciding when to terminate the process and/or to transfer the product, (5) estimating the mass and thermal balances, rates of reaction, kinetics, and yields, and (6) monitoring contamination.

Based on certain basic measurements other process data/performance are derived. For example,

1. From pH measurements the acid product produced can be determined.
2. From dissolved oxygen in the broth, oxygen in the exhaust gas and gas flow rate, the oxygen transfer and uptake rates (OTR and OUR), and respiratory coefficients can be estimated.
3. From the concentration of carbon dioxide in the exhaust stream and gas flow rate its evolution rate can be estimated.
4. From sugar level in the broth, sugar feed rate, and carbon dioxide evolution rate the yield and cell density can be determined.

The process control strategy consists of manipulating the control variables based on certain measurements and process scheme. For example, the fermentor temperature can be controlled by changing the flow of the heating or cooling medium; the pH, by changing the amount of acid or alkali added; and foaming, by the addition of antifoaming agent. The temperature is measured by resistance thermometer, the pH using pH electrode, and the foam level by conductivity probe. The liquid level inside the vessel is maintained constant by manipulating the feed rate. The dissolved oxygen concentration in the fermentation broth is a critical factor, which needs to be maintained tightly by manipulating individually or simultaneously agitation rpm, gas flow, gas composition, and fermentor pressure (Fig. 9.4).

Modern control strategies for fermentors aim towards maximizing productivity and minimize energy usage. This can be achieved by (1) tightly controlling the operating parameters and overcoming perturbations and

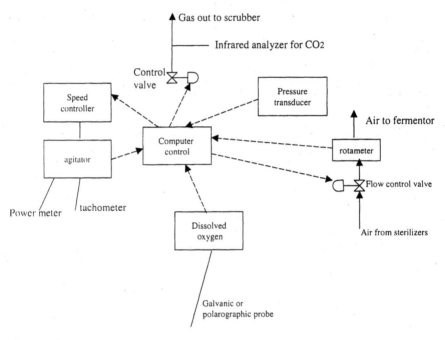

Figure 9.4 Control of dissolved oxygen.

disturbances, (2) identifying optimal operating conditions online, (3) improving product consistency, and (4) foresee problems a priori.

There are many control strategies that can be followed and a few of them are described below.

Feedback controller (Fig. 9.5a): Where the process variable that is being controlled is measured, compared with a set point, and the difference (or error) is sent to the controller, which sends a signal to the controlling element. If the fermentor temperature needs to be controlled, the controlling element could be a control valve, which operates the hot or cold water or steam flow rates. The control action takes place only after the controlled variable shifts away from the set point (desired value). No control action is taken on the disturbances that enter the process.

The control signal (I_s) is a function of the error (e) as shown below:

$$I_s = k_c e + \int \frac{e}{\tau_I} dt + \tau_c \frac{de}{dt} \tag{9.15}$$

The first term on the right-hand side of the equation indicates that the controller action is proportional to the error signal (then such a controller

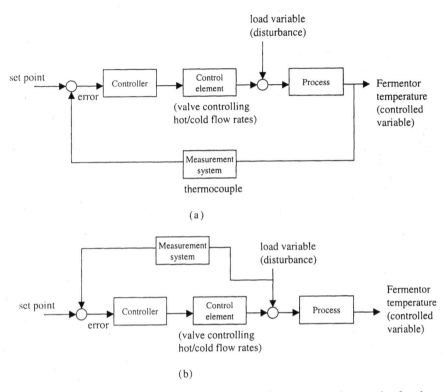

FIGURE 9.5 A typical (a) feedback and (b) feedforward control strategies for the control of fermentor temperature.

is called *proportional controller* and the inverse of k_c is called the *proportional gain*). The second term indicates that the controller action is an integral function of the error signal. This controller action is called *integral control* and τ_I, the *integral time*. The third term indicates that the controller action is a differential function of the error signal and is known as the *differential control*. τ_c is called the *differential time*. Controllers are designed to have a combination of these actions, i.e., a simple proportional control, proportional-integral control, or proportional-integral-derivative control.

Feed forward control (Fig. 9.5b): Here the disturbances are measured before they enter the process, and the controller sends signal to the controlling element to account for this load. Since there is no feedback action, no correction is done if there is a drift in the process variable. Hence feed forward control strategy is never used alone, but generally coupled with feedback strategy as well. In both these strategies the controller settings are fixed.

Soft-sensor control scheme: This is an equation based approach. Some variables like yield or cell productivity cannot be directly measured (inaccessible) but can be estimated (derived) by using equations relating variables that can be measured (like temperature, pressure, flow, pH, etc.) and those that cannot be measured. Such variables are called soft variables. If such variables need to be controlled, then the control algorithm is designed so that they tune the manipulative variables in such a way (based once again on the relationship between the soft variables and manipulative variables) to get the desired set point on the soft variables (Fig. 9.6). These schemes are based on model equations.

Adaptive control: Here the parameters of the controller are estimated on-line as the process proceeds based on the current state of the process and measured disturbances (Fig. 9.7). The software compares the current state of the process as measured from the plant and the model-based state and tunes the controller settings. This strategy is useful if the controllers have to adapt to the changes in the biocatalyst activity (may be due to deactivation) during the course of the reaction.

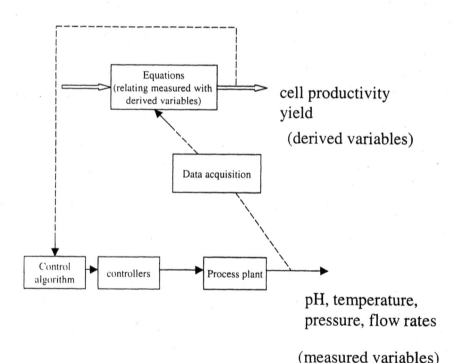

cell productivity
yield

(derived variables)

pH, temperature,
pressure, flow rates

(measured variables)

FIGURE 9.6 Soft sensor control scheme.

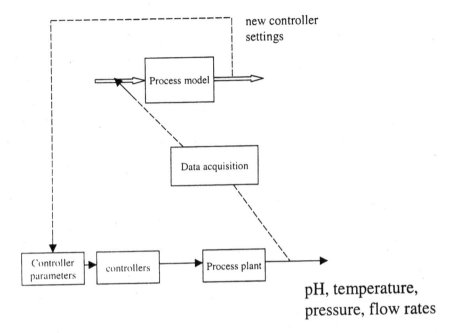

FIGURE 9.7 Adaptive control strategy.

Artificial intelligence (Fig. 9.8): This is a rule-based approach, where the on-line process computer manipulates the process and also gives advice to the operators based on in-built rules, knowledge database, and mathematical model. This approach can identify future process malfunctions and has the capability to self-learn from old information stored inside. This approach is also known as *expert system control*.

PROBLEMS

1. Consider a population of 10^4 cells/ml at 1 pm, which reaches 10^7 cells/ml at 9 pm. What is the time needed for doubling?

2. A solution has a residual yeast cells at 10^3 cells/ml. Assuming that the time for doubling of the cells is 1 day, how long it will take for the solution to have 10^6 cells/ml?

3. A solution has 10^{-12} g of cells. What will be the cell weight after 7 h if the doubling time is 5 h.

4. If a *log phase* culture goes from two cells to four cells during a 20-min interval, then the culture will go from four cells to eight cells during the next how many minutes?

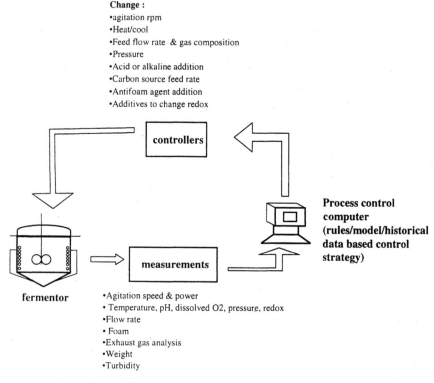

Change :
•agitation rpm
•Heat/cool
•Feed flow rate & gas composition
•Pressure
•Acid or alkaline addition
•Carbon source feed rate
•Antifoam agent addition
•Additives to change redox

controllers

Process control computer (rules/model/historical data based control strategy)

measurements

fermentor

•Agitation speed & power
• Temperature, pH, dissolved O2, pressure, redox
•Flow rate
• Foam
•Exhaust gas analysis
•Weight
•Turbidity

FIGURE 9.8 Artificial intelligence of fermentor control.

5. If during *exponential phase* there are 10 cells present at time 0, and 100 cells present at time 200, then at time 400 there will be how many cells present?

6. If at time 0 you had 10 million cells and 2 h later had 40 million cells, what is the generation doubling time?

7. Staring with 10 bacteria, a doubling time of 30 min, and given exponential growth, how many bacteria will you have after 3000 min (a little over 2 days) of continuous exponential growth?

8. Derive the growth equation for a mold assuming spherical shape and constant increase in the diameter of the pellet with time.

9. In the sequential type of mechanism determine the maximum amount of B formed if k_1 is 10 times that of k_2 (assume the reaction to be first order for both the steps, $C_{i0} = 1$ M).

10. In a batch fermentor if the initial substrate concentration and the initial rate are as given in the following table, comment on the mechanism and order of reaction.

Initial substrate concentration (mmol/l)	1	2	3	4	5
Initial rate (mmol/l/min)	0.023	0.046	0.069	0.092	0.115

11. What will be the time taken by a pellet to double its mass if $\alpha = 0.2$ and initial mass $= 0.25$ mg.

12. In a simple reaction, if the initial concentration is 1 M and the reaction follows a second order, determine the amount of product formed after 2 h if rate constant is 0.01/M/min.

NOMENCLATURE

D	Agitator diameter
N	rpm
P	Power
c_p	Product concentration
k_d	decay constant
r	rate of reaction
C_i, c_i	Nutrient concentration
C_{i0}, c_{i0}	Initial nutrient concentration
I_s	Controller signal
K_i	Equilibrium constant
M	Mass of a mold/spherical pellet
M_0	Initial mass
e	Error (difference between set point and the actual value)
n	Number of cells
n_0	Initial number of cells
t	Time
μ	Exponential growth factor
μ_{max}	Maximum growth rate
ρ	density of fluid

10

Reactor Engineering

10.1 INTRODUCTION TO REACTOR ENGINEERING

A reactor is a confined space used for transformation of reacting entities from one chemical form into another in a certain time period. This broad definition of the reactor space is equally applicable to bioreactors where the chemical transformation is achieved with assistance from either a living and growing population of microbial species or a biocatalyst in its pure or crude form, in a dissolved condition or in an immobilized form on a suitable support. The objective of the bioreactor could be conversion of a complex raw material into a value-added product, as in fermentation of molasses into alcohol or into relatively expensive drug intermediates or antibiotics such as penicillin, which may have to undergo further derivatization into semisynthetic drugs. A growing mould or bacterial colony in immobilized form on a solid support, wherein the nutrient carrying medium flows over the colony is another form of bioreactor. The medium can be well defined as that encountered in enzymatically catalyzed isomerization of glucose or separation of *dl*-amino acids mixtures, and it can be very complex, as in fermentation of corn syrup. In the former case the reactor resembles the packed bed operation in the conventional chemical industry.

The major objective of a reactor is to bring all reactants in intimate contact with each other. Unlike the conventional chemical reactors, in a bioreactor with an expanding number of bioparticles such as cells or micellium, the operation is difficult to be considered at steady state as the variables are continuously changing and so are the requirements of the growing population. The adequacy of reactor facilities to cater to the needs of evolving population needs to be, therefore, ascertained at the design stage itself. Another common characteristic of a bioreactor is the dynamic variation in the physicochemical properties of the growing population of the microbes and consequently, in the properties of the medium. The reactor must be able to meet the needs of all cells adequately despite the increasing and varying demands of the microbial population.

Three types of bioreactors are most common in practice; stirred vessels, tower reactors or air-lift reactors, and packed bed. The stirred reactors are the main workhorse of the biochemical industry, while tower reactors are being used for very large scale operations with low-viscosity media. The packed beds are used mainly for enzyme-catalyzed reactions.

In a stirred bioreactor, the nutrient medium is inoculated, and the biotransformation is allowed to proceed with intermittent addition, if necessary, of the nutrients at prespecified times, depending on the operating conditions. The contents of the stirred vessel are kept in mixed conditions with the help of an agitator, at a predetermined speed of agitation. In aerobic bioreactors gas, usually sterile air, is introduced into the liquid. The agitator not only serves to keep uniform conditions in the liquid phase but is also needed to break air bubbles into smaller ones to ensure availability of large interfacial area for mass transfer of oxygen from gas phase into the liquid phase.

In a tower bioreactor, the liquid is kept in circulation by imparting momentum using a stream of gas sparged in the reactor at the bottom. In air-lift reactor, which is a modified form of tower bioreactor, the liquid circulation is set up through an internal or external loop to avoid excessive relative velocity difference between the liquid and gas phases and to reduce the shear on the cells suspended in the liquid. The packed bed operation involves immobilization either of biomass or of enzymes on suitable supports, which are packed in a cylindrical vessel and the medium with reacting substrates or nutrients is passed over the particles.

A typical bioreactor invariably has a heterogeneous appearance with a mixture of the liquid medium, cells as suspended solids and air dispersed in the form of fine or big bubbles. The heterogeneity of the system makes an interplay of transfer processes with biochemical kinetics unavoidable. In the discussion that follows, it will be apparent that transfer processes, both mass and heat, have a significant effect on the performance of the bioreactors.

In the design of any reactor the main objective is to achieve the maximum production of desired chemical species and to minimize the cost of operation. For bioreactors, the selectivity towards the desired product assumes an additional significance. As a manufacturer, one would not like to be at the mercy of the microbial species that would like only to proliferate in their own number without converting expensive nutrients into products useful to the manufacturer. Unlike the nonbiological chemical reactors where the course of reaction does not deviate substantially, the change in environmental factors may induce the organism to adapt to newer environment producing entirely different products or to produce nothing at all of the desired product. It is, therefore, essential that the conditions of uniformity for all cells be maintained the same while they are in the reactor. The design aspects of a reactor, therefore, consist of adequate supply of nutrients to all growing population in the confines of the reactor. Different contacting methods have, therefore, been devised to bring about the proper fluid mixing.

10.2 MASS TRANSFER BASICS

A aerobic bioreactor is characterized by supply of oxygen from air bubbled in the fermentation medium. The resistance to mass transfer of oxygen from the gas phase, through an interface between the phases, is assumed to exist in two films on either sides of the interface [1]. The bulk phases are assumed to exist with uniform conditions, but the convective current dies out once the molecule reaches inside the film where molecular diffusion is considered as the sole mechanism of transfer of the solute to or from the interface (Fig. 10.1).

The Fick's equation for diffusion of species A, within the diffusion film, is

$$J_A = -D_m \frac{dC_A}{dX} \tag{10.1}$$

Integration of Eq. (10.1) using the boundary conditions at the two ends of a diffusion film gives a linear concentration profile across the film if the diffusion coefficient (D_m) of A is assumed to be constant.

$$J_A = \frac{D_m}{\delta}(\Delta C_A) = k_m \, \Delta C_A \tag{10.2}$$

The mass transfer coefficient (k_m), defined here for a film on one side of the interface, as the ratio of molecular diffusivity to the film thickness can be defined in the same manner for the second phase on the other side of the interface. The hydrodynamic conditions within the phase are represented by thickness of the film (δ). Highly agitated conditions means a thinner film near the interface. Because the hydrodynamics is usually not known accurately, the mass transfer coefficient is a lumped parameter that takes into account all phenomena related to mass transfer. The interface is assumed to provide no resistance to the mass transfer process, which may be difficult to accept for an interface loaded with surface active molecules, proteins, and in some cases

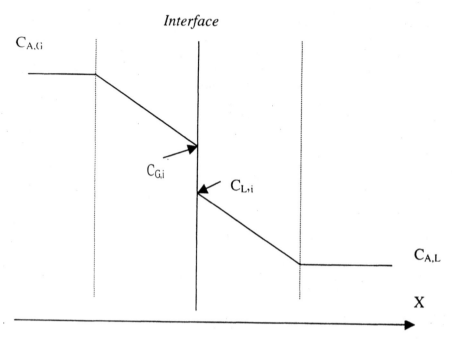

FIGURE 10.1 Two-film theory of mass transfer.

with biocells. In these cases, the interfacial resistance can be significant and comparable if not higher than the film resistance. In most reports on mass transfer, the mass transfer coefficients are reported with clean systems. For the bioreactors with complex media and in the presence of several surface-active compounds, which are part of the media or products from the cells, the mass transfer coefficient can be dependent on the type of the fermentation media.

For a typical mass transfer across liquid phase film, molecular diffusivity is of the order of 10^{-9} m^2/s, while the liquid film thickness is of the order of 10^{-5} m, giving a typical value of liquid side mass transfer coefficient, ~10^{-4} m/s. On the gas-phase side, the diffusivity is of the order 10^{-6} m^2/s giving a three orders of magnitude higher gas-phase mass transfer coefficient.

The equation for mass transfer flux in the gas phase is

$$J_A = k_g(C_{A,G} - C_{G,i}) \tag{10.3}$$

while the equation for mass transfer flux on the liquid side is

$$J_A = k_l(C_{L,i} - C_{A,L}) \tag{10.4}$$

Since the interface reaches the equilibrium almost instantaneously, the interfacial concentrations on the gas and liquid sides, $C_{G,i}$ and $C_{L,i}$, are related; therefore, through an equilibrium relationship

$$m = \frac{C_{G,i}}{C_{L,i}} \tag{10.5}$$

Eliminating the interfacial concentrations from Eqs. (10.3) and (10.4), the mass transfer flux can be written in terms of the bulk phase concentrations

$$J_A = K_{OL}(C_A^* - C_{A,L}) \tag{10.6}$$

where C_A^* is the liquid-phase concentration in equilibrium with the gas-phase concentration and the overall mass transfer coefficient, K_{OL} is defined as

$$\frac{1}{K_{OL}} = \frac{1}{k_l} + \frac{1}{mk_g} \tag{10.7}$$

For a sparingly soluble gaseous solute, such as oxygen, in aqueous media the second term is usually negligible because of a large m value, and the overall mass transfer coefficient can be considered to be equal to liquid-side mass transfer coefficient as the resistance on the liquid-side film is the controlling resistance. For oxygen, $m \sim 32$ in air-water system at $25°C$, and the diffusivity in the gas side is at least two orders of magnitude higher making the first term in Eq. (10.7), i.e., the liquid-side resistance, much higher than the second term.

The flux Eq. (10.6), when multiplied by the area available between air and liquid phases, gives the oxygen uptake rate (OUR). The above equation and the definition of mass transfer coefficient gives the dependence of the OUR on physicochemical properties of the solute such as diffusion coefficient, viscosity of solution, temperature through its effect on diffusivity and viscosity, and hydrodynamic conditions that decide the thickness of the liquid film (δ) surrounding the bubble.

Another model is proposed by Higbie [2] that considers the mass transfer as an unsteady process between a fluid element arrived at the interface and the gas phase. The fluid element spends a certain contact time (t_c), at the interface depending upon the flow conditions in the bulk. During this time, the solute diffuses into the fluid element as if it is diffusing into a stagnant medium of infinite depth before the fluid element is replaced by a new fresh fluid element from the bulk phase. During this diffusional process, the size of the eddy is considered to be much larger than the molecular size, so that for the diffusing solute molecule, the eddy is an infinite medium. A partial differential equation [Eq. (10.8)] describes the transient behavior of the diffusional process at any instant at X in the fluid element.

$$\frac{dC_A(X, t)}{dt} = D_m \frac{d^2C_A(X, t)}{dX^2} \tag{10.8}$$

With conditions,

$$X = 0 \qquad t > 0 \qquad C_A(0,t) = C_A^*(= \frac{C_{G,A}}{m})$$

$$X > 0 \qquad t = 0 \qquad C_A(X,0) = C_{L,0} \qquad\qquad (10.9)$$

$$X = \infty \qquad t > 0 \qquad C_A(\infty,t) = C_{L,0}$$

With these conditions the concentration profile of A at any instant t (Fig. 10.2) and the instantaneous flux are given as

$$C_A(X,t) = C_{L,0} + (C_A^* - C_{L,0})erfc\left(\frac{X}{2\sqrt{D_m t}}\right) \qquad (10.10)$$

$$J_A = -D_m\frac{dC_A}{dX}\bigg|_{x=0} = \sqrt{\frac{D_m}{\pi t}}(C_A^* - C_{L,0}) \qquad (10.11)$$

For a bubble of diameter d_b, rising through a liquid phase without much entrainment of the liquid phase, the contact time is approximated as diameter

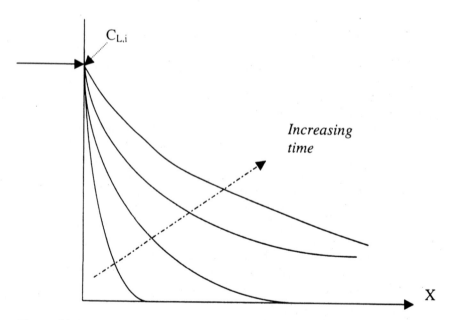

FIGURE 10.2 Concentration profiles in a stagnant liquid with time.

of the bubble divided by the bubble rise velocity. The integration of the flux over the time $t = 0$ to $t = t_c$, gives mass transfer coefficient as

$$k_m = 2\sqrt{\frac{D_m}{\pi t_c}} \qquad (10.12)$$

The mass transfer coefficient dependence on molecular diffusivity in the Higbie's penetration theory is square root type, unlike its direct dependence on D_m in the two-film theory. The hydrodynamic conditions are represented by the contact time, i.e., highly turbulent or circulation currents will renew the surface between the phases more often. Thus on increasing turbulence the contact time is reduced, maintaining higher concentration gradient at the interface and, therefore, higher mass transfer coefficient. The penetration theory assumes that all fluid elements, i.e., eddies, spend the same time at the interface. The assumption was relaxed by Danckwerts [3] in his surface renewal theory by considering a distribution of residence times for fluid elements at the interface. Although the film theory's postulate of existence of two stationary films lacks in reality, it, however, is the most widely used theory because of its simplicity and its ability to explain most of the mass transfer results. It has been able to provide consistent correlation of data in industrial reactors [4].

Small bubbles tend to move more like rigid solid bodies with little or no internal motions and thus show lower values of mass transfer coefficients. Larger bubbles, however, have a much more mobile surface and depending upon the liquid properties the mass transfer coefficient can be somewhat higher than the estimated values. Most of the literature data show a large scatter in mass transfer coefficient values considering the random fluctuations that must be existing at such mobile surfaces.

The main influence of temperature on gas–liquid mass transfer coefficient is because of its effect on diffusion coefficient. The diffusion shows a variation with temperature following the relation

$$D_m \propto T^{0.5-1.0} \qquad (10.13)$$

A similar effect of temperature will be reflected on the mass transfer coefficient.

10.3 MASS TRANSFER IN BIOREACTORS

The transfer of nutrients, including that of oxygen to the thriving microbiological population in an aerobic fermentor, may become an important aspect of the operation of the fermentation unit. At the laboratory scale, it is far easier to maintain homogeneous conditions throughout the liquid phase. At the plant level, however, when units as big as 100–1000 m^3 are handled, proper supply of all nutrients to each of the cells may not be achieved. In most fermentation processes, under aerobic conditions, the supply of oxygen is a crucial factor. Oxygen is sparingly soluble in a aqueous medium (1.25×10^{-6} mol/cm^3/atm), and it has to pass through a series of resistances in order to get consumed. A number of steps are involved in the process as shown in Fig. 10.3. The overall mass transfer process consists of

Transfer from bulk gas phase to gas–liquid interface
Transfer through interface
Transfer from interface to the bulk liquid phase
Transfer through liquid phase film around the single cell or around the floc of cells
Transfer through pellet (or floc of cells)
Transfer through cell membrane
Transfer of oxygen within a cell, which can be combined with microbial consumption inside the cell

The flux of oxygen transfer can be expressed in terms of the overall mass transfer coefficient or liquid-side mass transfer coefficient.

$$J_{0_2} = K_{0_2}(C_{0_2}^* - C_{L,0_2}) = k_{l,0_2}(C_{0_2}^* - C_{L,0_2}) \tag{10.14}$$

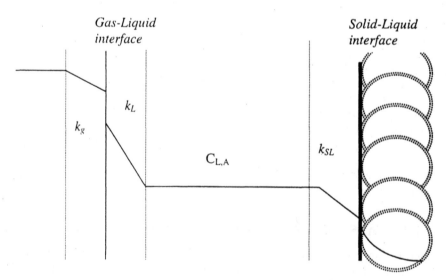

Gas-Liquid interface

Solid-Liquid interface

k_L

k_g

k_{SL}

$C_{L,A}$

Bulk liquid

FIGURE 10.3 Mass transfer steps involved in bioreactors.

For a sparingly soluble gas, like oxygen, the overall mass transfer coefficient is replaced by liquid-side mass transfer coefficient. This resistance is characterized by diffusivity of oxygen in the liquid phase and would also be dependent on the physical characteristics of the medium such as viscosity and ionic strength. The biological flocs or fungal micellium show highly non-Newtonian behavior and high viscosity. The diffusivity of oxygen, therefore, could be lowered in highly viscous media to ~1×10^{-10} m^2/s, and the mass transfer coefficient could still be lower as the hydrodynamic conditions are not as vigorous as in water alone. The maximum rate at which oxygen is made available to the microbial population would depend on the local conditions and physicochemical conditions of the medium and when its concentration in the bulk phase is reduced to negligible value,

$$\text{Maximum oxygen transfer rate} = k_l a \, C^*_{0_2} \tag{10.15}$$

The maximum oxygen utilization rate by a microbial medium of cell density x is related to its maximum growth rate (μ_{max}) through the oxygen yield factor, y_{o_2}

$$\text{Oxygen utilization rate} = \frac{x\mu_{max}}{y_{o_2}} \tag{10.16}$$

The yield factor (y_{o_2}) is defined as moles of carbon (biomass formed) per mole of oxygen consumed. The relative rates of the two steps decide whether the supply of oxygen is sufficient for growth of the microbes.

If $k_l a \, C^*_{o_2} \gg x\mu_{max}/y_{o_2}$, sufficient oxygen will be supplied to the microbial population as its consumption by the microbes for growth and sustenance will be at a lower rate than its supply, and the process is biochemical reaction limited. On the other hand, if $k_l a \, C^*_{o_2} \ll x\mu_{max}/y_{o_2}$, the consumption of oxygen by the microbial population is at a faster rate than the rate at which it can be supplied by the mass transfer processes. The process is then mass transfer limited. Under such conditions oxygen-deficient conditions can prevail in the reactor or within the cells, which may be detrimental for the growth of the cells or for the production of desired products. For the intermediate conditions where the rate of mass transfer of oxygen and biochemical reaction are comparable

$$k_l a \left(C^*_{o_2} - C_{L,0_2} \right) = \frac{x\mu}{y_{o_2}} \tag{10.17}$$

or in terms of Monod kinetics

$$k_l a C^*_{o_2} \left(1 - \frac{C_{L,0_2}}{C^*_{0_2}} \right) = \frac{x\mu_{max}}{y_{o_2}} \frac{C_{L,0_2}}{K_m + C_{L,0_2}} \tag{10.18}$$

Solving for C_{L,o_2}, one gets the critical concentration of oxygen for survival of the bioparticles in the unit, in terms of oxygen concentration in the gas phase, mass transfer coefficient, and the microbial kinetic parameters. If the concentration of oxygen falls below this level, then the mass transfer of oxygen becomes the limiting factor. The critical oxygen concentration for most microorganisms lies in the range 0.1 to 10% of the oxygen solubility. The oxygen demand may decide the type of the reactor and/or the operating conditions depending upon the value of mass transfer coefficients that can be achieved. The other factors which may decide the oxygen requirements include the growth phase of the microbial population (i.e., in exponential phase or stationary phase), carbon nutrients or type of substrate (i.e., glucose is metabolized more rapidly than other substrates), pH, and nature of desired microbial process (whether it is substrate utilization or for biomass production or secondary metabolite product yield). The fermentation broth contains many a times foaming substances that accumulate at the gas–liquid interface and provide additional resistance to the oxygen transfer.

The oxygen transfer in free-rise bubble reactors (bubble columns, airlift reactors, tower fermentors, etc.) and high turbulence reactors could be significantly different. In microbial cultures, air bubbles rise in a highly contaminated water. The medium has in some cases highly elevated non-Newtonian viscosity (500 times that of water). In these cases, the free-rise velocity of the bubble, except for large bubbles, is greatly reduced. Bubbles of very small size with very low rise velocity of 0.1 cm/s get equilibrated very quickly with the medium and after that would not contribute significantly even when the gas holdup is very high. Only those bubbles that have life span smaller than time required to absorb their oxygen contents would be useful for the effective mass transfer. With media containing bacteria and relatively low concentrations of extracellular long-chain molecules the viscosities are of the same magnitude as water. However, the presence of surface-active substances may make the surface of the small bubbles very rigid effectively reducing the mass transfer rate. The effect of surface rigidity is particularly marked on the liquid mass coefficient as the bubble size decreases. The common relation for mass transfer coefficient in agitated conditions is

$$\frac{k_l a}{\phi_G} = 0.4 \tag{10.19}$$

For cultures rendered viscous by polysaccharides and fungi, it is recommended [5]

$$\frac{k_l a}{\phi_G} = 0.4 \mu^{-0.5} \tag{10.20}$$

In addition to decreased diffusivity, the vastly increased viscosity increases the contact time of boundary layer around the bubbles because of the reduced rise velocity in viscous media and consequently the mass transfer coefficient is lowered. Relatively large bubbles must, therefore, be employed in very viscous cultures.

10.4 GAS HOLDUP

The area available for mass transfer at the gas–liquid interface depends on the amount of the dispersed phase and also on the degree of dispersion. In bioreactors, the gas phase is usually the dispersed phase in a large quantity of liquid. The volume of gas in the reactor is usually expressed in terms of fractional volume of the reactor occupied by the dispersed phase.

$$\text{Fractional gas holdup}(\phi_G) = \frac{V_G}{V_G + V_L} \tag{10.21}$$

The gas dispersion can also be written in terms of velocities of the two phases in flow conditions. The specific surface area offered by the dispersed phase depends on the size of the dispersed bubble or particle. For an average bubble diameter (d_b), specific area is given as

$$a = \frac{6\phi_G}{d_b} \tag{10.22}$$

Since the volume of the dispersed gas phase at any time is not very accurately known, but the amount of the liquid phase is known with certain degree of accuracy from the known amount of liquid added to the reactor, sometimes the specific surface area can be expressed per unit of liquid volume.

$$a' = \frac{6\phi_G}{d_b(1 - \phi_G)} \tag{10.23}$$

For the estimation of the area, the diameter of the bubble (d_b) must be known, which usually is the difficulty in accurate estimation of a. The bubble diameter depends strongly on properties of the medium. For coalescing media with high surface tension, the diameter of the bubble can be large. In noncoalescing liquids, like those with high viscosity and low surface tension, finer bubbles can be present in the liquid which rise very slowly in the medium and provide a large holdup. However, large holdup of very fine bubbles need not always mean higher mass transfer rate as many of fine bubbles spent long times and have already exhausted of their oxygen content. Tiny bubbles also have low surface mobility. Since for a highly viscous medium, large bubbles are preferred to maintain appreciable rise velocity, the available interfacial area gets limited. Because of the difficulty of knowing d_b accurately, usually values of overall mass transfer coefficient are measured for different operating conditions and different equipments. Depending upon the type of the microorganism and the operating conditions, the actual value of $k_l a$ can vary. Since the fermentations media have characteristics between those of coalescing and noncoalescing media and also show variation with time depending upon the age of culture and the presence of foaming or antifoaming agents, it is advisable to measure these values experimentally.

10.5 SOLID–LIQUID MASS TRANSFER

If respiring biomass particles are relatively large, mass transfer resistances from the bulk fermentation broth to the outer surface of the biomass particles may occur. If the density difference between the particle and the fluid becomes nil, then solid–liquid mass transfer coefficient (k_{sl}) can be approximated by a limiting value, i.e.,

$$\frac{k_{sl}d_p}{D_m} = Sh = 2$$

or

$$k_{sl} = \frac{2D_m}{d_p}$$

(10.24)

where d_p is the diameter of the particle. The liquid–solid mass transfer coefficient not only depends on the physicochemical properties of the liquids but also is influenced by hydrodynamic conditions, impeller speed, and its geometry. In general, liquid–solid mass transfer coefficients are correlated by an expression like

$$\frac{k_{sl}d_p}{D_m} = 1 + c\left(\frac{v_l}{D_m}\right)^n \left(\frac{u_s d_p}{v}\right)^{m'} = 1 + cSc^n Re^{m'}$$

(10.25)

where c, n, and m are constants and v is the kinematic viscosity. The slip velocity (u_s) in the definition of Reynolds' number (Re) is usually difficult to estimate. Therefore, it is common practice to compute Reynolds number on the basis of Kolmogoroff's theory for agitated conditions in terms of specific power input, which gives

$$Re = c\left(\frac{\varepsilon d_p^4}{v^3}\right)^p$$

(10.26)

The exponent p is dependent on the ratio of the particle size to the microscale of eddies. The following correlation closely represents the industrial fermentation systems [6].

$$Sh = 2 + 0.545Sc^{1/3}\left(\frac{\varepsilon d_p^4}{v^3}\right)^{0.264}$$

(10.27)

The relative importance of mass transfer resistance at the liquid–biomass interface (i.e., $1/k_{sl}a_s$) depends mainly on the size of the bioparticles. For instance, in yeast fermentations the particle sizes are in the range of micrometers. For typical fermentation conditions, $k_{sl}a_s \gg k_l a$, hence the oxygen transfer resistance is still located at the liquid–gas surface. On the other hand, in fermentations of *penicillium chrysogenum* in bubble columns, under special conditions biomass particles of 0.3 to 2 mm diameter can be grown. Here, the oxygen mass transfer at the liquid–solid interface (and possibly pore diffusion) should be considered as a major resistance as $k_{sl}a_s \ll k_l a$.

The area available for mass transfer at the solid–liquid interface can be estimated from diameter of the particle (d_p) and the solid suspension density (m_s).

$$a_s = \frac{6m_s}{\rho_s d_p} \tag{10.28}$$

10.6 MIXING AND SHEAR

Agitation is provided in a bioreactor to maintain homogeneity of all nutrients throughout the entire volume of the reactor. The nutrients include oxygen which must be supplied to all living cells at such a rate as to maintain a certain growth of the living organisms. Since the supply of oxygen is from air, it must be dispersed throughout the medium in an appropriate manner so as to increase the interfacial area for the mass transfer of oxygen. The medium characteristics may become important to maintain homogeneity throughout the bulk fluid and also for the dispersion of air. The medium nature has a significant effect on the dispersion characteristics such as fractional gas holdup, mass transfer coefficient, ease of dispersion, power input in creating homogeneous conditions, and also subsequently the cellular growth. However, the benefits of high shear agitation will have to be weighed against the possible detrimental effects on the bioparticle stability. Even for the enzyme-catalyzed reactions, the sensitivity of the catalyst to high speed of agitation needs to be ascertained.

Between two fluid layers, which are flowing with different velocities, the velocity gradient leads to the momentum transfer. The higher velocity fluid drags the slow moving fluid along with it, while the slow moving fluid element tries to slow down the fast moving fluid. The resultant shear between them is best visualized in the form of two planes of fluids flowing with a certain velocity gradient (Fig. 10.4). For a Newtonian fluid, a linear relation is given between the shear stress and the velocity gradient.

$$\tau = -\mu \frac{du}{dy} = \mu \dot{\gamma} \tag{10.29}$$

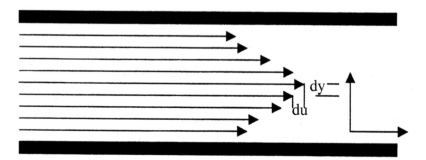

FIGURE 10.4 Velocity gradient in flowing fluid.

where μ is the viscosity and $\dot{\gamma}$ is the shear rate. The shear stress is taken positive in the direction of decreasing velocity. The viscosity of a Newtonian fluid is constant and independent of the shear rate. However, many a times, particularly for bioreactor media with either the filamentous growth or suspension of solids, the medium may show a strong viscosity dependence on the shear rate. For pseudoplastic fluids, such as a medium showing micelial growth, the viscosity decreases with shear rate, while for dilatant fluids the viscosity increases with shear rate as observed with densely packed bacterial suspensions.

For a non-Newtonian fluid the viscosity depends on the shear rate, time, elasticity, etc. The most common way of representing the non-Newtonian behavior is known as the *power law model* (Fig. 10.5)

$$\tau = K\dot{\gamma}^n = \left(K\dot{\gamma}^{n-1}\right)\dot{\gamma} = \mu_a\dot{\gamma} \tag{10.30}$$

The apparent viscosity μ_a of the fluid depends on the shear rate. For a dilatant fluid, the power law index (n) is greater than unity, while for a pseudoplastic fluid it is less than unity. The constant K is called *consistency index*. For some of the fluids, a certain yield stress is needed to set the fluid in motion. For such fluids the following Bingham plastic equation is used with τ_o as the yield stress:

$$\tau = \tau_o + K\dot{\gamma}^n \tag{10.31}$$

The rheological characteristics of the medium affect the shear and momentum transfer between different fluid elements flowing with different velocities. The rheological behavior, in turn, depends on the nature of material, such as whether the medium is a suspension of bacterial cells, suspension density, nature of cells, and filamentous growth, and even on proteinaceous and carbohydrate materials secreted by the growing population of cells in the medium. The viscosity can often originate from interaction between different

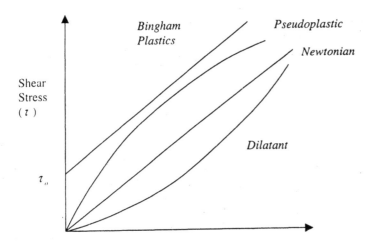

FIGURE 10.5 Fluid characteristics.

strands of the growing population and also from culture age, temperature, etc. and are unique for given fermentation broth under specified conditions. For a multispecies population, it would be difficult to have any unified correlation between these widely different parameters for prediction, and actual viscosity values have to be obtained experimentally.

At the beginning of the fermentation, the viscosity of the medium is dependent only on the medium composition because of a very low cell suspension density. However, as the fermentation proceeds, because of the increasing cell suspension density and changing structures of the cells and fermentation broth, the viscosity shows a nonlinear time dependent behavior. For micellial growth where the filaments of mycelium can intertwine and interact with each other, the consistency index shows a strong and nonlinear dependence on the concentration of the cells while the power law index gradually reduces below unity. The behavior is clearly pseudoplastic in nature. For most micelial growth the power law index (n) lies between 0.2 and 0.8 [7–11]. Concentrated micelial broth may show, in addition, the presence of yield stress. In a stirred vessel, the shear rate is very high near the impeller because of high tip velocity, while the velocity drops significantly as the fluid moves away from the impeller region. The region near the impeller may, therefore, show a low-viscosity behavior while near the reactor wall the apparent viscosity could be very high. In pseudoplactic fluids, as shown by micelial broth of moderately high cell suspension density or where the product is polymeric in nature, the rheological behavior varies within the vessel itself. With tip speeds of 2–10 m/s and reactor diameter up to 10 m, the shear rate ranges from 0.2 to 100 s^{-1} in the same reactor at different locations. The medium in production of polymers such as Xanthan and Pullunan shows a very high viscosity as the product polymer concentration builds up. Usually these solutions show a yield stress and time dependency. For bacterial cell suspensions, the viscosity is greatly affected only when the cell suspension density is increased to appreciable levels such as 15–30% by weight. But such densities are uncommon in commercial fermentors. At moderate density values, the cell suspension may show a dilatant behavior [12]. For a high suspension density broth, the apparent viscosity may reach to very large values making the objective of achieving homogeneity throughout the bulk fluid in the entire volume of reactor more difficult. In case of such gross inhomogeniety, particularly if it affects the growth of the microorganisms or stability of biocatalyst, it may be necessary to dilute the broth at the cost of production rate in the given volume.

The effect of viscosity is more pronounced for shear sensitive cells and biocatalysts, because of the effect of shear stress. It is not always advisable to operate the reactor of widely different rheological behaviors with a high speed of agitation to ensure homogeneity throughout the reactor as the increased shear stress could be detrimental to the growth of the cells or to the activity of biocatalyst, either free or immobilized. The shear force experienced by a cell or biocatalyst originates from the velocity gradient between the streamlines of flow, high-speed impact with impeller blades and reactor walls. For a rational design of a bioreactor, and also for its commercial viability, the sensitivity of the cells, their growth rates under different conditions of shear, the product formation, and growth rate kinetics, overall productivity must be known for the cells under consideration. The data would have to be obtained for the species under consideration as the cell sensitivity is unique for each species.

Systematic experiments are the only way to ascertain the cell viability under differing shear conditions in the reactor and for scale-up from laboratory scale to commercial plants. The same is also applicable to biocatalysts including immobilized enzymes. It may be necessary to isolate the effects of each parameter that can affect the cell viability. The animal cells attached to a support are more prone to shear sensitivity due to lack of protective cell wall and also because of relatively large size of ~20 μm. Being attached to a large-size support, they also cannot orient very easily with the flow of the fluid and bear the brunt of the shear stress generated at the surface because of high shear rate. The velocity of fluid at the solid surface can be assumed to be almost negligible. Since the cells cannot rotate freely and align with the streamlines, the net force and torque experienced by the cells can be enormously high in high shear intensity field such as in the impeller region.

In a tower bioreactor where the homogeneity or mixing is achieved by sparging a gas, the shear is generated at the surface of a rising gas bubble

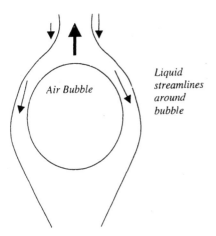

FIGURE 10.6 Relative velocity between gas and liquid phases.

depending upon the relative velocity between the rising bubble and the fluid (Fig. 10.6). The rise velocity depends on the characteristics of the broth such as its density, viscosity, and the size of the bubble; larger bubbles rise very fast while smaller bubbles rise very slowly. For bubbles of size 1 mm to 1 cm with rise velocity of 0.1 to 50 cm/s, the shear rate varies from 1 to 100 s^{-1}.

10.7 NONIDEAL FLOW CONDITIONS IN BIOREACTORS AND BACKMIXING

The mixed flow and plug flow are the two extremes of ideal reactor flow behavior. In the plug flow, every element entering into the reactor is considered to spend exactly the same amount of time before moving out of the reactor space. In the mixed flow conditions, the reactant added to the reactor is mixed completely with the reactor content. For a batch stirred reactor, the mixing will have no influence on the output of the reactor if the reaction is faster as all the ingredients are added to the reactor at the beginning. For a semibatch or continuous stirred reactor, however, the uniform distribution of added reactant stream may be important for maintaining quality of the product. Even for a well-stirred continuous flow reactor, fluid elements added into the reactor would spend different times in the reactor giving rise to a distribution of residence time for different fractions of the exit stream.

A real flow bioreactor would have a residence time distribution between those of the plug flow reactor and a mixed flow reactor. The ideal plug flow, in

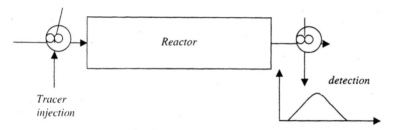

FIGURE 10.7 Stimulus response technique.

fact, would have no distribution at all. The residence time distribution has the consequence of different conversions in different fractions of the exit stream usually with an average conversion between those obtained with ideal plug flow and completely mixed flow. The extent of residence time distribution can be decided by stimulus-response technique using a tracer, which can be injected into the reactor and then is detected in the exit stream with respect to time (Fig. 10.7). Knowing the time when the tracer was injected into the reactor, its detection over time gives its distribution, which is converted into the fraction of the tracer injected into the system.

The residence time distribution function of a fluid in a reactor is given by $E(t)$ function which represents the distribution of age of different fractions of exit stream. The typical $E(t)$ function is shown in Fig. 10.8. The residence time distribution is defined as [13]

$$E(t)dt = \text{fraction of the exist stream}$$
$$\text{with residence time between } t \text{ and } t + dt \tag{10.32}$$

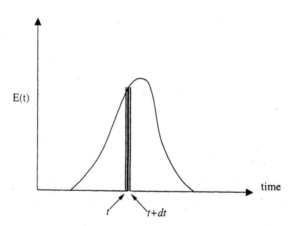

FIGURE 10.8 Exit age distribution.

The $E(t)$ curve can be obtained by injecting a tracer into a flow system and by measuring its concentration with respect to time in the exit stream. If the concentration of the tracer is $C(t)$, then the total amount of tracer is $\int_0^\infty C(t)\,dt$ or $\Sigma\, C(t)\,\Delta t$.

The $E(t)$ function can be then estimated as

$$E(t) = \frac{C(t)}{\displaystyle\int_0^\infty C(t)dt} \tag{10.33}$$

By its definition it is clear that

$$\int_0^\infty E(t)\,dt = 1.0 \tag{10.34}$$

It is often desirable to know the spread of the distribution which is quantified by variance of the exit age distribution about the mean time of the distribution. These two terms are defined as

$$\theta = \int_0^\infty tE(t)dt \tag{10.35}$$

and

$$\sigma^2 = \int_0^\infty (t - \theta)^2 E(t)\,dt \tag{10.36}$$

The residence time distribution (RTD) can be used as an diagnostic tool to identify the deviation of the reactor flow from an ideal flow behavior. However, for most flow conditions the exit age distribution is not sufficient information, and actual flow behavior must be known for quantifying mixing taking place at *micro*level. The RTD information when supplemented with information at microlevel mixedness, becomes a powerful tool in evaluation of the performance of the reactor. The mixing of fluid elements of different ages within the reactor is termed *backmixing*, which should decrease the yield of the reactor.

Two models, most commonly used to represent the nonideal flows or micromixedness are the dispersion model [14] (Fig. 10.9), which is mostly

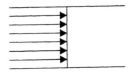

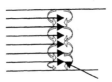

Dispersion imposed on plug flow

Plug flow

FIGURE 10.9 Dispersion model.

applied to tubular reactors, and the tanks-in-series model [15]. In ideal plug flow conditions, every element entering in the reactor spends the same time as any other fluid element and there is no residence time distribution (Fig. 10.10). If, however, backmixing takes place between different elements entered into the reactor at different times then the effective concentration of the reacting species decreases. The mixing in axial direction is by diffusion of species or movement of eddies of different sizes, which increases the residence time distribution. The process is akin to molecular diffusion, but unlike molecular diffusion, which is dependent on molecular properties, the dispersion is affected by the flow conditions in the vessel. If turbulent flow prevails in the reactor, the random velocity fluctuations in all directions will cause mixing of elements at different positions, both forward and backward directions. The dispersion imposed on bulk flow is represented by an equation similar to that used to represent the molecular diffusion flux (Fig. 10.11).

$$J_d = -D_d \frac{dC_A}{dX} \tag{10.37}$$

In a tubular reactor, this flow is imposed on the net flow along the axial direction and therefore, is termed *axial dispersion model*. However, the dispersion can exist in radial direction and with an order different from that in the axial direction. For a plug flow with dispersion (Fig. 10.11), Eq. (10.37) can be written as

$$-u \frac{dC_A}{dX} + D_d \frac{d^2C_A}{dX^2} - r_A = 0 \tag{10.38}$$

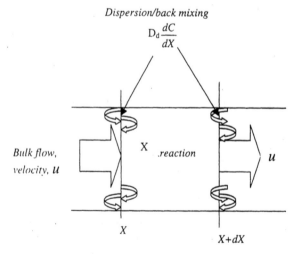

FIGURE 10.10 Plug flow with dispersion.

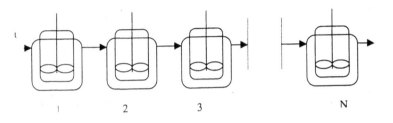

residence time, $\theta = V/u$

FIGURE 10.11   Tanks-in-series model.

When a rate expression is known for r_A and if the dispersion coefficient (D_d) is known from RTD studies, Eq. (10.38) can be integrated to obtain the concentration profile of A as a result of reaction and dispersion. It is more common to write the above equation in dimensionless variables, $z \,(= X/L)$ and $C \,(= C_A/C_{Ao})$

$$-\frac{dC}{dz} + \frac{D_d}{uL}\frac{d^2C}{dz^2} - \frac{r_A L}{uC_{Ao}} = 0 \qquad (10.39)$$

The dimensionless number (uL/D_d) is called *Peclet number* (Pe). For a flow approaching true plug flow, the value of the Peclet number approaches to ∞, while for a completely mixed flow conditions it should be zero. An intermediate value of Peclet number indicates flow conditions between plug flow and mixed flow. For the use of the above equations the axial dispersion coefficient must be known. From experimental exit age distribution, Pe and, therefore, dispersion coefficient can be estimated from the variance and mean time of the $E(t)$ function.

For the dispersion model, the relationship between the distribution characteristics and Peclet number is

$$\frac{\sigma^2(t)}{\theta^2} \approx \frac{2}{Pe} = 2\frac{D_d}{uL} \qquad (10.40)$$

In general, the axial dispersion coefficient is a function of the flow conditions (decided by Reynold's number) and molecular diffusivity of the solute (decided by Schmidt number). A large amount experimental data available from literature in the form of correlations can be used to evaluate the performance of the reactor. Most correlations are of the form

$$Pe = f(Re, Sc)$$

Axial mixing in reactors with mainly axial flow can also modeled by a system of N well-stirred tanks in series. Each of the tanks is treated in a perfectly mixed conditions. The residence time distribution of the tanks-in-series model with a total residence time of θ is given as [14]

$$E(t) = \frac{t^{N-1}N^N}{\theta^N(N-1)!} \exp\frac{-Nt}{\theta} \qquad (10.41)$$

This equation can be used for modeling a real reactor with a nonideal residence time distribution data obtained with a tracer-response experiment. For the tanks-in-series model, the number of tanks (N) in series can be obtained from the variance and mean time of the distribution.

$$\frac{\sigma^2(t)}{\theta^2} = \frac{1}{N} \qquad (10.42)$$

The number of tanks in series once obtained can be used along with equations for ideal mixed flow reactor for each reactor to evaluate the performance of the real reactor. This approach, however, does not need any detail information about the actual flow pattern inside the reactor and its application for design on larger scale may become questionable. The tanks-in-series model also does not permit backmixing beyond the size of individual reactor in the series.

When two phases, say a liquid phase and another gas phase, are brought together in a reactor, a certain amount of reactant must be transferred from one phase to another. This can be ensured only if there is intimate contact between two phases. In a bioreactor, it is the supply of oxygen from gaseous phase into the aqueous phase where the "reaction" takes place. The two process phases enter into reactor at different points and at least one of them may leave the process or both may pass through the reactor. The flow of each phase may fall either in mixed flow or plug flow or plug flow with super-imposed dispersion. A variety of combinations will be possible within the reactor. In addition, if a third phase is also present, it can either remain within the continuous phase or can get associated with the dispersed phase. For example, solid particles are carried in the wake of gas-phase bubbles through the liquid continuum. When agitation is at very high speed, both continuous as well as dispersed phases are well mixed. Usually, because of sufficient density difference between the gaseous and liquid phases, separation takes place in the reactor phase, and two phases move at different velocities in the reactor.

In the case of heterogeneous systems the situation becomes more complex because the degree of nonideal flow behavior and segregation/mixedness must be known for both phases. Particularly for the dispersed phase, such as a gas phase in stirred contactor or tower bioreactor, the segregation depends on the bubble breakup and coalescence. Any physico-chemical parameter that can influence these processes will also have effect on the degree of mixedness or segregation in that phase and consequently on the mass transfer processes.

10.8 HEAT TRANSFER

The transport of heat plays an important role in the operation of bioreactors if the heat effects become very large. Although, extreme temperatures are uncommon in actual operation of bioreactors, sterilization of medium might require elevation of medium temperature beyond $100\,^{\circ}$C. The supply of heat to medium under such conditions or removal of heat from the fermentation medium to maintain temperature below the critical temperature for survival or for optimum performance of microbial species needs careful analysis of heat transfer techniques.

Three different mechanisms according which heat can be transported from higher temperature to lower temperature can be distinguished.

10.8.1 Conduction

In stagnant media, the heat is transported by molecular movement in the presence of temperature gradient, according to Fourier law, i.e.,

$$q_h = -\lambda A_h \frac{dT}{dX} \tag{10.43}$$

10.8.2 Convection

In a flowing medium, the heat transfer is result of heat entrained with the flow of the medium. The flow could be forced one or set by other external forces, such as convective currents because of density difference, etc. The rate of heat transfer near the heat transfer area is described by an equation analogous to mass transfer, i.e., using a concept of stationary film near a surface offering major resistance to the heat transfer and using a concept of heat transfer coefficient (h).

$$q_h = hA_h\Delta T \tag{10.44}$$

10.8.3 Radiation

This mode of heat transfer by electromagnetic radiation is unlikely to be important in bioreactors as it becomes appreciable only at high very temperatures.

The former two modes of heat transfer are often present together. For example, the heat transfer through the fluid film near the heat transfer surface will be by convective transport and is characterized by h while the heat transport through the wall takes place by conduction only. Since for heat transfer, another fluid flows on the other side of the wall, the heat transfer again is characterized by heat transfer coefficient for that fluid. The two resistances to heat transfer on both fluids and the wall itself are combined in an overall heat transfer coefficient (U) with overall driving force (ΔT) for heat transfer between two fluids (Fig. 10.12):

$$q_h = UA_h \Delta T \tag{10.45}$$

where the overall heat transfer coefficient (U) is related to all resistances to heat transfer

$$\frac{1}{U} = \frac{1}{h_i} + \frac{\Delta x}{\lambda_{\text{wall}}} + \frac{1}{h_o} \tag{10.46}$$

where the subscripts i and o indicate the parameters associated with an inner fluid and outer fluid, respectively. The individual values of heat transfer coefficients are related to the flow conditions on two sides and the physicochemical properties of the fluids.

Fermentation processes are accompanied by heat effects due to heat dissipation by the mechanical agitator in the liquid, heat absorbed from hot gas from compressor, which compresses air adiabatically, and because of aerobic and anaerobic fermentation processes. The heat may be lost because of expanding gas bubbles on their way up in the reactor or for vaporization of part of medium if dry air is used as the feed. If the amount of the heat generated is more than the heat lost, the temperature of the broth may rise. Higher temperatures are not always useful for viability of the organisms and it may be necessary to remove the heat through heat transfer area, either through reactor wall or immersing cooling coils in the reactor. The equations

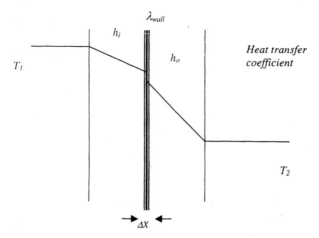

Figure 10.12 Heat transfer by convection and conduction.

given above can be used to estimate the necessary heat transfer area, if the heat generation terms and heat removal terms are known. The energy dissipation by impellors has been discussed separately in the next chapter. The heat energy input because of agitation is of the order of 0.1 to 5 kWm^{-3}. The heat absorption from incoming air depends on its feed temperature and flow rate and the temperature of the broth. The heat generation because of metabolic activity in a number of fermentation systems has been correlated to the specific oxygen uptake rate [16].

$$Q_m = 465 \times 10^3 (\text{J/mol of oxygen})$$
$$\times OUR(\text{mol of oxygen/m}^3/\text{h}) \times V(\text{m}^3) \tag{10.47}$$

One can estimate from this equation that heat generation in bioprocesses has the same order of magnitude as chemical reactions. For instance, in 3000 m^3 fermentor of the Hoeshst-UHDE SCP process the reported OUR is 10 kg $O_2/m^3/\text{h}$ giving heat generation of 4.36×10^8 kJ/h. In case of bioreactors, the heat is dissipated in vessels which are larger by one or two orders of magnitude than conventional chemical reactor. In addition, the temperature level is low, therefore, latent heats cannot be used for heat removal. Owing to a small temperature difference between reaction and cooling media the heat transfer area must be large. The calculation of heat transfer coefficients is also important for biological processes, considering the possibility of slime formation and adherence of biomass particles to the heat transfer area, which might reduce the heat transfer considerably.

The overall heat transfer coefficient has to adjusted if the medium shows strong fouling tendency. The fouling layer, which could be a growing mass of cells on the heat transfer area, provides additional resistance (R_f) to heat transfer. Since the fouling is a time dependent phenomena, the resistance to the heat transfer builds up over a period and needs attention for cleaning if it becomes appreciable.

$$\frac{1}{U} = \frac{1}{h_i} + \frac{\Delta x}{\lambda_{\text{wall}}} + \frac{1}{h_o} + R_f \tag{10.48}$$

Usually inside the cooling coils the flow of fluid is faster than the flow conditions in the fermentor making the heat transfer coefficient on the fermentation side the controlling parameter. In some instances, the heat exchange area is provided outside the fermentation by recirculation of medium, provided the microbes can withstand higher shear region in the pump and contamination is avoided. In addition, the residence time in heat exchanger is kept minimum to avoid oxygen-deficient conditions for the microbes during the recirculation.

10.9 FOAMING IN BIOREACTORS

In aerobic fermentors, generation of excessive foam by gas bubbles escaping from the fermentation medium may cause decrease in effective reactor volume and product loss and/or contamination if the foam overflows. A number of reviews about foaming are known [17–19]. Foaming liquids are characterized by low surface tension because of dissolved surface active materials that could be present in the medium itself or are generated by the microbial population during growth. Proteins particularly have the tendency because of their amphiphilic nature to reside at the air-water interface.

Proteins are always present in the fermentation media. Many fermentations are directed to production of proteins or cells produce them as extracellular products. Proteins as biopolymers have amino acid's side chains forming hydrophobic parts and ionic residues forming hydrophilic parts. The flexibility of a protein molecule in orienting itself at the surface with continuous change from its three-dimensional structure helps in its adsorption at the air–water interface in a such a way that hydrophobic chains try to move away from the bulk water while ionic groups anchor themselves at the surface. The conformation of an adsorbed protein is such that the hydrophilic parts form a chain at the surface while hydrophobic parts form loops and tails. The process is dynamic as a protein molecule takes some time in uncoiling itself and adsorbing at the interface. The process of adsorption of a single protein molecule is slow because it involves continuous uncoiling and adsorption of a large number of segments from the structure of the protein. Because a large number of segments of a single molecule is involved in the process even a small amount of the protein (1 ppm) could be sufficient to reduce the surface tension substantially. Once adsorbed, however, it is difficult to dislodge the molecule because of the multipoint attachment. They can be, however, be dislodged by another small molecular weight surfactants. The adsorption of surface active molecules reduces the surface tension of the fluid. Since the protein adsorption is slow, the equilibrium is reached after a long time, sometimes in hours. The rate of change and the ultimate change in surface tension is not necessarily decided by the amount of the adsorbed protein but how a protein is adsorbed. And the entire process is random unlike the surface tension experienced in the presence of conventional surfactants.

Another important consequence of protein adsorption on the surface is the sustenance of surface-active gradients on the surface for a long time. Since these gradients act against the shear stress exerted because of liquid flow or gravity, the liquid velocity can be slowed down considerably, almost to zero. The local viscosity experienced by solvent also could be high because of the presence of entangled chains and loops of protein segments at the surface. When two bubbles collide and remain in contact with each other with a thin film between them, their coalescence depends upon the rate of drainage of the liquid between them. As the bubbles rise and if the drainage is not fast, the bubbles may not coalesce. On one hand this may help in maintaining higher surface area for mass transfer. It may also provide a cushioning effect on the cells by opposing the shear created at the interface because of the fluid motion. On the other hand, when such bubbles enter in the foam zone, the film drainage could be a slow process giving rise to stable foam on the liquid surface. The foam mainly consists of gaseous phase encapsulated within a structure with a very low holdup of liquid. The liquid films give rise to a honeycomb structure that breaks if the surface tension forces cannot support the liquid against the gravitational forces. In the protein-stabilized films, distribution of proteins molecules is also not as fast as compared to that observed with conventional small molecular weight surfactants, and because of the protruding chains and loops the protein stabilized film is thicker and more stable.

The foaminess because of proteins is also dependent on those characteristics of the medium, which can affect their structures. Since pH has direct effect on the structure of the proteins, particularly the charged groups, their numbers, and type of charge, the protein adsorption, and subsequently the foaminess depend on pH, and the effect is maximum at the isoelectric point of the protein where its net charge is zero and a protein molecule will have the highest tendency to move out of the liquid medium. In addition, if the protein precipitates at pI, the foaminess increases. The presence of ionic salts in the medium has a similar effect. The increased counterion association with the charged groups in the presence of salts can decrease effective charge on a protein molecule, making it more hydrophobic and more surface active. Increase in temperature decreases foaming, probably due to decreased viscosity of the medium and facilitating film drainage. At higher temperatures denaturation or precipitation of proteins can affect the foam stability.

For the fermentation systems the foam control may become essential. Any defoaming technique, however, should not hamper the mass transfer processes as defoamers can decrease the mass transfer area. The most widely employed method is the use of chemical defoamers as relatively a small amount (<1%) is needed, and the method is reliable. Typical defoamers include oils, fatty acids, esters, polyglycols, and siloxanes. Oily materials basically aim at replacing the film stabilizing proteins molecules from the surface. The presence of hydrophobic oil droplets breaks the aqueous film by creating holes by exploiting the contact angle between air and water and the oily phase. Small molecular weight surfactants displace the protein from the interface. But the addition of defoamers also increases the possibility of contamination and affects the acceptability of the process for products of pharmaceutical end uses. A number of other techniques such as a high-speed rotating disks, ultrasonication, and temperature shocks are available without changing the nature of the fermentation medium. High-speed rotating disks, however, on larger scales become too expensive because of massive constructions, high energy costs, and limited efficiency, and for highly foaming liquids their capacity can be very low.

PROBLEMS

1. In what respects does a bioreactor resemble a chemical reactor and differ from it?

2. Why is it necessary to maintain uniform environment for each cell in a bioreactor?

3. Show that by reducing a bubble diameter the area available for mass transfer per unit volume of the gas phase increases.

4. Why is a typical bioreactor inherently a heterogeneous reactor?

5. What differences do you expect in diffusional rates in the following conditions:

    a. (i) A glucose molecule and (ii) protein molecule of MW 150 kD in aqueous solution.

    b. Glucose molecule through (i) aqueous salt solution and (ii) glycerol solution

    c. Glucose through (i) aqueous solution and (ii) cell membrane.

6. Show that for a sparingly soluble gas the controlling resistance lies on liquid side. What will be the controlling resistance for a highly soluble gas?

7. How does the penetration theory differ from the film theory of mass transfer?

8. What is the difference between penetration theory and surface renewal theory of mass transfer?

9. What is the difference between mass transfer coefficients from film theory and penetration theory?

10. Stirred reactors can break a gas bubble into very small size bubbles, but will it be always advantageous?

11. What is a critical oxygen concentration?

12. Estimate a solid–liquid mass transfer coefficient for a particle of size 100 μm in aqueous solutions.

13. Why do mycelia broths show pseudoplastic behavior?

14. What is a RTD? What kind of RTD do you expect for a laminar flow in tubular reactor?

15. What are the possible sources of stress that a cell can experience in (a) stirred reactor and (b) bubble column.

16. Estimate heat generation because of metabolic activity in a fermentor having an average OUR of 25 kg O_2/m^3/h.

17. What are defoamers? Why is their use essential in fermentation?

18.   If mass transfer coefficient and a $(k_l a)$ $=0.05/s$, equilibrium concentration of oxygen in solvent is 0.25 mol/l, cell density $x = 0.25$, $\mu = 0.12$,
and $Y_{o_2} = 0.4$, then is the reaction kinetic or mass transfer controlled?

19.   What is the total heat transferred if the heat transfer area is 20 m^2,
heat transfer coefficient is 0.01 kcal/m^2/C/s and hot fluid temperature is 150°
and cold fluid temperature is 50°C?

20.   Estimate the heat transfer area for transferring heat of 150 kcal/s if
the heat transfer coefficient is 0.0144 kcal/m^2/C/s and hot fluid temperature is
160° and cold fluid temperature is 45°C.

21.   What should be the driving force to achieve a oxygen transfer rate
of 0.5 gmol/s of oxygen if the mass transfer coefficient and area $(k_l a)$ is 0.01/s.

22.   Determine the OTR if the driving force is 0.01 mol/l, yield factor is
0.5, and $k_l a = 0.009/s$.

23.   An organic liquid (T_h) is condensed at 115°C on the outside of a
tube through which cooling water is flowing at an average temperature (T) of
25°C. The inside and outside diameter of the tube is $D_I = 15$ mm and $D_0 = 20$
mm, respectively. The heat transfer coefficients for water $(h_i) = 2270$ W/m^2°C
and for the organic liquid $h_o = 2840$. If you neglect the resistance offered by
the tube wall estimate the tube wall temperature. Note: For heating $T_w = T +
\Delta T_i$ and for cooling $T_w = T + \Delta T_i \cdot \Delta T_i = (\Delta T/h_i)/(1/h_i + D_i/D_o\ h_o)$ and
$\Delta T = T_h - T$.

24.   Alcohol is flowing inside the inner pipe of a double-pipe heat
exchanger having an inner (D_i) and outer tube diameter (D_o) of 2.6 cm and
3.1 cm respectively. Water is flowing outside. The thermal conductivity of the
tube (k_w) is 40 W/m °C. What is the over all heat transfer coefficient based on
the outside area (U_o) of the inner tube. Note the following: Alcohol coefficient
$(h_i) = 1000$ W/m^2 °C, water coefficient $(h_o) = 1650$ W/m^2 °C, inside fouling
factor $(h_{di}) = 5500$ W/m^2 °C, outside fouling factor $(h_{do}) = 2900$ W/m^2 °C.
The equation for the overall coefficient is given as

$$U_o = \frac{1}{(D_0/D_i h_{di}) + (D_0/D_i h_i) + (x_w/k_w)(D_0/D_L) + (1/h_o) + (1/h_{d0})}$$

x_w is the wall thickness and D_L the logarithmic mean diameter.

25. What should be the area for heat transfer if the volume of fluid that
needs heating $= 1000$ lt to a $T = 150°C$ using steam at 190°C. The $\rho\ C_p$ of the
fluid is $=1.5$ and the heat transfer coefficient is $=0.01$.

NOMENCLATURE

A	Specific surface area of gas–liquid interface, m^{-1}
A	Diffusing species
A_h	Heat transfer area, m^2
a_s	Specific surface are of solid particles, m^{-1}

$C(t)$	Concentration at time t, mol/m^3
C_A	Concentration of A, mol/m^3
$C_{A,G}$	Concentration of A in gas phase, mol/m^3
$C_{A,L}$	Concentration of A in liquid phase, mol/m^3
C_A^*	Concentration of A in liquid phase in equilibrium with the gas phase concentration, mol/m^3
D	Diameter of impeller, m
d_b	Diameter of bubble, m
D_g	Dispersion coefficient in gas phase, m^2/s
D_l	Dispersion coefficient in liquid phase, m^2/s
D_m	Molecular diffusivity of solute, m^2/s
d_0	Orifice diameter, m
D_s	Dispersion coefficient in solid phase, m^2/s
d_s	Diameter of bubble at the orifice, m
$E(t)$	Exit age distribution
F	Pumping rate of impeller, m^3/s
H	Height of tank, m
H_d	Width of impeller blade, m
J	Diffusion flux, $mol/m^2/s$
J_A	Mass transfer flux of A, $mol/m^2/s$
J_d	Dispersion flux, $mol/m^2/s$
K	Consistency index [Eq. (10.30)]
k_g	Gas side mass transfer coefficient on gas side, m/s
k_l	Gas side mass transfer coefficient on liquid side, m/s
k_m	Mass transfer coefficient, m/s
K_{OL}	Overall mass transfer coefficient, m/s
k_{sl}	Solid–liquid mass transfer coefficient, m/s
L	Characteristic length, m
M	Equilibrium concentration ratio
m_s	Solid suspension density, kg/m^3
n_{cr}	Critical speed of agitation for suspension of solids, rps
$n_{cr,s/g}$	Critical speed of agitation for suspension of solids in presence of gas, rps
N_p	Power number of impeller
n_q	Flow number of impeller
n_s	Speed of agitation, rps
P	Power input, W
P	Pressure of gas, N/m^2
q_h	Heat transfer flux, $J/m^2/sec$
Q_m	Heat generation because of metabolic activity, J/sec
r_A	Rate of reaction, $mol/m^3/sec$
Re	Reynolds's number

R_f	Fouling resistance
T	Tank diameter. m
t	Time, s
t_c	Time of contact, s
t_{cir}	Circulation time, s
t_m	Mixing time, sec
U	Overall heat transfer coefficient, $J/m^2/K/s$
U	Velocity, m/s
u_s	Slip velocity between liquid and solid phrases, m/sec
V	Volume, m^3
X	Distance, m
X	Cell suspension density, number/m^3
Y	Yield factor
Z	Dimensionless distance
Δx	Thickness of wall, m
ν	Kinematic viscosity
θ	Mean time of exit age distribution
σ^2	Variance of exit age distribution
σ	Surface tension, N/m
δ	Film thickness, m
ε	Energy input density, W/m^3
γ	Shear rate, 1/s
λ	Heat conductivity, $J/m/K/s$
λ_κ	Size of eddy, Kolmogoroff scale, m
μ	Viscosity, Ns/m^2
μ_{max}	Maximum specific growth rate of cells, 1/s
ρ	Density, kg/m^3
τ	Shear stress, N/m^2

REFERENCES

1. Lewis, W.K.; Whitman, K.R. Ind Eng Chem 1924, *12*, 1215.
2. Higbie, R. Trans Am Inst Chem Eng 1935, *31*, 365.
3. Danckwerts, P.V. Gas-Liquid Reactions; McGraw Hill: New York, 1970.
4. Doraiswamy, L.K.; Sharma, M.M. Heterogeneous Reactions. John Wiley: New York, 1984; Vol. II.
5. Andrew, S.P.S. Trans I Chem E 1982, *60*, 3.
6. Sanger, P.; Deckwer, W.D. Chem Eng J 1981.
7. Roels, J.A.; Van den Berg, J.; Vonchen, R.M. Biotechnol Bioeng 1974, *16*, 181.
8. Charles, M. Adv Biochem. Eng 1978, *8*, 1.
9. Blakebrough, N.; McManamey, W.J.; Tart, K.R. J Appl Chem Biotechnol 1978, *28*, 453.

10. Manchanda, A.C.; Jogdand, V.V.; Karanth, N.G. J Chem Technol Biotechnol 1982, *32*, 660.
11. Henzler, H.J.; Schafer, E.E. Chem Ing Tech 1987, 59, 940.
12. Van Suydam, J.C.; Dusselje, P.C.B. Concept of apparent morphology as a toll in Fermentation. In *Physical Aspects of Bioreactors*; Crueger, W., Ed.; Dechema, VCH: Wainheim, 1987; 107 pp.
13. Danckwerts, P.V. Chem Eng Sci 1953, *2*, 1.
14. Levenspiel, O.; Bischoff, K.B. Adv Chem Eng. Academic Press: New York, 1963; Vol. 4.
15. Roemer, M.H.; Durbin, L.D. Ind Eng Chem Fundament 1967, *6*, 121.
16. Roels, J.A. Energetics and Kinetics in Biotechnology; Elsevier Biomedical Press: Amsterdam, 1983.
17. Bikerman, J.J. Foams: Theory and industrial Publications; Reinhold: New York, 1953.
18. Hall, M.J.; Dickinson, S.D.; Pritchard, R.; Evans, J.I. Prog Ind Microbiol 1973, *12*, 169.
19. Viestures, U.E.; Kriscapsons, M.Z.; Levitans, E.S. Adv Biochem Eng 1982, *21*, 169.

11

Stirred Bioreactors

11.1 STIRRED BIOREACTORS

Stirred vessels form the major workhorse in biochemical industry, many of them are operated batchwise and at a very large scale. Whether in a catalytic reaction involving enzymes or in a fermentation process, in order to make the process possible, the components have to be brought in contact at the molecular scale. In a typical reactor, the reacting species are added separately into the vessel in which reaction takes place, either as miscible or immiscible phases. The components from different streams are to be brought in intimate contact with each other if the reaction between them is to proceed. The reactor is, therefore, a three-dimensional space where these components are contacted with each other under either controlled conditions of temperature, pressure, etc., or are left to react under uncontrolled conditions. When the reactants are miscible with each other or are present in the soluble form in a solvent, the reaction is said to *homogeneous*. On the other hand, when the reactants are present in different phases, the system is *heterogeneous*. For a soluble enzyme-catalyzed reaction in aqueous phase where the substrate is also soluble, the reaction system is homogeneous. For an aerobic fermentation process, where oxygen transfer from gaseous phase to the liquid medium is necessary for the survival and growth of the microorganisms, the system is heterogeneous with distinct liquid, solid, and gaseous phases. Since in such

241

heterogeneous conditions, the existence of transport processes across an interface characterizes the overall process, the relative rates of kinetic and transfer processes determine the rate-determining step.

11.2 MIXING IN STIRRED CONDITIONS

Mixing in everyday uses means blending of two or more materials. The scale of mixing is decided by the smallest size of a sample from the mixture that shows inhomogeniety or difference in properties from those of its surrounding. In practice, the scale of mixing is much higher than the molecular scale and much smaller than the system itself. For homogeneous mixtures, the scale of mixing is almost at molecular level given sufficient time, for example, when a water-soluble substrate, such as a buffer salt, is dissolved in water. The mixing in such a solution can be defined at the molecular level of the salt and water. The mixing in the solution is said to be complete if a sample taken from any part of the solution has the same concentration of the dissolved salt. This state is achieved only after sufficient time is given for the salt to dissolve and if the solution is agitated with an appropriate device. The terms *sufficient time* and *agitation* are important here as they decide the scale of operation.

If a certain amount of solute is added to a large amount of water in a vessel, the distribution of the solute molecules will not be spontaneously uniform throughout the solution without an external aid. If the solution is left unperturbed, it will have a region high in concentration in the dissolved solute where the solid is dissolving, while the region away from the solute may will be completely devoid of the solute. The only mechanism by which the solute distributes itself throughout the liquid bulk phase will be by molecular diffusion. In a liquid phase, the diffusion process is a somewhat slow process with a typical value of 1×10^{-9} m^2/s. Even in a laboratory beaker, the uniform distribution of the solute will take days if not years, if left unagitated. In manufacturing processes, one cannot be at the mercy of the slowly diffusing species, particularly if we are interested in bringing two or more components in contact with each other. The rate of the reaction between any two species will be decided by the number of molecules of two species coming in intimate contact with each other in a given space in a given time. If a large number of different reacting species are brought together in the given space, the number of them undergoing the reaction will be expectedly higher. This also implies that the dispersion can also reduce the number density of the molecules, which in turn can decrease the local concentrations of the reacting species and, therefore, the rate of the reaction. So if the reaction has a stronger dependence on concentration, the mixing is not always advantageous. But if we need to intimately contact two separate streams of different reacting species for the reaction to occur, we have to live with the decreased rates. The backmixing effects can be expressed by either dispersion model or tanks-in-series model. The extent the concentration deviates from ideally expected value defines the intensity of mixing, or rather demixing.

The mixing time, in general, can be determined experimentally by a tracer experiment, in which the tracer is followed after its injection into the system by some means. The time the tracer takes to reach within 5% of the ultimate value is taken as the mixing time. Of course, the mixing time should depend upon the degree of mixedness and intensity of mixing. A typical experimental mixing time response is shown in Fig. 11.1.

Let us first, however, concentrate on the major ways of bringing the mixing process. The most common way of mixing is using a mechanical device, which provides energy for the mixing process. The device can be an agitator or a blender or a pump. The mixing device functions by generating velocity gradients and shear stresses. The velocity gradient is because of the different velocities of different segments of the fluid and momentum transfer across them. In an agitated system, the mechanical energy is transferred to different fluid elements in the vicinity of the impeller blade setting them in motion with respect to the rest of the fluid. As a fast moving fluid element shears the rest of the fluid during the circulation set up in the vessel, it transfers its momentum to the neighboring fluid elements. For larger size vessels the circulation will be over a longer distance before the same fluid element comes back to the impeller space. The high shear region can also break the fluid elements or 'eddies' into smaller eddies. The degree of the breakup decides the level of turbulence in the vessel. The circulation current set up in the reactor

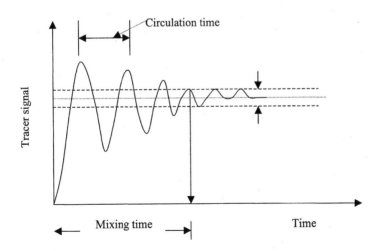

FIGURE 11.1 Experimental tracer study.

results in macromixing, while the shear and elongational flow, i.e., thinning of an eddy because of the shear, causes the micromixing. The thinning of an eddy, which ultimately results in its breakup in smaller fluid pockets, ensures more uniformity and mixedness.

    The mixing of liquids can be achieved with various types of impellers (Fig. 11.2). The shafts of the impeller are mounted vertically in the axis of a cylindrical vessel. The impeller blades are mounted usually at a height between 0.25 and 0.5 of the height of the liquid layer. For vessels having H/D ratio of 2–3, multiple impellers are used with spacing such that for a section of vertical height approximately equal to the diameter of the vessel, one im-

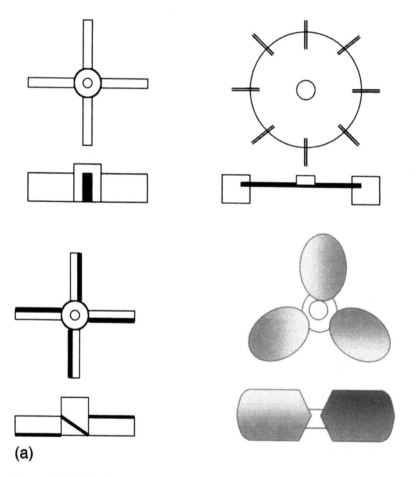

(a)

FIGURE 11.2   Various agitator designs.

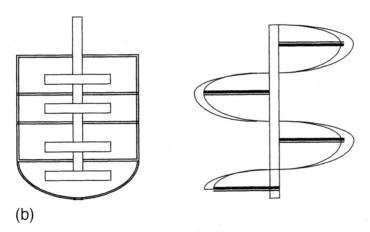

(b)

FIGURE 11.2 Continued.

peller is available (Fig. 11.3). The diameter of the impeller commonly varies from 0.2 to 0.5 of the tank diameter, a diameter ratio of 1/3 is often preferred. The tanks are also provided with, most often four, baffles, i.e., flat plates of width of about 0.1 of the tank diameter, mounted vertically on the reactor walls. The baffles serve to reduce the formation of a vortex and promote vertical circulation of the fluid in the vessel. Turbine impellers pump the fluid radially so that in the upper part of the vessel the fluid moves upward along the wall of the vessel and downward in the center of the vessel, while in the lower section of the vessel, the fluid moves downward along the wall and upward at the center. For a propeller-type impeller, depending upon whether it is downward or upward pumping, the fluid moves downward or upward at the center of the vessel and the other way around along the wall. For a pitched-blade turbine where the impeller blades are at an angle with the plane of the impeller, the flow can be either downward or upward at the center with a radial component, usually a small one, also accompanying the flow of the fluid motion. The flow induced by the impeller sets the fluid in circulatory motion throughout the vessel. The circulation velocity largely decided by the impeller dimension, its speed, and the size of the vessel itself. Typical circulation patterns are shown in Fig. 11.4.

For mixing viscous fluids, encountered in special cases, a good circulation is necessary to move the liquid in far corners of the vessels. The flow pattern in highly viscous fluids is difficult to maintain in highly turbulent conditions. The mixing is, therefore, achieved by elongation of fluid elements and shear using gate or anchor type of impellers that rotate almost the entire volume of the vessel and create considerable shear near the walls. The flow

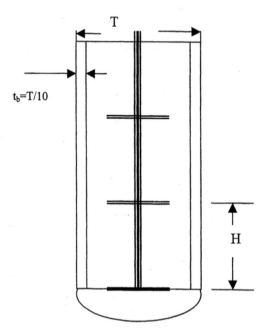

FIGURE 11.3 Typical multiple-impeller stirred vessel system.

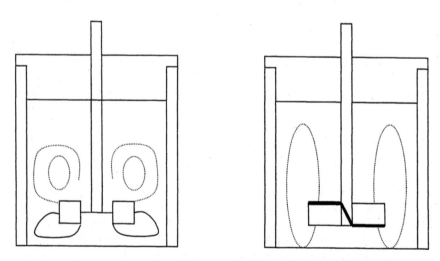

FIGURE 11.4 Circulation patterns in an agitated vessel.

pattern of such fluids is, however, circulatory, and little mixing in vertical direction can be achieved. A better construction for acquiring both shear and circulation is using a helical impeller. These impellers are, however, characterized by the heavy consumption of power despite low speeds of agitation and heavy mechanical constructions.

The purpose of mixing is to reduce the concentration difference in the reactor at different points. It is also important to know how fast this homogeneity is achieved. The time required for such a complete mixing as compared to that required for consumption of the substrate can decide the relative importance of hydrodynamics and kinetics. For a reaction which is slow as compared to the mixing process, the emphasis should be to improve the reaction kinetics, while for a relatively fast reaction as compared to the transport process, efforts to improve further kinetics, say, by improved catalyst or reaction conditions will not be fruitful unless they have other advantages such as improved selectivity.

For a process where the hydrodynamic effect is significant, understanding mixing becomes important. The mixing can be characterized by the degree of mixing and the rate of mixing. Usually the latter is quantified in terms of mixing time, i.e., the time taken by the system to reduce the difference in a measurable property below the detectable value. The degree of mixing is strongly decided by the hydrodynamic conditions, the bulk movement pattern set up in the reactor, physical properties of the fluids such as density or density difference, viscosity, homogeneous or heterogeneous conditions, presence of suspended solids, etc. When a stream is added continuously into a mixed reactor where it is miscible, it causes local concentration difference at the point of addition. For a reaction to occur uniformly throughout the reactor, uniform distribution of this reactant may be necessary. If the reaction is slow, the time required for uniform distribution may be relatively insignificant. On the other hand, however, for a faster reaction the distribution time will affect not only the reaction rate and thus production rate but can also affect the final product distribution at the end of the reaction.

The uniformity of mixing depends on overall circulation brought about by either pumping action of an impeller or a pump itself, shearing action of the impeller blades or that of rising gas bubbles, on transfer of momentum from eddies to each other when there is appreciable velocity difference between the neighboring eddies and molecular diffusion within the eddies or between the eddies when the relative velocity is negligible between them. The relative sizes of the fluid eddies, depending upon the degree of turbulence or shear experienced by them, decide the degree of each of the above factors. For small eddies with sizes below 0.1 mm and normally observed for highly turbulent conditions, the diffusion within the eddy element is faster while for a highly viscous fluid the circulation currents could be faster than the diffusional mixing within the eddy elements mainly because of their size. Overall, however, even the circulation patterns could be slower than those achieved in low-viscosity fluids. Most solutions, even if with a low-viscosity solvent, because of the presence of suspended particles including cells, behave as a fluid with high viscosity.

The mixing within a small fluid element is a faster process, but the mixing of eddies may become slower depending upon the type of flow pattern. If the flow is laminar, the eddies may spend a long time maintaining their identity. The segregation for longer duration could be detrimental for efficiency of the operation. Particularly when one reactant is continuously added to the reactor, its contact with other reactant(s) will be insufficient when segregation takes place. The larger is the size of such eddies, worse would be the performance of the reactor. For high-viscosity liquids the performance of the reactor could not be acceptable. The most common means of improving effective contact is then the vigorous agitation. The energy dissipation in the mixture would do the job of breaking the fluid elements to smaller eddies. However, as the scale of operation increases the energy dissipation would be nonuniform, particularly for high-viscosity fluids or with liquids with high solid loading.

For a fully developed turbulent flow in an agitated vessel, the energy delivered by an impeller is fully transformed into kinetic energy of the fluid pumped by the impeller. For a radially pumping impeller the pumping rate is the volume swept by the impeller in unit time

$$F = \pi d^2 n_s H_d \sim k n_s d^3 \tag{11.1}$$

and the power input is given as

$$P = F\frac{\rho u^2}{2} = N_p \rho n_s^3 d^5 \tag{11.2}$$

where

N_p = the power number
ρ = density of the medium
n_s = rotational speed (s^{-1}), for an impeller of diameter d (m)
u = the impeller tip velocity
H_d = height of the impeller blade

For the fully developed conditions, as normally used in the industry, the power number of the impeller is constant. However, in the lower range of Reynolds number (Re), which characterizes the flow conditions for a given geometrical conditions, it shows an inverse relationship with Re (Fig. 11.5).

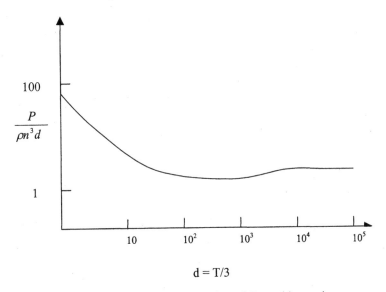

$$d = T/3$$

FIGURE 11.5 Power number as a function of Reynolds number.

The intensity of flow conditions can be gauged by the Reynolds number (Re), which is defined as

$$Re = \frac{\rho n_s d^2}{\mu} \tag{11.3}$$

For values $Re > 10^5$ the flow is considered turbulent, while below $Re < 100$ the flow is laminar. The intermediate region is the transition region. For an impeller diameter of 0.5 m and speed of agitation of 2 s^{-1} in a fluid of density 10^3 kg/m^3 and viscosity of 1×10^{-3} Nm^{-2}s, the Reynolds number is 5×10^5, which is clearly in turbulent region. Increase in diameter of the vessel will maintain the turbulent conditions. At relatively large scale, the local scale of operation can be turbulent in a medium of low viscosity. The energy input density is defined as

$$\varepsilon = \frac{P}{\rho_l V} = N_p n_s^3 d^2 \tag{11.4}$$

This value of energy input density is an average value as the velocities and velocity gradients vary locally within the stirred vessel as one moves away from the impeller region toward the wall. Obviously the shear rates and local energy dissipation would also vary in the same vessel from point to point. If the energy dissipation is the same throughout the medium, the process is isotropic and in the scaled-up version the same flow conditions can be expected if energy input density is kept the same. However, it is difficult to assume to the same energy input throughout the bulk phase if the process is nonisotropic, and keeping the same energy dissipation rate may not ensure the same flow conditions in the scaled-up version.

The liquid flow in most stirred vessels is at least locally turbulent, specially at large scale and higher impeller tip velocity. For sufficiently high bulk flow Reynolds number, the smallest eddies exist in a state of isotropic equilibrium. The smallest size of eddie can be estimated by the Kolmogoroff length scale λ_κ(m)

$$\lambda_k = \left(\frac{v^3}{\varepsilon}\right)^{1/4} \tag{11.5}$$

where ε is the power dissipation rate per unit mass. The size of the smallest eddy decreases with increase in the power dissipation rate per unit volume (or mass). During scale-up the constant energy dissipation rate can be considered as one of the criteria. It is assumed that if the energy dissipation density is kept constant at the higher scale of operation, then the mixing effects will be the same. As discussed above, in the absence of uniform energy dissipation this criteria becomes questionable.

The pumping capacity, i.e., the rate of liquid discharge from the impeller, decides the local flow rate in the vicinity of the impeller. The pumping rate of an impeller is related to the speed of impeller and the area swept by the impeller blades in every revolution.

$$Q = N_q n_s d^3 \tag{11.6}$$

When the flow of fluid discharges from the turbine impeller blade, it also carries additional fluid from the area surrounding the impeller without entering into the turbine's area. Therefore, the discharge from the turbine impeller can be greater than that estimated simply from the speed of the impeller and its geometrical configuration. The circulation rate within the vessel, therefore, is more than that estimated from the pumping rate of the impeller. The ratio of circulation rate to the pumping rate can be as high as 4 for turbine impeller in unbaffled tanks. In baffled tanks this number could be reduced to 2, while for pitched blade turbine the ratio is between 1.2 to 1.6. The circulation rate decides the time taken by the same fluid element to arrive again at the impeller space. Thus, the mixing time in a vessel is proportional to the linear dimension of the vessel.

For a vessel of diameter T and height H the circulation volume is the volume of the vessel

$$V = \frac{\pi}{4} T^2 H \tag{11.7}$$

and since the pumping rate for turbine impeller is

$$F = \pi d n_s H_d \tag{11.1}$$

the circulation time is

$$t_{cir} = \frac{(\pi/4)T^2 H}{\pi d n_s H_d} = \frac{T^2 H}{4 T n_s H_d} = \frac{TH}{4 n_s H_d} \tag{11.8}$$

Since H_d is proportional to d (or T), and H is usually the same as T or proportional to T, the mixing time t_m is proportional to the linear dimension of the vessel. It also indicates that mixing time increases with the dimension of the vessel. Depending on the actual pumping capacity, the mixing time can be estimated. Since for a turbine impeller in a baffled vessel the pumping capacity is double the discharge rate of the impeller, the circulation time will be half of the value calculated using the equation.

If the pumping rate is considered as a mixed flow through a capacity of the vessel as a pure first-order system for its dynamic characteristics, the mixing time for almost 98% mixing should be about four times the circulation time

$$t_m = 4 t_{cir} \tag{11.9}$$

In practice, however, the experimental value is at least 1.6–2 times that calculated from Eq. 11.9. For highly viscous media, Reynolds number may be low ($<10^4$), which increases the mixing time considerably as enough turbulence would not exist in the vessel. For larger reactors, viscosity beyond 1 Nsm^{-2} should be avoided. Since the product $n_s t_m$ is constant on larger scales, as n_s decreases for constant power input density ε, t_m increases considerably. For such conditions a constant ε criteria for scale-up may not be advisable.

Stirring is also required for suspension of solids, either of cells or pellets or immobilized enzymes. A minimum speed of agitation can be estimated from the energy imparted by impeller to the liquid and the settling velocity of the particles. Agitation also helps in liquid–solid mass transfer. A particle in a stirred vessel experiences (1) gravitational force because of density difference, (2) viscous and inertial drag from liquid because of velocity gradients, and (3) frictional forces with other particles. The performance of the impeller is specified in terms of critical speed of the impeller required to lift the particles from the bottom of the vessel or at which no particles will settle at the bottom for more than, say, 1 s. A correlation for critical suspension speed was given by Datta and Pangarkar [1] for unaerated conditions.

$$n_{cr} = 3.5 v^{0.1} \left(\frac{g \Delta \rho}{\rho_l} \right)^{0.42} \frac{m_s^{0.15} d_p^{0.18} T^{0.58}}{d^{0.15}} \tag{11.10}$$

where m_s is the weight percentage of the solid in suspension.

On sparging a small amount of gas into a two-phase (solid–liquid) system with particles in just suspended conditions, the particles start settling down. For aerated systems, therefore, the critical speed of agitation increases for suspension of solid, and the change in the critical speed is proportional to the gas flow rate. The induced gas accumulates in the lower pressure region behind the blades in the gas-filled cavities. The increased pressure behind the blades results in a lower pressure drop and reduced drag and subsequently lower power consumption. Any decrease in power input and pumping capacity reduces all the parameters that cause solid suspension, i.e., drag force, local energy dissipation, and associated turbulent eddies. Even a minor change in impeller type, d/T ratio, type and design of sparger, etc., can affect the solid suspension. The same authors [1] have suggested the following correlation for critical speed for solid suspension in aerated conditions.

$$n_{cr,s/g} \propto d_p^{0.18} \Delta\rho^{0.42} m^{0.146} d^{-1.23} \tag{11.11}$$

The solid–liquid mass transfer coefficient ($k_{s/g}$) in aerated conditions further has been correlated by Pangarkar et al. [2] to the critical impeller speed of agitation for suspension of solids by following equation:

$$k_{s/g} = 1.8.10^{-3} \left(\frac{n_s}{n_{cr,s/g}} \right) Sc^{-0.53} \tag{11.12}$$

The ratio of actual speed of agitation n_s to critical speed of agitation for suspension is a significant operating parameter as evident from Eq. (11.12). It may also be noticed that the dependence on molecular diffusivity is square-root type, as predicted by penetration theory. This correlation for mass transfer coefficient is the most generalized form of the equation and should be useful for any system involving three phases.

A stirred vessel with aeration has been a preferred choice for several fermentation systems including those of highly viscous media where other types of bioreactor may not operable. The preference is due to higher heat and mass transfer coefficients, good liquid–air interfacial area, good mixing capability, and a wide range of possibilities of liquid and gas residence time distributions. In aerated bioreactors, involving a gas phase sparged in the fluid, the mixing time increases almost by a factor of 2 from the value for homogeneous systems under otherwise identical conditions [3–5]. Most of the gas, under the conditions of very high Reynolds number ($>10^5$), show a cavity formation behind the impeller blades. The pumping capacity of the impeller, therefore, decreases and so does the power number of the impeller. The mixing time in single- and two-phase conditions are related, most experimental data showing a difference of 2: i.e., the two-phase mixing time is almost double of the single-phase mixing time. This difference may become very important on the larger scales as the mixing efficiency decreases with the scale.

For heterogeneous conditions the mixing in the gas phase may also become a point of concern. In most well-agitated vessels the gas phase can be considered to be well mixed in the same way as the liquid phase. When the gas is sparged in the eye of the impeller, which is usually the practice, the impeller blades break up the bubbles into smaller bubbles. The larger bubbles rise in the fluid phase faster, but smaller bubbles rise slowly and very slowly in high viscosity fluids. If the impeller speed is high enough the gas bubbles, having diameters too small to escape from the liquid phase, will be continuously recirculated with the incoming fresh gas phase. As a result one can assume a complete mixing of the gas phase under well-agitated conditions. If the mixing mechanism is the same as the liquid phase, the mixing time for gaseous phase also can be approximated by the liquid-phase mixing time.

The mixing pattern in stirred vessel in the presence of air is different than that in the absence of it. At higher impeller speed and low aeration rate, the pattern is more or less as that in unaerated vessel. On increasing air flow rate, however, the flow pattern gets increasingly dominated by air flow. At other extreme of high air flow rate, the air is not dispersed at all with a reversal of circulation flow at the center. In this condition the impeller is said to be flooded where energy dissipation by air is higher than the energy dissipation by the impeller. On that basis the proposed criteria for the absence of flooding is [6]

$$\frac{F}{n_s d^3} < 0.3 \frac{n_s^2 d}{g} \qquad (11.13)$$

Single-impeller stirred vessels, however, are criticized particularly in large-scale applications because of enormous energy requirements and uneven distribution of the energy in the vessel. Multiple impellers are increasingly being used due to efficient air distribution, increased gas holdups and highger gas residence time, better flow characteristics and lower power requirements. Height-to-diameter ratios of 2 or more are now common which also provide higher heat transfer area per unit volume, redistribution of air bubbles by impellers in each impeller zone and better utilization of air dispersed into the reactor. Since oxygen supply in aerobic fermentations is of prime importance, anything that can increase the mass transfer rate is welcome. An excellent review on multi-impeller systems as bioreactor has been presented by Gogate et al. [7].

The circulation pattern set up by multiple impellers depends on the type of the impellers, the speed of agitation, geometry of the vessel, and the distance between impellers. For impellers spacing higher than diameter of the vessel, however, the flow patterns generated by the impellers are quite independent of each other. For spacing equal to impeller diameter the flow patterns can either merge with each other or work against each other. For a disk turbine and pitched-blade turbine dual-impellers system a maximum in radial velocity has been reported that also has a higher hydraulic efficiency in energy utilization in fluid transport and better pumping capacity [8]. The higher radial velocity also means bubbles generated at the impeller tip can travel longer toward the wall of the vessel dispersing in a better way, which should help in improving mass transfer. With smaller diameter multiple-impeller systems, the distance between the impeller zone and wall is reduced considerably.

For stirred vessels with multiple impellers, compartmentalization of fluids in different sections within the reactor is apparent at lower speeds of agitation. For Re <50, the mixing between different sections is almost non-existent as transport of fluids from one section to another is reduced considerably. In the case of unhindered transport from one compartment to another, the total circulation time will be a sum of circulations times of individual compartments. Thus the mixing time also will be a sum of individual sections's mixing times. The system of multiple impellers can be modeled as a number of mixed zones with some backmixing allowed between them. If circulation loop promoted by a single impeller is considered as a single tank, the exchange flow rate between the circulation loops become the controlling factor for mixing time. The exchange flow rate in turn is decided by the type of the flow pattern generated by individual impellers. If mixing efficiency is defined as reciprocal of the product of mixing time and power consumption for multiple-impeller systems, the mixing efficiency has been observed to increase with impeller spacing. In general, for multiple impellers the mixing time increases over the single-impeller system for the given volume [7].

Abardi et al. [9] have given the following relation for estimation of mixing time in multiple impeller system

$$n_s t_m = 0.5 \left(\frac{T}{d}\right)^{2.3} \ln \frac{2}{M} \tag{11.14}$$

where M is for degree of homogenization of the liquid. While for a dual-impeller system Vasconcelos et al. [10] have correlated the mixing time for 95% mixing with the power input per unit volume:

$$t_{m,95\%} = 1500 \left(\frac{P}{V}\right)^{-1/3} \tag{11.15}$$

For the two-phase conditions in multi-impeller systems, the effect of the air sparged in the fluid on mixing time depends on the operation conditions. At lower impeller speeds of agitation the mixing time increases marginally or stays constant while at higher speeds of agitation the mixing time increases with the superficial gas velocity. The critical speed for suspension of solids in multi-impeller systems depends on the number of impellers, and it increases with the number of impellers. For any multi-impeller system, the lower impeller plays a pivotal role. Any interference in its role by impeller above it will affect the solid suspension behavior. Any dependence on the number of impellers, therefore, will be decided by the interimpeller spacing. If the impeller spacing is more than the impeller diameter, the critical speed of lower impeller remains unchanged. If the impeller spacing is less than the impeller diameter, there would be increase in critical speed as the upper impellers interfere in the flow pattern generated by the lower impeller. Aeration of solid–liquid suspension results in an increased value of critical speed of agitation for suspension of particles in the same manner as in single-impeller systems. The difference is dependent on properties of the system and gas flow rate. Since in a bioreactor every cell or pellet needs its oxygen supply, a uniform suspension of cells is desired. At the same time the critical speed of agitation must be minimum to avoid adverse effect of shear on the cell viability.

The power consumption in multiple-impeller systems depends on the impeller spacing. When the impellers spacing is less than the impeller diameter, the total power consumption has been observed to be less than sum of power consumptions of individual impellers. If the impeller spacing is equal to the vessel diameter, the power consumption is equal to sum of power consumption of individual impellers. The power consumption in presence of air, is always less than that under unaerated conditions, the reduction being more for the lowest impeller. Most of the gas sparged into a multi-impeller system goes through the bottom impeller with recirculation of part of the gas below the bottom impeller while the above impeller may not see any part of the gas that has been recirculated by the bottom impeller.

Although the multiple-impeller system is at some disadvantage because of increased critical suspension velocity in solid–gas–liquid systems, additional impellers help in effective distribution of solids throughout the vessel. For bioreactors, where uniform suspension of all particles is essential, this definitely is an advantage with additional benefit of increased mass transfer coefficient at the solid–liquid interface.

11.3 GAS DISPERSION IN STIRRED BIOREACTORS

The purpose of agitation is to create gas holdup by producing small bubbles that, however, because of a low free rise velocity escape slowly from the liquid. Consequently the area available for mass transfer should increase; however, the mass transfer coefficient does not change much. In fact, it may show a decrease for very fine bubbles that behave like rigid solid particles.

The gas holdup in a stirred vessel is controlled by the energy input and gas sparging rate. At low speeds of agitation the gas holdup will be independent of stirrer speed and is mainly controlled by sparger design. Above a certain speed of agitation, however, the gas bubbles are dispersed by the stirrer and with breakup of bubbles into smaller ones, their residence time is changed and so also is their recirculation in the impeller region. The speed of agitation should be above a minimum speed of agitation (n_m), which is decided by type of impeller, clearance from bottom, type of sparger, and gas superficial velocity. For multiple impellers this value is also decided by gas flow rate, unlike its independence in single-impeller system. Above a certain speed of agitation, however, the gas holdup becomes independent of the speed of agitation where the breakup rate of bubbles because of shearing action of impellers matches the coalescence rate of bubbles.

High-speed agitators are frequently used as means of creating dispersion where air is injected into liquid stream entering the eye of the impeller. The main gas dispersing process takes place in the immediate vicinity of the impeller. On introduction of air in stirred vessel, it is entrained in the vortex behind the impeller blade. Its dispersion takes place at the tip of the cavity where dispersed gas enters into liquid phase. At lower air flow rates gas remains at the center of the vortex, but with increasing air flow rate the cavity is increased occupying the entire vortex at higher air flow rates and which allows gas to cling to the impeller blades. All the air introduced into the vessel along with that being circulated in form of fine bubbles is entrained into this cavity. This also ensures mixing in the gas phase even for noncoalescing liquids as all bubbles originate from this cavity. Since the air is continuously sucked into the cavity and then distributed depending upon the speed of the impeller, the gas sparger design should have no influence on the degree of dispersion and also on its quality. If the liquid is noncoalescing, collisions between gas bubbles may not lead to their coalescence. These bubbles would, therefore, have smaller size and retain their identity without mixing with other bubbles during circulation cycles. Coalescing liquids, on the other hand, promote the formation of bigger bubbles which also are easily broken down by average turbulence in the vessel. The air phase in such liquids undergoes continuous coalescence and breakup cycles leading to a characteristic bubble size distribution.

The presence of surface active compounds, ionic salts, and polymers affects the nature of the medium. If the medium shows non-Newtonian characteristics, the flow behavior and gas dispersion get affected substantially. In a shear thinning liquid, because shear is very high in the impeller region, the viscosity is low but away from it the viscosity increases. The gas from the vortex from the impeller blade will be not then dispersed easily into the liquid phase outside the impeller zone.

For geometrically similar impellers at low gas voidage the volume pumped is proportional to $H_a d^2$ and the bubble size leaving the impeller is determined by the balance between the Newtonian forces on the bubble due to the shear near the impeller tip and the surface-tension-induced strength of the bubble [11]. Thus,

$$\pi d_b \sigma = K\rho(n_s d)^2 d_b^2 \tag{11.16}$$

As the bubble disruption is very rapid, surface tension agents have little influence on the process and one can safely consider the surface tension of water as a representative value. The above expression can be rearranged in terms of power consumption

$$P \propto \rho d^5 n_s^3 \tag{11.17}$$

that is,

$$P \propto \rho d^2 (dn_s)^3 \tag{11.18}$$

that is,

$$P\rho^{-1}d^{-2} \propto (dn_s)^3 \tag{11.19}$$

or

$$(n_s d)^2 \propto (P\rho^{-1}d^{-2})^{2/3} \tag{11.20}$$

Substituting in Eq. (11.16) for $(n_s d)^2$

$$d_b \propto \sigma\rho^{-1/3} P^{-2/3}d^{4/3} \tag{11.21}$$

The equation indicates that the size of the bubble would decrease with an increase in the power input. In high-viscosity solutions the coalescence of these bubbles is difficult, and the area for mass transfer will be provided by these dispersions. If the volumetric gas flow rate is V_g, then the holdup of the gas is proportional to

$$\phi_G \propto \frac{\mu V_g}{d^2} \tag{11.22}$$

or

$$\phi_G = \mu V_g \left(\frac{V}{H_a d}\right)^{4/3} P^{4/3} \tag{11.23}$$

This expression shows that the small $H_a d$ high-speed impellers make effective use of power to create gas holdup in agitated system. Microbiological reactors typically place the impellers close to the air inlet at the center at the bottom of a relatively tall cylindrical tank often with a draft tube to lead the slowly circulating medium down.

The common relation for mass transfer coefficient in agitated conditions is [11]

$$\frac{k_l a}{\phi_G} = 0.4 \tag{11.24}$$

For cultures rendered viscous by polysachharides and fungi, it is recommended that it should be assumed that [11]

$$\frac{k_l a}{\phi_G} = 0.4\mu^{-5} \tag{11.25}$$

In the highly turbulent region in agitated systems, $k_l a$ is no longer constant but increases with power input. The transition from free bubble regime into turbulent regime occurs at P/V of between 1 and 2 kW/m^3. In highly turbulent regime, the whole contents of the tank are close to the impeller. The bubbles produced by the impellers, instead of floating to the surface of the liquid, are circulated throughout the fluid phase with frequent distortions and coalescence and breakup. If the characteristic time is considered between the distortions or breakup at the impeller region, then it is inversely proportional to the speed of agitation ($\sim 1/n_s$) and equivalent to the contact time for mass transfer when the bubble surface is renewed.

$$k_l \propto (D_m n_s)^{1/2} \tag{11.26}$$

$$\frac{k_l a}{\phi_G} \propto \frac{1}{d_b} \sqrt{D_m n_s} \tag{11.27}$$

or

$$\frac{k_l a}{\phi_G} \propto \sigma^{-1} \frac{1}{d} \sqrt{D_m \rho^{1/6}} \left(\frac{H_a d}{V}\right)^{-5/6} d_b^{-1/2} P^{5/6} \tag{11.28}$$

Experimentally, $(k_l a/\phi_G)$ is found to be proportional to $P^{0.8}$ at a laboratory scale [11]. In addition, the ratio is inversely proportional to square root of viscosity. The gas holdup increases until limited by the reducing efficiency of the impeller, which becomes flooded with gas. Values of ϕ_G in this region usually do not exceed 30%. The value of 20% can be used for estimation. As virtually all microbial systems contain added antifoam, it is recommended that $(k_l a/\phi_G)$ be reduced by 40% relative to pure liquids. Gas holdup in a multi-impeller system has been reported to be higher than that for single-impeller systems at the same stirrer speed and gas flow rate.

11.4 CORRELATIONS FOR GAS HOLDUP IN STIRRED BIOREACTORS

A number of correlations are available from numerous experimental studies for estimation of holdup of gas phase in stirred conditions and most of them are in the form

$$\phi_G = c \left(\frac{P^a}{V} \right) u_G^b \tag{11.29}$$

The constants a, b, and c depend on the type of impeller and combination of impeller in the case of multi-impeller systems. Feijen et al. [6] gave a correlation relating pressure difference across the unit to correct for superficial gas velocity

$$\phi_G = 0.13 \left(\frac{P}{V} \right)^{1/3} \left(u_G \frac{P_{surface}}{P_{stirrerlevel}} \right)^{2/3} \tag{11.30}$$

An other correlation has been suggested by Hughmark [12], however, from an extensive number of results on turbine stirrer in pure liquids.

$$\phi_G = 0.74 \left(\frac{F_g}{n_s V} \right)^{0.5} \left(\frac{n_s^2 d^4}{g H_a V^{2/3}} \right)^{1/2} \left(\frac{0.0025 \rho n_s^2 d^4}{\sigma V^{2/3}} \right)^{1/4} \tag{11.31}$$

A few studies have been conducted with multi-impellers systems, and the correlations can be presented for P/V range of 1–5000 W/m^3 and u_G range of 1–2.5 cm/s [7].

$$\phi_G = (0.151 - 0.328) \left(\frac{P}{V} \right)^{0.195 - 0.244} u_G^{0.566 - 0.695} \tag{11.32}$$

11.5 CORRELATIONS FOR MASS TRANSFER COEFFICIENTS

Since the correlations for holdup are different in different impellers and also in multi-impellers of different combinations, it is not surprising to see different correlations for volumetric mass transfer coefficients as well. Calderbank and Moo-Yang [13] proposed the following correlation:

$$Sh = 2 + 0.31 \left(\frac{d_b \Delta \rho g}{\mu D_m} \right)^{1/3} \tag{11.33}$$

Equation (11.33) shows that the mass transfer coefficient depends on size of the bubble. For very small bubbles (<1 mm), the surface is rigid because of surface tension force. Larger bubbles (>5 mm), on the other hand, have more mobile surface and follow contact time equal to size of the bubble divided by its rise velocity:

$$k_l = 2\sqrt{\frac{D_m}{\pi t_c}} = 2\sqrt{\frac{D_m u_r}{\pi d_b}} \qquad (11.34)$$

Volumetric mass transfer coefficients also have been correlated in usual form in single- and multi-impeller systems

$$k_l a = c\left(\frac{P}{V}\right)^a u_G^b \qquad (11.35)$$

The exponent a in most multi-impeller experimental investigations is in the range 0.6–0.68 for a clean system involving air and water, but in the presence of ionic salts the exponent changes to higher values 0.82–1.17 [7]. The dependence on superficial gas velocity, however, is not influenced so much by the salts in aqueous medium. The values of exponent – are in the range 0.5–0.59 for clean systems, while in the presence of salts it is in the range 0.4–0.55. The constant c, however, strongly depends on the combination of the multiple impellers. Since the presence of salts shows a strong effect on the mass transfer coefficient, these correlations cannot be applied directly to fermentation medium. The effect of viscosity can be considered by adding viscosity ratio to these equations.

$$k_l a = c\left(\frac{P}{V}\right)^a u_G^b \left(\frac{\mu}{\mu_w}\right)^d \qquad (11.36)$$

For example, Kawase and Moo-Young [14] gave the following correlation for volumetric mass transfer coefficient

$$k_l a = 0.675\left(\frac{P}{V}\right) u_G^{0.5} \left(\frac{\mu}{\mu_w}\right)^{-1.25} D_m^{0.05} \sigma^{-0.6} \qquad (11.37)$$

The value of d varies between 0 and −1.3. Hence, the viscosity effect cannot generalized for any system. Since viscosity also affects other properties, such as rise velocity, diffusivity, and even coalescence of the bubbles, it is not easy to generalize its effect. However, it is clear that increased viscosity reduces the mass transfer coefficient. In general the diffusivity dependence on viscosity can be represented by

$$D_m \mu^n = \text{constant} \qquad (11.38)$$

For small molecular weight solute in solutions the exponent of viscosity is unity but for large molecules it is equal to −2/3. In the case of non-Newtonian media the dependence on agitation speed is less, but dependence on superficial gas velocity is increased due to its direct effect on the overall mixing. For highly viscous medium the following correlation is available for bubbles of hindered surface flow [15]:

$$k_l a = 0733 \left(\frac{P}{V}\right)^{0.903} u_G^{0.457} D_m^{0.33} \mu^{-1.6} \sigma^{-0.6} \rho^{0.533} \qquad (11.39)$$

Other correlations have the (P/V) ratio multiplied by $(1 - \phi_G)$ giving equations implicit in ϕ_G. It is difficult to derive an unique equation to represent holdup data as estimations from these correlations show a wide variation, probably because they have been obtained with different impellers and different combinations. The properties of the medium, such as ionic salts have shown significant effects on the holdup and clearly their applications to fermentation systems should be with caution even if the equations are dimensionless and have been obtained with units of geometric similarity. The presence of components affecting coalescing tendency of the bubbles would give rise to a very different holdup. To apply a particular correlation, care should be exercised for the type of sparger, power dissipation rate, number and type of impellers, and properties of the medium.

Most biological reaction systems produce foam stabilizing substances, which may stabilize the gas–liquid dispersions by reducing the rate of drainage from the liquid film. The lack of bubble coalescing ability does help in increasing gas holdup and subsequently $k_l a$, but a tradeoff is necessary between the carryover of foam and reduced value of $k_l a$ as many a time long aged small bubbles will be useless as far as transfer of oxygen is concerned. As the superficial gas velocity increases, the uncertainty in predicting the gas holdup also increases, particularly in bubble column. At low velocities the rise velocity of bubbles approximates to the free rise velocity, and in liquids similar to water it is about 20–30 cm/s. However, as superficial velocity increases the relative rise velocity becomes uncertain.

11.6 EFFECT OF SHEAR ON CELLS

Since microbial cells are affected adversely by shear, high shear conditions in stirred vessel could be detrimental to the viability of cells. It may be necessary to consider the magnitude of the shear experienced by the cells. The highest shear will be definitely in the vicinity of the agitator while it should decrease as the fluid moves away from the impeller. Croughen et al. [16] defined an integrated shear factor (ISF) as a measure of shear field strength between an impeller and vessel wall.

$$\text{ISF} = \frac{2\pi n d}{T - d} \qquad (11.40)$$

The experimental observations showed that above ISF of 19 s^{-1}, the growth of animal cells droped to zero. Another parameter suggested by the same authors [16] is time averaged shear (τ_{avg} for turbulent regime):

$$\tau_{avg} = \frac{113.1 n_s d^{1.8}(T^{0.2} - d^{0.2})(d_f/d)^{1/8}}{T^2 - d^2} \quad (11.41)$$

where

$$d_f = \frac{0.625 \text{ Re}}{625 + \text{Re}} \quad (11.42)$$

Croughen et al. [16] also showed that if one considers the hydrodynamic forces solely arising from spatial velocity gradients, then the maximum shear on the surface of a particle is given as

$$\tau_{max} = 3\mu\gamma \quad (11.43)$$

The maximum shear is proportional to the tip velocity of the impeller blade.

$$\gamma = cu_{tip} = c\pi n_s d \quad (11.44)$$

For a flat-blade turbine $c \sim 40$ m^{-1}. For the laminar region the maximum shear rate is in the trailing vortex of the impeller. For an average shear rate of 7 s^{-1} and for water $\mu = 0.9$ mNsm^{-2} the estimated shear is about 18.9 mNm^{-2}. For animal cells the critical stress has been reported as 0.65 Nm^{-2}, above which the cells are damaged. In highly viscous medium where the clearance between the impeller blade and vessel wall is small, the shear rate generated could be large enough to kill the cells. For such conditions Croughen et al. [17] estimated the maximum tangential shear rate from the following equation, considering the flow profile between two concentric cylinders:

$$\tau_{max} = \frac{4\pi n_s T^2}{T^2 - d^2} \quad (11.45)$$

If the smallest eddy size is larger than the biocatalyst particles, the particles will follow the local flow patterns since in most cases the bioparticles have densities closer to that of the medium. The relative velocity between the two will be almost nonexistent, and so will the effect of the shear on the activity of the cells. For particles of much higher density, however, the effect cannot be ruled out. For turbulent and isotropic conditions if the eddy size is similar or smaller than the bioparticle size, the effect on the activity could be significant. Detrimental effects of turbulent conditions become more apparent when the Kolmogoroff eddy size drops below 100 μm. Excessive agitation thus lead to smaller eddy size with large enough energy to damage the cells.

Croughen et al. [16] correlated the energy dissipation density to the death rate of the cells as

$$q_d = k_e(\lambda_k)^n \qquad (11.46)$$

where k_e is the specific death rate constant, which depends on the cell and support properties and is reasonably constant for geometrically similar systems. Although these authors conclude that "if the turbulent eddy model is valid, scale-up at constant power input per unit mass (or volume) should not lead to detrimental hydrodynamic conditions," it may not be always possible to have isotropic conditions in the large-scale reactor, particularly handling non-Newtonian fluids where the energy dissipation is nonuniform. The increase in scale also increases the impeller diameter and the tip velocity proportionately. Near the impeller blades the turbulent conditions should be more intense and thus may still lead to detrimental effects on the cell viability.

As per the Kolmogoroff's theory [18], the angular velocity of Kolmogoroff's fluctuations is proportional to $(\varepsilon/v)^{0.5}$. If the hydrodynamic blows of these fluctuations or eddies are responsible for the death of the cells, then

$$q_d \propto \left(\frac{\varepsilon}{v}\right)^{0.5} \quad or \quad q_d = k_1\left(\frac{\varepsilon}{v}\right)^{0.5} \qquad (11.47)$$

If k_1 were constant, q_d should be proportional to $\varepsilon^{0.5}$. Since the experimental observations indicate that q_d is proportional to $\varepsilon^{0.75}$, k_1 is dependent on $\varepsilon^{0.25}$ or inversely proportional to λ_k, i.e., Kolmogoroff's scale. Thus

$$q_d\left(\frac{v}{\varepsilon}\right)^{0.5} = k_1 = k_2\frac{d_p}{\lambda_k} = k_2 d_p\left(\frac{\varepsilon^3}{v^5}\right)^{1/4} \qquad (11.48)$$

The diameter of the particle is used to make the eddy size, i.e., the Kolmogoroff scale, dimensionless. Since the energy dissipation rate depends on the impeller and vessel diameter along with the properties of the medium and speed of agitation, the effect of system parameters on the bioparticle viability can be estimated. Increased viscosity, reduced agitation speed, smaller particle size, and smaller impeller diameter would reduce the detrimental effect on the cells. The effect of speed of agitation and viscosity is the most prominent. It must be borne in mind that to maintain a homogeneous conditions in the reactor, a certain speed must be maintained. At least the circulation time for bioparticles throughout the reactor must be smaller than the biological response time.

For single particle settling velocity in the range 10^{-7} to 10^{-4} m/s, maximum shear stresses are ~0.01 Nm^{-2}, which are far below the stress to damage the cells. Keeping the cells just in suspension should not be detrimental at all. For maintaining homogeneity of the system for oxygen transfer needs liquid velocities in turbulent conditions, i.e., in excess of 5 cm/s. At larger scales, even if the liquid velocity is kept constant, the increased scale of operation increases the degree of turbulence in the bulk phase. If the velocities have to be increased to ensure sufficient turbulence for oxygen transfer, it may not be advisable to operate the same in a stirred conditions, and other option of employing a tower bioreactor can be exercised.

11.7 HEAT TRANSFER IN STIRRED VESSELS

The heat generated in stirred vessels by metabolic activity of microbes can be removed by heat transfer fluid via jacket, cooling coils, or an external heat exchanger. Since the flow rate of heat transfer fluid through coils or jackets can be maintained at high values and mostly clean fluids are used, the main resistance to heat transfer lies on the fermentation fluid side. The resistance is further compounded by fouling tendency of the fermentation medium and tendency of the cells to grow on the heat transfer area.

Heat transfer in stirred vessels is strongly dependent on the flow pattern produced by the stirrer. A turbine stirrer generates a radial flow that on reaching the wall moves in the axial direction. The thickness of the boundary layer formed near the wall depends on the energy of eddy reaching to the surface and thus on the Re of the stirrer. The mean velocity near the wall should be proportional to the velocity generated by the stirrer, i.e., $n_s d$.

The general form of heat transfer correlations in similar hydrodynamic conditions is

$$\text{Nu} = \text{constant} \left(\frac{\rho n d^2}{\mu}\right)^a \left(\frac{c_p \mu}{\lambda}\right)^b \left(\frac{\mu}{\mu_w}\right)^c \tag{11.49}$$

For relatively high Re numbers the momentum transport between the core of eddy approaching the surface boundary layer determines the thickness of the layer and its heat transfer characteristics. For turbine type impellers that induce axial flow near the wall, the experimental value of exponent of Re varies between 0.5 (low Re, Re $< 10^4$) and 0.7 to 0.8 (for very high Re number). For Re number in the region $10^3 < \text{Re} < 10^6$, the Re exponent varies from 0.65 to 0.70. For example, the following relation has been suggested by Henzler [19]:

$$\text{Nu} = 0.6 \left(\frac{\rho n d^2}{\mu}\right)^{2/3} \left(\frac{c_p \mu}{\lambda}\right)^{1/3} \tag{11.50}$$

If the medium has a very low viscosity, the last term in Eq. (11.49) can be taken as unity. The application of such correlations to fermentation media, however, should be suspect because these correlations are available from studies with clean systems, while fermentation media are quite complex and show a wide range of behavior, even for the same medium over the period of fermentation. Further, in the presence of sparged conditions the flow patterns are modified. For low gas flow rate in low-viscosity medium the heat transfer coefficient shows an increased heat transfer coefficient. With further increase in air flow rate the impellers gets more and more flooded with gas with reduction in gas dispersion and in heat transfer coefficient. The effect of aeration on heat transfer characteristics depends on liquid flow patterns and, therefore, on the gas flow rate and the speed of agitation. For a micellium growth that shows a pseudoplastic behavior the heat transfer may become quite a problem as the maximum shear is possible only in the region close to impeller region. The region away from impeller shows increased viscosity and therefore very poor heat transfer characteristics. The heat transfer within such a region will reduce to conduction instead of convection. This kind of media are best handled using impellers that can kept the entire mass in circulation, such as helical ribbon or paddle type impellers.

In aerated conditions the effect of aeration on heat transfer is a function of gas velocity and speed of agitation. It depends on the liquid circulation patterns set up in the presence of gas bubbles and impeller action. It is expected that the heat transfer characteristics should be similar to the mixing time behavior in the stirred reactors in the presence of gas [20].

The data on multiple impeller systems heat transfer are scarce and limited to dual-impeller systems. Nearly 50% increase in heat transfer efficiency has been reported by Streak and Karoz [21] and Steiff [22], which has been attributed to intensification of heat transfer in the region above the lower impeller and circulation velocities generated by the top impeller. The heat transfer coefficients are higher with multi-impeller systems for the same power input; probably the circulation currents set up by the impellers help in intensification of heat transfer processes. Since greatest turbulence is created in the plane of impeller, increase in the number of planes definitely assists in improving the heat transfer coefficient. It may, however, decrease in the presence of air, since the increased gas hold up in the vessel can decrease the heat transfer characteristics of the reactor. Karcz [23] has shown that heat transfer coefficient decreases with increase in the gas flow rate in the agitated vessel. An average heat transfer coefficient is given by a correlation

$$\mathrm{Nu} = 0.769 \mathrm{Re}^{0.67} \mathrm{Pr}^{0.33} \left(\frac{\mu}{\mu_w} \right)^{0.14} \tag{11.51}$$

For a single-impeller system the correlation constant is 0.6, indicating increased heat transfer coefficient for multiple-impeller systems. In a multi-impeller system the distance between the heat transfer area and impeller blade can be reduced by keeping higher H/T ratio. This also provides a higher surface area for heat transfer per unit volume and, better control must be possible on heat transfer processes.

    Example 11.1   A large scale fermentor ($T = 4.0$; $H/T = 2$) needs to maintain the oxygen uptake (OUR) of 0.028 mol/m^{3}s. The gas is available at the velocity of 2 cm/s at the total pressure of 3×10^5 N/m^2 near the impeller.

    The gross composition of oxygen in air, at atmospheric condition is

$$C_{o_2,gas} = 0.25 \text{ mol/m}^3$$

At a pressure of 3×10^5 N/m^2, the gas phase concentration is

$$C_{o_2,gas} = 3 \times 0.25 \text{ mol/m}^3 = 0.75 \text{ mol/m}^3$$

The corresponding equilibrium concentration at the liquid–gas interface on the liquid side, considering the ratio of oxygen concentration in air to that in water to be 32 at 30°C, is

$$C_{o_2} = \frac{1}{32} \times 3 \times 0.25 \text{ mol/m}^3 = 0.0234 \text{ mol/m}^3$$

Consider that in the liquid phase we have to maintain at least, say, a 10% of the saturation solubility.

$$C_l = 0.1 \times \frac{1}{32} \times 3 \times 0.25 \text{ mol/m}^3 = 0.00234 \text{ mol/m}^3$$

The required $k_L a$ can be estimated as

$$\text{OUR} = k_L a \cdot (0.0234 - 0.00234)$$

$$0.028 = k_L a (0.234 - 0.00234)$$

$$k_L a = 1.327 \text{ s}^{-1}$$

Can the stirred reactor give the required $k_l a$?

    Example 11.2   A stirrer bioreactor of diameter 4.00 m is equipped with an agitator of diameter 1.0 m. For air dispersion the tip velocity should be greater than 2.5 m/s. Estimate the gas flow rate for the absence of flooding of the impeller.

    Since for the dispersion of air-bubbles a minimum stirrer speed is to be maintained, let us consider the impeller tip velocity

$$v_{tip} = 3 \text{ m/s}$$

For an agitator of diameter d,

$$n_s = \frac{v_{tip}}{\pi d} = \frac{3.0}{3.14 \times 1.0} = 0.955 \text{ rps}$$

For the absence of flooding, the criteria is

$$\frac{F_{air}}{n_s d^3} < 0.34 \left(\frac{n_s^2 d}{g} \right)$$

Therefore,

$$F_{gas} < 0.34 \left(\frac{n_s^3 d^4}{g} \right) = \frac{0.34(0.955)^3 (1.0)^4}{9.8} \text{ m}^3/\text{s}$$

$$< 0.030 \, \text{m}^3/\text{s}$$

For different tip velocity values, i.e., speeds of rotation, corresponding flow rates of air can be estimated.

Example 11.3   Estimate the minimum stirrer speed for particle suspension of 5% (w/v) of solid particles of size 50 μm each. The density of the medium is $1.05 \times 10^3 \text{ kg/m}^3$ and that of particle is $1.2 \times 10^3 \text{ kg/m}^3$. A propeller of diameter 0.75 m is used in a reactor of 3.0-m diameter. The medium has viscosity of $0.4 \times 10^{-2} \text{ N.s.m}^{-2}$.

Zwietering [24] gave the following correlation for the critical speed of agitation for suspension of particles.

$$n_{s,\min} = c v_2^{0.1} d_p^{0.2} \left(g \frac{(\rho_s - \rho_l)}{\rho_l} \right)^{0.45} X^{0.45} d^{-0.85}$$

For a propeller $c = 6.4$,

$$n_{s,\min} = 6.4 \left(\frac{0.4 \times 10^{-2}}{1.05 \times 10^3} \right)^{0.1} (50 \times 10^{-6})^{0.2}$$

$$\times \left(\frac{9.8 \times (1.2 - 1.05) \times 10^3}{1.05 \times 10^3} \right)^{0.45} (0.05)^{0.45} (0.75)^{-0.85}$$

$$n_{s,\min} = 0.098 \, s^{-1} \approx 6 \, \text{rpm}$$

Example 11.4   What is the particle settling velocity in the above example?

For very small particles, in the Stokes region

$$v_s = \frac{d_p^2(\rho_s - \rho_l)g}{18\mu_l}$$

$$v_s = \frac{(50 \times 10^{-6})^2(1.2 - 1.05) \times 10^3 \times 9.8}{18 \times 0.4 \times 10^{-2}}$$

$$v_s = 5.1 \times 10^{-5} \text{ m/s}$$

The settling velocity is very low because of highly viscous nature of the fluid and low density difference.

Example 11.5   A commercial fermentor of 200 m^3, the height of the ungassed volume is twice that of diameter and the stirrer diameter is 0.4 times the fermentor diameter. The fermentor contains a filamentous broth at a concentration (C_x') of 2 kg/m^3. For most filamentous broth, the behavior of medium is pseudoplastic with power index between 0.3 and 0.6. For a shear rate of 10 s^{-1} what is the viscosity of the medium?

Let us take $n = 0.5$. From Van't Riet and Tramper [25], the consistency index can be related to the concentration of the biomass.

$$K = 0.07(C_x')^{1.5}$$

For a non-Newtonian fluid the power law model can be used as a stress–strain relationship

$$\mu_a = (K\gamma^{n-1}) = 0.07 \times (2)^{1.5} \times (10)^{-0.5} = 0.062 \text{ Nsm}^{-2}$$

The viscosity of the fermentation broth is not constant and will change with time as biomass concentration changes with time. Increasing the concentration of the biomass will increase the apparent viscosity.

Example 11.6   Croughan et al. [16] showed that hydrodynamic forces because of time-averaged spatial gradients in fluid velocity, give rise to shear stresses on the surface of a particle. The maximum stress is $\tau_{max} = 3\mu_l \gamma$, while average stress is $\tau_{avg} = 0.5\mu_l \gamma$. For a medium of viscosity twice that of water and for an average shear rate of 7s^{-1}, what are the maximum and average shear stresses?

At 30°C, the viscosity of water is 0.86×10^{-3} Nsm^{-2}. The maximum stress is

$$\tau_{max} = 3 \times (2 \times 0.86 \times 10^{-3}) \times 7$$

and the average stress is

$$\tau_{avg} = \frac{1}{2}\mu_l\gamma = 0.5 \times (2 \times 0.86 \times 10^{-3}) \times 7$$

$$= 0.036 \ N/m^2$$

$$= 6.02 \times 10^{-3} \ N/m^2$$

Example 11.7 For a flat-blade turbine impeller of 0.6 times the tank diameter (2.0 m), estimate the power consumption at a speed of 30 rpm. The tank is fully baffled. What is the size of a turbulent eddy in the agitated condition if the density of the fluid is $1.05 \times 10^3 \ kg/m^3$ and viscosity $= 1.1 \times 10^{-3} \ Nsm^{-2}$?

For a turbine impeller, power number in turbulent conditions (N_p) is 3.0. The power input (P) is given as

$$P = N_p \rho_l \ n_s^3 d^5$$

$$= 3.0 \times 1.05 \times 10^3 \times \left(\frac{30}{60}\right)^3 (2.0 \times 0.6)^5$$

$$= 979 \ W$$

The energy density is power input per unit weight of the process fluid,

$$\varepsilon = energy \ input \ density$$

$$= \frac{P}{\rho_l \cdot V_l}$$

$$= \frac{979}{(1.05 \times 10^3) \times \frac{\pi}{4}(2.0)^2(3.0)} = 0.099 \ W/kg$$

for the vessel with H/T ratio of 3.0.

The Kolmogoroff eddy size depends on the energy density input.

$$Eddy \ size = \lambda_k = \left(\frac{\nu^3}{\varepsilon}\right)^{1/4} = \left[\frac{\left(\frac{1.1 \times 10^{-3}}{1.05 \times 10^3}\right)^3}{0.099}\right]^{1/4}$$

$$= 5.84 \times 10^{-5} m \approx 58.4 \mu m$$

The turbulent eddy size is comparable to most biocells at the given conditions and can have substantial damaging effect on the cells.

Example 11.8   Estimate mixing time in a fermentor of diameter 3.0 m equipped with a six-blade turbine impeller running at a speed of 30 rpm. Also estimate the power requirement of the agitation.

Let us assume $H/T = 1.0$, diameter of impeller $= d = 1/3T = 1.0$ m, and $H_d =$ blade height $= 0.2 \times d = 0.2$ m, as for most standard impellers. For a radially pumping impeller, the pumping capacity $= v \cdot dH_d$, where

$$v = \text{impeller tip velocity} = \pi d n_s = 3.14 \times 1.0 \times \frac{30}{60} \text{ m/s} = 1.57 \text{ m/s}$$

The pumping capacity of impeller $= F_c \text{pd}$

$$F_c = 1.57 \times 3.14 \times 1.0 \times 0.2 = 0.98 \text{ m}^3/\text{s}$$

With an assumption that volume swept during one circulation equals the volume of the vessel,

$$V_c = \frac{1}{4} \pi T^2 H = \frac{1}{4} \pi (3.0)^2 (3.0) \text{ m}^3 = 21.2 \text{ m}^3$$

$$\text{Circulation time} = t_c = \frac{V_c}{F_c} = \frac{21.2}{0.98} = 21.64 \text{ s}$$

The mixing time is 4 times the circulation time. Therefore, mixing time is

$$t_m = 4 \, t_c = 86.55 \text{ s}$$

In terms of T/d, n_s, and power number of impeller (N_p), an experimental correlation is

$$t_m = \frac{3.00 \, \left(T/d\right)^3}{n_s \quad N_p^{1/3}} \qquad \text{for Re} > 10,000$$

$$= \frac{3.00}{0.5} \frac{(3.0)^3}{(3.0)^{1/3}} = 112 \text{ s}$$

$$\text{Re} = \frac{d(dn_s)\rho}{\mu} = \frac{1.0(1.0 \times 0.5) \times 1.0 \times 10^3}{0.9 \times 10^{-3}} = 5.55 \times 10^5$$

The value of Re clearly shows the flow to be turbulent in the unit.

In the presence of air sparged in the fermentor, mixing time is twice that in the absence of aeration because of decrease in pumping capacity of the impeller due to cavity formation behind impeller.

$$t_{m,\text{two phase}} = 2 \times 112 = 224 \text{ s}$$

Power consumption by impeller is

$$P = N_p \rho n_s^3 d^5 = 3.0 \times 1.0 \times 10^3 \times (0.5)^3 (1.0)^5 \text{ W} = 375 \text{ W}$$

Specific power input within the unit is

$$\epsilon = \frac{P}{\rho_e V} = \frac{375}{1.0 \times 10^3 \times (\pi/4)(3.0)^2 \times (3.0)} = 0.076 \frac{W}{kg}$$

PROBLEMS

1. What are the differences in flow patterns induced by flat-blade turbine, pitch-blade turbine, and propeller-type impellers?

2. Discuss shear induced by different types of impellers. If one needs an impeller with low shearing action, what type of impeller would you recommend?

3. What are the typical dimensions of impellers in single-impeller and multiple-impeller units for a tank of diameter T?

4. How does non-Newtonian character of fluids commonly encountered in fermentation affect the choice of the impeller?

5. Define *power number* and *flow number* of an impeller.

6. Why is the power number of a propeller much lower than that of a flat turbine impeller?

7. What is the significance of energy input density in stirred bioreactor?

8. What are the criteria for scale-up of a stirred bioreactor?

9. Why do particles from near suspension conditions settle on introduction of gas into a solid–liquid suspension?

10. What will be the effect on the mixing time of sparging gas into a solid–liquid suspension and why?

11. Show that small dH_d high-speed impellers make effective use of power to create holdup in agitated system.

12. Estimate average and maximum shear stresses experienced by a cell in a reactor with average shear rate of (a) 1.0 s^{-1} and (b) 10 s^{-1} and with a viscosity of solution 5 times that of water.

REFERENCES

1. Dutta, N.N.; Pangarkar, V.G. Chem Eng Commun 1995, *73*, 273.
2. Pangarkar, V.G.; Yawalkar, A.A.; Sharma, M.M.; Beenackers, A.A.C.M. Ind Eng Chem Res 2002, *41*, 4141.
3. Einsele, A.; Finn, R.K. Ind Eng Chem Proc Des Dev 1980, *19*, 600.
4. Einsele, A. Proc Biochem 1978, *7*, 3.

5. Middleton, J.C. Proceedings of Third European Conference on Mixing.
6. Feijen, J.; Heijem, J.J.; van't Riet, K. Proceedings of the Symposium on Mixing and Dispersion Processes, Institute of Chemical Engineers, Delft Technology University.
7. Gogate, P.R.; Beenackers, A.A.C.M.; Pandit, A.B. Bichem Eng J 2000, *6*, 109.
8. Machon, V.; Vleck, J.; Shrivaneck. Proceedings of the European Conference on Mixing 1985, *5*, 155.
9. Abardi, V.; Rovero, G.; Sicardi, S.; Baldi, G.; Conti, R. Proceedings of the European Conference on Mixing 1988, *6*, 63.
10. Vasconcelos, J.M.T.; Orvalho, S.C.P.; Rodrigues, A.M.A.F.; Alves, S.S. Ind Eng Chem Res 2000, *39*, 203.
11. Andrew, S.P.S. Trans I Chem Eng 1982, *60*, 3.
12. Hughmark, G.A. Ind Eng Chem Proc Des Dev 1980, *19*, 638.
13. Calderbank, P.H.; Mo-Young, M.B. Chem Eng Sci 1961, *16*, 39.
14. Kawase, Y.; Moo-Young, M.B. Chem Eng Res Dev 1988, *66*, 284.
15. Linek, V.; Vacek, V.; Benes, P. Chem Eng J 1987, *34*, 11.
16. Croughen, M.S.; Hamel, J.F.; Wang, D.I.C. Biotechnol Bioeng 1987, *29*, 130.
17. Croughen, M.S.; Sayre, E.S.; Wang, D.I.C. Biotechnol Bioeng 1989, *33*, 862.
18. Levich, V.G. *Physicochemical Hydrodynamics*; Englewood Cliffs, NJ: Prentice-Hall, 1962.
19. Henzler, H.J. Chem. Ing. Tech. 1982, *54*, 461.
20. Pandit, A.B.; Joshi, J.B. Chem Eng Sci 1983, *38*, 1189.
21. Streak, F.; Karoz, J. Proceedings of the European Conference of Mixing 1988, *6*, 375.
22. Stieff, A. Proceedings of the European Conference of Mixing 1985, *5*, 209.
23. Karcz, J. Chem Eng J 1999, *72*, 217.
24. Zwietering, Th.N. Chem Eng Sci 1958, *8*, 244.
25. Tramper, J.; Van't Reit *Basic Bioreactor Design*; Marcel Decker: New York, 1991.

12

Tower Bioreactors

12.1 TOWER BIOREACTORS

The aerobic bioreactor is a device that brings three phases, liquid, air, and solid cell suspension, in intimate contact with each other and ensures adequate transfer rates of nutrients from the gas phase to the solid phase of cells. The mass transfer rates of necessary ingredients for growth and multiplication of cells are largely dependent on the area available for mass transfer and hydrodynamic conditions existing in the reactor. The relative velocities between different phases decide the efficiency of the transfer, while dispersion of gas either in the form of small bubbles and that of cells can provide the necessary area for mass transfer.

The mixing can be achieved by passing a stream of gas bubbles at a very high velocity. The sparged gas, usually air in the case of bioreactor, breaks into bubbles of different sizes and rises in the liquid carrying a substantial amount of liquid in the wake. The fluid is set in circulatory motion either in the entire unit or in circulation cells of size equivalent to the diameter of the vessel. The overall effect is partial or complete mixing in the fluid phase. The circulations may also achieve a somewhat uniform energy distribution unless gross channeling of the gas phase in the form of large bubbles takes place. The fluid circulation may be set up with distribution of gas throughout the entire volume of the fluid as in a bubble column or the gas is sparged in only one limb of the circulation loop setting up an effective circulation pattern throughout the reactor, as in an air-lift reactor, without a large shearing effect on the biological species.

Two distinct flow regimes can be distinguished in the operation of the column. If all bubbles rise in the vessel without any circulatory currents in bulk liquid, the flow is considered as homogeneous. The mixing is provided by the liquid carried by the bubbles in their wake while rising through the liquid pool. The homogeneous regime is possible only at very low velocities (1–4 cm/s) and with a precise distribution of gas uniformly at the bottom of the column. The bubbles are created at the sparger openings at the bottom of the vessel. At the sparger, the bubble size depends on the gas velocity through the sparger pipe opening and the diameter of the sparger orifice (d_o). If the gas velocity is low (<1 cm/s), the size of the bubble can be estimated from the buoyancy force on the bubble at the sparger plate and the surface tension force, which tries to retain the bubble at the sparger hole.

$$\pi d_o \sigma = \frac{\pi d_b^3}{6}\left(\rho_l - \rho_g\right)g \tag{12.1}$$

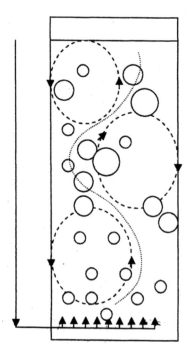

Figure 12.1 Time-averaged circulatory flows in bubble column.

In heterogeneous flow regime the circulatory flow is dominant. The jets of gas coming out of the orifices split into a stream of bubbles. The distribution of the gas phase into bubbles is nonuniform over the cross section of the column and that causes nonuniform flow patterns in the liquid and gas phases. The circulation loops change dynamically with an approximate loop size equal to the diameter of the vessel (Fig. 12.1). The overall appearance is of the gas flow through the center of vessel and downward flow of liquid near the wall.

The air-lift reactor, which is a modified form of bubble column, consists of a two vertical tubes connected at both ends (Fig.12.2). Air is sparged at the bottom of one of the tubes, called the *riser*. The bubbles rise through the riser and disengage from the liquid in the top section. The presence of the gas phase in the riser reduces the bulk density of the fluid in the riser, and the density difference in the two tubes sets the fluid in circulatory motion; upward in the riser and downward in the other tube, the *downcomer*. Another form of loop reactor, where an external pump recirculates fluid in the reactor, can allow mixing of several streams in the pump assembly itself. Centrifugal pump is very effective for such mixing where the liquid is fed into the reactor with sufficiently high velocity.

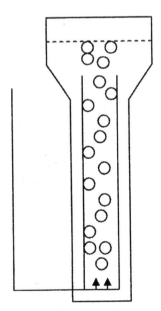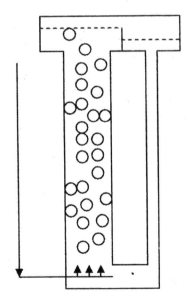

FIGURE 12.2 Air-lift reactor.

12.2 MIXING IN TOWER BIOREACTORS

The purpose of mixing is to reduce the concentration difference in the reactor at different points. It is also important to know how fast this homogeneity is achieved. The time required for such a complete mixing as compared to that required for consumption of the substrate can decide the relative importance of hydrodynamics and chemical kinetics. For a reaction that is slow as compared to the mixing process, the emphasis should be to improve the reaction kinetics, while for a relatively fast reaction as compared to the transport process, efforts to improve further kinetics say by improved catalyst or reaction conditions will not be fruitful unless they have other advantages such as improved selectivity.

The uniformity of mixing depends on overall circulation brought about either by pumping action of an impeller of a pump itself, shearing action of the impeller blades, or rising gas bubbles, which transfer momentum from eddies to each other when there is appreciable velocity difference between the neighboring eddies and molecular diffusion within the eddies or between the eddies when the relative velocity is negligible between them. The relative sizes of the fluid eddies, depending upon the degree of turbulence or shear experienced by them, decide the degree of each of the above factors.

The mixing in a bubble column is the result of rising bubbles in the fluid. In the homogeneous bubble flow regime, the liquid is entrained in the wake of the rising bubbles. The homogeneous flow regime is possible if the gas distribution is uniform at the bottom of the vessel and the gas velocities are low enough to have uniform bubbles generation. At higher gas velocities the local differences in the gas velocities cause uneven liquid velocities and distribution of gas phase across the cross section of the column. Nearly all commercial fermentors show heterogeneous regime of operation. If the sparger holes are not uniformly distributed at the sparger level in the column, the local differences in the dispersion lead to uneven gas distribution and local holdups that further facilitate different coalescence rates and lead to heterogeneous regime at all gas velocities. The stabilized flow regime with its dynamic variation shows upward flow at the center of the column and downward flow along the wall of the column. In tall columns this circulation is divided into loops.

The circulatory liquid velocity(u_l) in heterogeneous regime at the center of the column is related to the superficial gas velocity (u_g) [1].

$$u_l = 0.9(gTu_G)^{1/3} \qquad (12.2)$$

The average upward liquid velocity is $0.8\, u_l$. Introducing the tank volume, the circulation time for a given column ($H/T = 1$ or for a circulation cell) is

$$t_{cir} = 2.8 \left(\frac{gu_G}{T^2} \right)^{-1/3} \qquad (12.3)$$

Considering that the mixing time is about 4 times the circulation times, for a bubble column with H/T ratio different from 1,

$$t_m = 11 \frac{H}{T} \left(\frac{gu_G}{T^2} \right)^{-1/3} \tag{12.4}$$

For the homogeneous flow regime in a bubble column, the gas-phase flow is regarded as plug flow. For the heterogeneous flow regime, the liquid circulatory flow entrains the gas bubbles and certain mixing will be allowed in the gas phase as well. For large and not too tall columns the gas phase also can be considered as mixed.

For the air-lift reactors the liquid circulation velocity and so the mixing time are decided by the gas holdup and the friction in the reactor. The driving force for the liquid circulation is the density difference between riser and downcomer. The gas holdup in downcomer, in any case, is much smaller than that in riser. The circulation velocity is decided by the driving force, i.e., gas holdup (ϕ_G), and the hydrodynamic resistance

$$\rho_l \phi_G g H_r = \frac{1}{2} f \rho_l u_l^2 \tag{12.5}$$

$$u_l = \left(\frac{2\phi_G g H_r}{f} \right)^{0.5} \tag{12.5a}$$

where f is the frictional coefficient, u_l is the superficial liquid velocity in downcomer, H_r is the riser height. With a criterion of good mixing, average mixing time is about 4–7 times the circulation time in the air-lift reactor. At low velocities, the bubble phase in the riser behaves more or less as that in the bubble column. At higher liquid velocities, however, the flow patterns in air-lift reactors should be more or less plug flow.

12.3 BACKMIXING EFFECTS IN TOWER BIOREACTORS

The global effects of mixing in tower bioreactors can be conveniently described by the dispersion coefficients of both phases. The liquid-phase dispersion mainly depends on gas velocity and column diameter and physicochemical properties, like viscosity and density, do not significantly affect the liquid-phase dispersion. For homogeneous flow, the liquid-phase dispersion coefficient is in the range [2]

$$D_l = 10^{-3} \text{ to } 10^{-2} \text{ cm}^2/\text{s} \tag{12.6}$$

while for the heterogeneous regime it can be related to the gas phase velocity
[2]

$$D_l = 0.33(gT^4 u_G)^{1/3} \tag{12.7}$$

In most tower bioreactors, the dispersion coefficient is given [3] by

$$D_l = 2.7T^{1.4} u_G^{1/3} \tag{12.8}$$

where D_l is in cm^2/s, T in cm, and u_G in cm/s. In a dimensionless form a
correlation is given as [4]

$$\text{Pe}_L = \frac{U_G T}{D_l} = 2.83 \left(\frac{u_G^2}{gT}\right)^{1/3} \tag{12.9}$$

For very small diameter bubbles from porous spargers, a major fraction of
liquid is attached with the bubbles at the gas–liquid interface and is carried
upward, which causes an underpressure and leads to violent eddies. At
medium values of d_b (1 to 3 mm) bubbly flow is assumed to prevail, which
yields low values of D_l. Thus the structure of gas–liquid dispersion may play
an important role in mixing and therefore decide D_l. For fermentation it is
the mixing time rather than the dispersion coefficient, which is used to characterize the global mixing effects. The mixing time can be obtained from the
transient solution of dispersion model and is usually defined for 90–95%
homogeneity.

Owing to slight density difference between fermentation liquid and biomass, the particles have the tendency to settle. Thus a biomass concentration
profile along the tower height may result. The pertinent model to account for
biomass concentration profiles is the *sedimentation–dispersion* model. This
model involves two parameters, the solid dispersion coefficient (D_s) and the
mean settling velocity of the biomass ($u_{p,s}$). The biomass dispersion coefficient
in similar to the liquid-phase dispersion coefficient. Most biomass particles
can be easily suspended in bubble column due to their small size and relatively
small density difference from the liquid medium. For a good dispersion the
minimum velocity in the bulk phase should be at least twice the settling
velocity of the solid particles. If the particles are very small and the particle
Reynolds number defined as

$$\text{Re}_{dp} = \frac{\rho_l u_{p,s} d_p}{\mu_l} \tag{12.10}$$

is less than 1, Stokes regime can be considered for estimation of the settling
velocity

$$u_{p,s} = \frac{d_p^2 (\Delta \rho) g}{18 \mu_l} \tag{12.11}$$

For an intermediate range ($1 < Re < 10^3$) the following relation can be used for estimation of the settling velocity with the knowledge of drag coefficient (C_D) depending on its size.

$$C_D Re_{dp}^2 = \frac{4 d_p^3 \rho_l \Delta \rho g}{3 \mu_l^2} \tag{12.12}$$

In contrast to liquid-phase dispersion coefficients, the data on gas-phase dispersions are sparse and in general, the measurements reveal considerable scatter. The bubble rise velocity in the swarm is a characteristic variable which mainly influences the gas-phase dispersion. For the homogeneous flow regime the gas phase can be regarded as a plug flow. For the heterogeneous regime the liquid circulation will entrain gas bubbles and certain amount of backmixing occurs. The following correlation may be used for an approximate estimation of the gas-phase dispersion coefficient [5].

$$D_G = 5 \times 10^{-4} u_G \phi_g T^{3/2}, m^2/s \tag{12.13}$$

Both these correlations show a strong dependence of the gas-phase dispersion coefficient on the tank diameter (T). For large and not too tall tanks ($H/T < 2$) the gas can be considered as ideally mixed.

Though gas-phase dispersion coefficients are high and often considerably larger than those of the liquid phase, the impact of gas-phase dispersion on reactor performance is seldom taken into account. Particularly no significant data are available for bioreactors, though owing to their usually large size gas-phase dispersion may be of significant influence.

12.4 GAS DISPERSION AND HOLDUP IN TOWER BIOREACTORS

A decisive mass transfer problem in the majority of fermentations constitutes the transport of oxygen from the air phase to the locale of the reaction, i.e., the biomass phase, which, in accordance to the chemical engineering, will be referred to as the solid phase. The major reason that oxygen transfer may play an important role in biological processes is the limited oxygen capacity of the fermentation broth due to the low solubility of oxygen.

The fractional gas holdup (ϕ_G) is an important parameter to characterize gas in liquid, which depends mainly on the gas throughput, the sparger design, and physicochemical properties. If the column diameter is large compared to the bubble diameter, say by a factor of about 40, the column diameter has no significant effect. This is commonly valid if $T > 10$ cm. At the sparger the bubble size depends on the sparger pipe opening and the gas velocity through it. On release from the sparger opening the bubbles are subjected to dispersion and coalescence processes. An equilibrium size of bubble is one where the dispersion and coalescence phenomena lead to a stable size of the bubble. The size of the bubble will not then be dependent on the sparger design. In a coalescing medium if the bubble size grows above the equilibrium size, the shear effect at the liquid–gas interface can break the bubble into smaller ones again. A typical size of bubble in a coalescing liquid is about 6 mm. In noncoalescing liquid, the bubble generated at the orifice would not undergo substantial changes in size because of the liquid properties unless they are bigger than the equilibrium size. For smaller bubbles, the dispersion will not be important, and the bubble size will be decided by the sparger design. If the bubbles size at the sparger exceeds the equilibrium bubble size, any type of sparger can be used in the column. The influence of the gas velocity on ϕ_G can be conveniently expressed by

$$\phi_G \propto u_G^n \tag{12.14}$$

At low gas velocities and if porous spargers are used, bubbly flow prevails. Then the exponent n may vary from 0.7 to 1.2. In churn-turbulent (heterogeneous regime), which occurs at higher gas velocities, and if single and multinozzle spargers ($d_o > 1$ mm) are used, n is in the range 0.4 to 0.7. For most commercial fermentors, with both coalescing liquids and non-coalescing liquids with bubble size at sparger bigger than the equilibrium size, the following equation can be applied for the fractional gas holdup [1].

$$\phi_G = 0.6u_G^{0.7} \tag{12.15}$$

The most important variable in bubble column could be viscosity of the medium. At viscosities above 10–100 mNs/m^2 the holdup will decrease with viscosity as large bubbles (0.1–1.0 m) are formed. However, for such high viscous solutions bubble column cannot be recommended because of very slow rise of bubbles through the medium. Considerably higher gas holdups for fermentations have been reported, particularly if more effective spargers such as porous plates and two-phase nozzles are used. At low velocities a homogeneous flow can be realized in these contactors. On increasing gas velocities the final regime will be slug flow after a transition range where the flow is heterogeneous. In this flow regime, with porous spargers the exponent is 0.85. If perforated plates are used, slug flow exists for all gas velocities and the exponent is 0.67. For a fully developed slug flow the holdup is independent of viscosity. In homogeneous regime, ϕ_G values can be obtained at low gas velocities, which can be obtained in the slug flow only at higher gas velocities. As the holdup can be measured easily by various methods, it seems, therefore, better to measure out ϕ_G in lab-scale column with $T > 10$ cm.

Akita and Yoshida [6] correlation gives a conservative estimate of the hold-up in bubble column.

$$\frac{\phi_G}{(1 - \phi_G)^4} = 0.2 \left(\frac{gT_c\rho_l}{\sigma}\right)^{1/8} \left(\frac{gT^3}{\rho_l}\right)^{1/12} \frac{u_G}{(gT)^{1/2}} \quad (12.16)$$

In a air-lift reactor, the gas bubbles are carried with the liquid and disengage from the liquid at the top. The holdup values in the air-lift reactor are lower than those in a bubble column under the same conditions. The holdup in the bubble column will depend on the liquid circulation velocity. It is also needed to ascertain the flow regime in the reactor before the estimation of the holdup. The upper limit on the holdup in a air-lift reactor can be estimated from the equation used for a bubble column.

12.5 MASS TRANSFER COEFFICIENTS

Various models are available to calculate liquid-side mass transfer coefficients k_l. The value of this hydrodynamic parameter and the equations that apply to its calculations largely depend on the bubble size and the constitution of the bubble surface. Large circulating bubbles with mobile surface yield k_l values, which approach the predictions of model of Higbie [7].

$$\frac{k_l d_b}{D_m} = 1.13 \left(\frac{u_B d_b}{D_m}\right)^{1/2} \quad (12.17)$$

where the contact time t_c is calculated from the bubble rise velocity u_B and the diameter d_b considering that during the passage of a bubble its surface is completely renewed when the bubble rises through a distance equal to its own diameter. If the bubbles are small, the surface mobility is decreased and they behave as rigid spheres. Therefore, the k_l values of small bubbles approach the limiting solution given by Levich [8].

$$Sh = 0.997Pe^{1/3} \quad \text{if Pe} \gg 1 \text{ and Re} \ll 1 \quad (12.18)$$

For practical purposes and if only estimates are needed the following correlations can be used [9].
For $d_b < 2.5$ mm

$$k_l Sc^{2/3} = 0.31 \left(\frac{\Delta\rho u_l g}{\rho_l^2}\right)$$

and for $d_b > 2.5$ mm

$$k_l Sc^{1/2} = 0.42 \left(\frac{\Delta \rho u_l g}{\rho_l^2} \right)^{1/3} \tag{12.19}$$

Thus, provided it is known whether the bubble diameter is greater or smaller than 2.5 mm, k_l can be calculated from physicochemical properties alone. If the bubbles do not interfere with each other in a swarm of bubbles, which should be true only for low gas hold-ups and for noncoalescing and high-viscosity liquids, the mass transfer coefficient of a single bubble can be extended to the conditions involving the swarm of bubbles. In bubbly flow regime it is particularly true where the gas velocities could be as low as 5 cm/s. At higher velocities however, these estimates could deviate considerably. Together with the fractional gas holdup the bubble diameter (volume-to-surface-area value) decisively determines the gas–liquid interfacial area, which is given by

$$a = \frac{6 \phi_G}{d_b} \tag{12.20}$$

In addition, the discussion of k_l for single bubble has indicated the importance of the bubble diameter which essentially influences the constitution of the gas–liquid interface and the mean bubble rise velocity. Initial bubble size (d_s) generated from sparger orifices can be estimated from force balance on the bubble detaching from the orifice diameter (i.e., the buoyant force is opposed by the surface tension force working at the rim of the orifice)

$$\pi d_o \sigma = \frac{\pi}{6} d_S^3 (\rho_l - \rho_g) \tag{12.21}$$

or from an empirical correlation [10]

$$\frac{d_s}{d_o} = 1.88 \left[\frac{u_o}{g d_o} \right]^{1/3} \tag{12.22}$$

where d_o is the orifice diameter and u_o is the velocity at the orifice.

The volume to surface mean bubble diameter d_{32} in gas–liquid dispersion can be estimated by a correlation [10]:

$$\frac{d_{32}}{T} = 26 \left(\frac{g T^2}{\sigma} \right)^{-0.5} \left(\frac{g T^3}{v_l} \right)^{-0.12} \left(\frac{u_g}{\sqrt{g T}} \right)^{-0.12} \tag{12.23}$$

The size of bubbles as they move upward in a bubble column depends on the initial size and on the coalescing tendency depending upon the nature of the liquid. The coalescence properties of the liquid are mainly dependent on the added salts and organic substances present. As a bubble is generated, the concentration of electrolyte at the interface and in the bulk liquid are equal at first. The ions have the tendency to move away from the interface, giving an enrichment of water there accompanied by an increase in surface tension. Since the transport of ions in the bulk liquid requires some time, the coalescence hindering action is only pronounced at short residence times of the bubbles. Therefore, large effects of added electrolytes on d_s (and hence on interfacial area and volumetric transfer effects) can be observed in stirred vessels and multistage columns).

12.6 VOLUMETRIC MASS TRANSFER COEFFICIENTS

Although it is most useful to know the influence of various physical and operating variables on the individual values of k_l and a, the model equations of a biological reactor involve only $k_l a$. An absorption enhancement because of fast reactions is not expected in biological reactors of practical importance. The operating variable that strongly affects $k_l a$ in gas–liquid dispersion is the gas flow rates. Kastanek [11] used the contact time from Higbie's [7] theory and isotropic turbulence from Kolmogoroff's theory to arrive at an equation

$$k_l = \frac{u_G(u_G + c)^{13/20}}{2u_G + c}$$ (12.24)

which can be reduced to

$$k_l a = bu_G^n$$ (12.25)

where exponent n varies from 0.8 to 1.2. This exponent is little affected by liquid flow rate and the kind of sparger used in the column. However, the constant b is largely dependent on the sparger design and liquid media. In the vicinity of the gas spargers of porous and plate type, higher turbulence intensities prevail, and therefore, higher $k_l a$ values are observed. In industry, porous plates are uncommon though the achievable mass transfer rates are favorable. The gas is either sparged by single or multiorifice distributors or by two compartment nozzles of various types. The interfacial area is dependent on the gas velocity because it is decided by hold up and size of the bubble. Since k_l, as decided by the bubble size, shows dependence on gas velocity the volumetric mass transfer coefficient also should show dependence on the gas velocity. For a coalescing medium the following correlation can be applied for estimation of $k_l a$ [1].

$$k_l a = 0.32u_G^{0.7}$$ (12.26)

For noncoalescing liquids, if the diameter of the bubble at the sparger is smaller than the equilibrium value of 6 mm, k_la should be greater than that estimated from the above equation. The influence of viscosity on k_la in bubble column is more pronounced. For example, Deckwer et al. [12] have given the dependence as

$$k_la = c\mu^{-0.84} \tag{12.27}$$

For a medium of viscosity 0.1 Nsm^{-2} the decrease in mass transfer coefficient is very large and the process may be strongly hindered by mass transfer limitations. A viscosity of 0.1 Nsm^{-2} can be considered as upper limit for bubble column operations. For the case of the less effective single and multi-orifice nozzles following correlation can be used [6], which, however, gives a conservative estimate of k_la

$$k_la\frac{d_d^2}{D_l} = 0.6\phi_G^{1.1}\left(\frac{v_l}{D_l}\right)^{0.5}\left(\frac{gT^2\rho_l}{\sigma}\right)^{0.62}\left(\frac{gT^3}{v_l^2}\right)^{0.31} \tag{12.28}$$

In the most important flow regime the measured data have been correlated as

$$k_la = 2.08 \times 10^{-4}u_G^{0.6}\mu_{eff}^{-0.84} \tag{12.29}$$

In an air-lift reactor, liquid circulation velocity is imposed on the circulation cells in the conventional bubble column. The transfer of the bubbles in the riser is faster than in bubble column. The effective volumetric mass transfer coefficient may show a decrease from the value from the bubble column. Results reported by Verlaan et al. [13] indicate a substantial decrease in the value of k_la, almost by factor of 3.

12.7 SOLID–LIQUID MASS TRANSFER

If respiring biomass particles are relatively large, mass transfer resistances from the bulk fermentation broth to the outer surface of the biomass particles may occur. If the density difference between the particle and the fluid becomes nil, then k_{sl} can be approximated by limiting value, i.e.,

$$Sh = 2 \quad \text{or} \quad k_{sl} = \frac{2D_m}{d_p} \tag{12.30}$$

The liquid–solid mass transfer coefficient (k_{sl}) depends not only on the physicochemical properties but also is influenced by hydrodynamic conditions, impeller speed, and geometry. In general, liquid–solid mass transfer coefficients are correlated by

$$\frac{k_{sl}d_p}{D_m} = 1 + c\left(\frac{v_l}{D_m}\right)^n\left(\frac{u_s d_p}{v}\right)^m = 1 + cSc^n Re^m \tag{12.31}$$

The slip velocity u_s is usually difficult to estimate. Therefore, it is a common practice to compute Reynolds number on the basis of Kolmogorof's theory, which gives

$$Re = c\left(\frac{\varepsilon d_p^4}{v^3}\right)^p \tag{12.32}$$

Here, the exponent p is dependent on the ratio of the particle size to the microscale of the eddies. In the case of tower bioreactor, the energy dissipation rate per unit mass ε can be simply calculated from

$$\varepsilon = u_G g \tag{12.33}$$

The power of the gas flow can be derived from the change of entropy of the gas as

$$P = \text{Flow rate of gas} \times RT \ln\frac{p_o}{p_{top}} \tag{12.34}$$

As a rule of thumb, each 1 cm/s of gas velocity corresponds to a dissipated power of 100 W/m^3 up to $H = 1$ m, above which it decreases gradually.

The following correlation has been proposed by Sanger and Deckwer [14] on systems which closely represents the fermentation systems in tower bioreactor.

$$Sh = 2 + 0.545Sc^{1/3}\left(\frac{\varepsilon d_p^4}{v^3}\right)^{0.264} \tag{12.35}$$

The relative importance of mass transfer resistances at the liquid–biomass interface (i.e., $1/k_{sl}a_s$) depends mainly on the size of the bioparticles. For instance, in yeast fermentations the particle sizes are in the range of microns. For typical fermentation conditions $k_{sl}a_s \gg k_l a$; hence the oxygen transfer resistance is located at the liquid-gas surface. On the other hand, in fermentations of *Penicillium chrysogenum* in bubble columns, under special conditions biomass particles of 0.3 to 2 mm diameter can be grown [12]. Here, oxygen mass transfer at the liquid–solid interface (and possibly pore diffusion) should be considered as a major resistance since $k_{sl}a_s \ll k_l a$.

12.8 HEAT TRANSFER COEFFICIENTS IN BUBBLE COLUMN

In a bubble column the liquid circulation is determined by the superficial gas velocity. Heat transfer properties also can be related to the superficial gas velocity. Heijenen and van't Riet [1] gave the heat transfer coefficient on the fermentation side as

$$h = 9.3919 \times 10^3 u_G^{0.25} \left(\frac{\mu_w}{\mu}\right)^{0.35} \tag{12.36}$$

The available data on heat transfer coefficients on tower reactors is best described by the correlation [15]

$$\frac{h}{\rho c_p u_G} = 0.1 \left[\frac{u_G^3}{gv}\left(\frac{v\rho c_p}{\lambda}\right)^2\right]^{-1/4} \tag{12.37}$$

One should consider, however, the possibility of slime formation and adherence of biomass particles at the heat transfer area, which might reduce the heat transfer considerably.

For air-lift reactor, for estimation of heat transfer coefficient the liquid circulation velocity must be known. When the average liquid velocity is known in downcomer, the conditions can be considered single phase turbulent flow for the estimation of heat transfer coefficient.

$$\frac{hd_{dc}}{\lambda} = 0.027 \left(\frac{\rho_l u_{l,dc} d_{dc}}{\mu}\right)^{0.8} \left(\frac{\mu c_p}{\lambda}\right)^{1/3} \tag{12.38}$$

Since the riser behaves as a bubble column the equation suggested for bubble column can be applied. If the bubbles do not contribute to the heat transfer very much the equation of downcomer can be applied for estimation of heat transfer coefficient in the riser as well.

12.9 EFFECT OF SHEAR

In a tower bioreactor where the homogeneity or mixing is achieved by sparging a gas, the shear is generated near the rising gas bubble depending upon the relative velocity between the rising bubble and the fluid. The rise velocity depends on characteristics of the broth such as its density and viscosity, the size of the bubble; larger bubbles rise very fast while smaller bubble rise very slowly;. For bubbles of size 1 mm to 1 cm with rise velocity of 0.1 to 50 cm/s, the shear rate value varies from 1 to 100 s^{-1}. For high suspension density broth the apparent viscosity may reach to very large values, making the objective of achieving homogeneity throughout the bulk fluid in the entire volume of reactor more difficult. In case of such gross inhomogeniety, particularly if it affects the growth of the microorganisms or stability of biocatalyst, it may be necessary to dilute the broth at the cost of production rate in the given volume.

For larger scales of operation where the power consumption may become prohibitively high in stirred contactor, the tower or air-lift reactors may become useful. Although there is no agitation, turbulence is still created by rising swarm of bubbles. Since the bubbles have a size distribution and also undergo coalescence and breakup during their way up, the rising velocities also show a distribution. The rise velocity is mainly decided by size of the bubble: smaller bubbles move with lower velocities, while bigger bubbles reach to very high velocities. The shear is experienced by bioparticles at the liquid–bubble interface. The bioparticles at the surface have exposure to liquid phase, which is more or less in stagnant conditions as compared to the velocity in air on the other side. The relative velocity and the subsequent shear experienced by the cell or bioparticles can indeed be very high. During the three phases of the bubbles, at the entry point of sparger, during rise, and at the exit point the bubbles can expose the associated bioparticles to shearing action.

At the orifice, the velocity can be calculated from gas flow rate and the sparger/nozzle diameter. For a velocity of $1/2$ m/s and characteristic length equal to the size of a cell ($\sim 10\,\mu m$) the shear stress is about $20\,Nm^{-2}$, which is an order of magnitude above the critical shear stress for the cells. The velocity of a rising air bubble is of the order of 25 cm/s. The corresponding stress on the cells is about $15\,Nm^{-2}$, which is still above the critical stress for the cell [2].

When the bubble reaches to the surface of the liquid, it first forms the liquid film dome. The film can retain the bubble below until the surface tension is sufficient to overcome the buoyancy force because of density difference. As soon as the surface tension becomes weak the film breaks and liquid falls back. The associated fluid velocity is estimated by the Culick equation [16],

$$u_{l,b} = 2\left(\frac{\sigma}{\rho t_f}\right)^{0.5} \tag{12.39}$$

If the surface tension is assumed to be of the order of 50×10^{-3} N/m and film thickness 40×10^{-6} m, with density similar to that of water, the velocity is estimated to be 2.5 m/s. Using a characteristic length for velocity gradient the stress experienced by cell could be as high as $140\,Nm^{-2}$, which is far higher than the critical stress value. If all cells experience such levels of stress, there would be no production at all as the death rate would be very high than the birth rate. However, how much time the cell experiences such a shearing force would be also important. If the contact time is estimated from the size of the bubble and its rise velocity, for the cells to survive the contact time must be smaller than that the rise time of the bubble to cover distance equal to its own size. On the other hand, the relative velocity of at the interface must be much smaller than that estimated rise velocity of the bubble. The loss of viability of cells can be taken as a first-order process

$$C_x(t) = C_x(o) \exp(-k_d t) \tag{12.40}$$

where k_d is the death rate constant. If V_k is considered to the volume associated with a rising bubble in which all viable cells are killed [2].

$$-V\frac{dC_x}{dt} = n_b C_x V_k = \frac{6F_b}{\pi d_b^3} C_x V_k \tag{12.41}$$

Integration of the equation and comparison with the first-order dependence of the cell concentration, gives

$$C_x(t) = C_x(0)\exp(-k_d t) \tag{12.42}$$

where

$$k_d = \frac{6F_b V_k}{\pi d_b^3 V} \tag{12.42a}$$

The experimental evidence, though limited, shows that the rising air bubbles can have negligible effect on the viability of the cells. It is possible that the most damaging effect of shear is either at the point of entry or during bubble bursting at the exit.

12.10 MODEL EQUATIONS FOR FERMENTORS

The design and scale-up of biochemical reactors, though they are definitely more complex, should be based on mathematical models and computational optimization procedures. The mathematical models combine the microscopic picture of mass transport and reaction with the macroscopic balance equations of the fermentor. With the exception of stirred vessels, biological reactors usually have some degree of dispersion. A plug flow model with dispersion in each of the phases can be a pertinent approach to describe fermentation. The differential equations of this model are obtained in the usual way by balancing over a volume element under consideration of those phenomena, which are thought to be of influence. In view of Fig. (12.3) the balance equations are as follows:

12.10.1 Gas Phase

$$\frac{d}{d_x}\left(\phi_G D_G \frac{dC_g}{dx}\right) - \frac{d}{dx}(u_G C_g) - k_l a\left(C_l^* - C_l\right) = 0 \qquad (12.43)$$

12.10.2 Liquid Phase

$$\frac{d}{d_x}\left(\phi_l D_l \frac{dC_l}{dx}\right) - \frac{d}{dx}(u_l C_l) + k_l a\left(C_l^* - C_l\right) - k_{sl}a_s(C_l - C_s) = 0$$

$$(12.44)$$

12.10.3 Biomass Phase (External Surface)

$$k_{sl}a_s(C_l - C_s) = \text{rate of consumption of oxygen} \qquad (12.45)$$

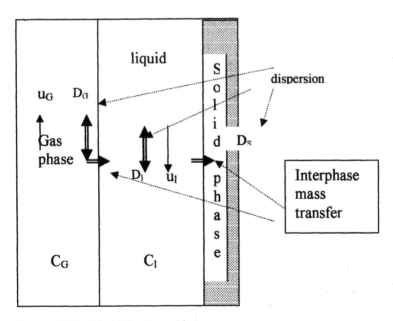

FIGURE 12.3 Model for tower bioreactor.

The rate term in the oxygen balance depends additionally on the local concentrations of the carbon source and the biomass, and corresponding balances for them have to be formulated. In the case of biomass, sedimentation may also have to be taken into consideration, which gives

$$\frac{d}{dx}\left(\phi_l D_s \frac{dC_B}{dx}\right) + (u_s - u_l)\frac{dC_B}{dx} + R = 0 \tag{12.46}$$

where R is biomass generation term and u_s is the settling velocity of the biomass particles in the swarm.

The distributed reactor model equations involve mixing or dispersion parameters(D_G, D_l, D_s) and mass transfer parameters (k_l, k_{sl}, a_s, a_p), which depend on the physicochemical parameters of the fluid, geometrical sizes, the gas distribution, phase velocities, and flow regime.

Hydrodynamically, bubble column is the simplest bioreactor used for gas-liquid transfer in which relatively large bubbles (>0.3 cm) are introduced at the base. The circulation of liquid is aided by an internal partition. The design is relatively simple as all that needs to be estimated is the residence time using free rise velocity. Considering that this type of reactor is used for relatively low viscosity, unstirred liquid and with large bubbles, the assumption of a rise velocity of 25 cm/s coupled with the ratio of $k_l a/\phi_G = 0.4$ is probably accurate provided the coalescence rate is kept low and voidage of gas is not allowed to rise above 5% [17]. A conservative estimate of this ratio is 0.25 if the designer is faced with a rich microbiological culture. If the viscosity is appreciable higher, then the viscosity correction can be also applied. The value of surface area in the bubble column can be estimated from the sparger orifice diameter, overall reactor information using chemical means of measuring a, or buy photographic data. If the bubble residence time in the reactor is T_b, the volumetric flow rate is F_o per orifice, and actual number of orifices is n_o, then

$$a = F_o T_b \frac{4\pi d^2 n_o}{\pi d^3/6V} \tag{12.47}$$

Small size bubbles find difficulty to disengage in viscous liquid. The use of multi hole sparger is an alternative where the size of the bubble is reduced by reducing the air flow per orifice. In relatively tall reactors the expression for holdup in terms of free rise velocity of the initially produced bubbles is only an approximation. In noncoalescing liquid the holdup can be as high as 40%, while in a strongly coalescing system it would be difficult to get holdup above 15%, typically close to 5% [17].

Most biological reaction systems produce foam stabilizing substances, which may stabilize the gas–liquid dispersions by reducing the rate of drainage from the liquid film. The lack of bubble coalescing ability does help in increasing gas holdup and subsequently $k_l a$, but a tradeoff is necessary between the carryover of foam and reduced value of $k_l a$ as many times long aged small bubbles will be useless as far as transfer of oxygen is concerned. As the superficial gas velocity increases, the uncertainty in predicting the gas holdup also increases, particularly in bubble column. At low velocities the rise velocity of bubbles approximates to the free rise velocity and in liquids similar to water it is about 20–30 cm/s. However, as superficial velocity increases the relative rise velocity becomes uncertain.

With gas-lift reactors the tendency of coalescence can be reduced by installing static baffles. The mixture velocities through orifices could be 50–200 cm/s, which are still lower than the typical impeller tip velocities. The bubbles can be redispersed using redistributors. Bubbles produced such a way will be in the millimeter range. The performance of gas–liquid mass transfer is not affected significantly with the size of the bubbles as the liquid backmixing is reduced and at the same time larger bubbles size is reduced in highly coalescing systems.

Example 12.1 Estimate the shear experienced by a cell at the gas–liquid interface of (a) rising bubble and (b) air bubble bursting at the gas–liquid interface, in a bubble column.

a. If it is assumed that the cell at the surface of a rising bubble experiences liquid velocity on one side and the air velocity through rise velocity of a bubble on another side, the velocity gradient will be velocity difference across the cell divided by its size. Let us consider, $d_p \sim 20\,\mu m$ of the cell and the velocity difference equals to the rise velocity of the bubble.

The rise velocity of the bubble, say of size 5.0-mm air bubble as encountered in a bubble column, can be estimated by the force balance for the bubble.

$$C_D \left(\frac{\pi d_b}{4} \right)^2 \frac{1}{2} \rho_l v_b^2 = \frac{\pi}{6} d_b^3 (\rho_l - \rho_g) g$$

Assuming $Re > 1000$, where the drag coefficient is more or less constant for spherical particles, $C_D \sim 0.44$. The force balance on a suspended particle is

$$0.44 \frac{\pi}{4} d_b^2 \frac{1}{2} \rho_l v_b^2 = \frac{\pi}{6} d_b^3 (\rho_l - \rho_g) g$$

Rearranging

$$v_b^2 = \frac{8}{6} \frac{d_b (\rho_l - \rho_g) g}{\rho_l \times 0.44}$$

$$v_b = \sqrt{\frac{4}{3} \frac{d_b(\rho_l - \rho_g)g}{\rho_l \times 0.44}}$$

$$= \left(\frac{4}{3} \times (5 \times 10^{-3}) \frac{(1.0 \times 10^3 - 1.0)}{1.0 \times 10^3} \times \frac{9.8}{0.44}\right)^{0.5} = 0.38 \text{ m/s}$$

The shear stress can be estimated from the velocity gradient and viscosity of the medium.

$$\text{Shear stress} = \tau = -\mu_l \frac{dv}{dx} = \mu_l \times \left(\frac{v_b}{d_p}\right)$$

$$= 0.89 \times 10^{-3} \times \frac{0.38}{20 \times 10^{-6}} = 16.91 \frac{N}{m^2}$$

b. A bubble escaping from a liquid medium first stretches a liquid film rising above the liquid surface. As soon as the surface tension of the film becomes weaker, the film breaks. The associated liquid velocity of the film can be estimated by Culick equation [16].

$$v_b = 2\left(\frac{\sigma}{\rho_l t_{film}}\right)^{0.5}$$

$$= 2\left(\frac{35 \times 10^{-3} N/m}{(1 \times 10^3 \text{ kg/m}^3)(40 \times 10^{-6} m)}\right)^{0.5} = 1.87 \text{ m/s}$$

$$\text{Shear stress} = -\mu_l \frac{dv_b}{dx} = 1 \times 10^{-3} \times \frac{1.87}{20 \times 10^{-6}} = 93.5 \text{ N/m}^2$$

Considering that for most animal cells the critical shear for cell viability is close to 0.65–1.0 N/m² as they do not have the protection of cell wall unlike the plant cells, the stress experienced by the cells is far in excess of the critical stress for cells' survival.

Example 12.2 The following data are available on viability of *spodoptera frugiperda* cells in a bubble column of 32 cm in height [18].

Time (h)	0.5	1.0	1.5	2.0	2.5	3.0	3.5
Fraction of viable cells (%)	90	87	82	81	80	80	79

If the deactivation is considered as a first-order process, the cell concentration at any time t can be related through deactivation constant $-k_d$.

$$C_x(t) = C_x \log e^{-kdt}$$

A plot of log $C_x(t)$ with respect to t can be used to estimate the death rate coefficient. (See Fig. 12.4, $k_d = -\text{slope} = 0.0181 \text{ h}^{-1}$.)The estimated death rate constant is 0.0181 h^{-1} or $5.03 \times 10^{-6} \text{ s}^{-1}$. In a bubble column, the cell death rate constant is proportional to the airflow and inversely proportional to volume of the bubble column. The estimation of shear stress associated with rising bubbles is relatively low compared to that associated with the injection of air bubbles into the medium and their bursting at the surface. The death rate constant, therefore, should be reduced by increasing the height of the vessels as the percent time spent by the bubbles in the vessel is increased as compared to the time for formation and collapse of bubbles.The death rate constant is related to volume of the reactor and gas flow rate (V_g), and specific hypothetical killing volume V'_k

$$k_d = \frac{F_g}{(\pi/4)/T^2 H} \times V'_k$$

If V'_k is known from small-scale laboratory experiments, the death rate constant for large reactor can be estimated. The value of k_d should be also compared with the growth rate constant. For the viability of the cells the growth rate constant must be higher than the death rate constant. For example, if minimal doubling time of microbes is 8 h, the growth rate constant can be estimated from the first-order kinetics, i.e.,

$$k_g = \frac{\ln 2}{8 \times 3600} \text{s}^{-1} = 2.406 \times 10^{-5} \text{s}^{-1}$$

If $k_g > k_d$, the cells will survive and grow, while when $k_g < k_d$, no growth will be possible in the reactor. Since in the present case, the death constant is smaller than the growth constant, cells should grow in the medium.

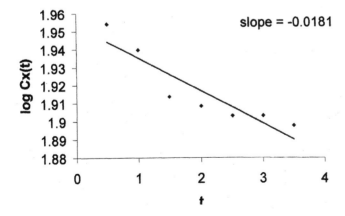

FIGURE 12.4 Deactivation.

Example 12.3 Check the backmixing in the gas phase in a bubble column with diameter 5.0 m and $H/T = 10$, for superficial gas velocity of 0.05 m/s. The average conventional velocity of the gas relative to the surrounding liquid can be estimated if we know the circulation velocity of liquid set in the bubble column. In addition, the bubbles tend to concentrate in the center of the column. These bubbles move with their rise velocity relative to the surrounding liquid. The liquid velocity at the center of the column is

$$v_{l,c} = 0.9(gTu_G)^{\frac{1}{3}}$$

$$= 0.9(9.8 \times 5.0 \times 0.05)^{\frac{1}{3}}$$

$$= 1.21 \text{ m/s}$$

Bubble velocity = rise velocity + $v_{l,c}$

$$= 0.25 + 1.21 = 1.46 \text{ m/s}$$

The gas-phase mixing can be estimated using the gas-phase dispersion coefficient

$$D_g = 78(u_G T)^{\frac{3}{2}} \text{ m}^2/\text{s}$$

$$= 78 \cdot (0.05 \times 5.0)^{\frac{3}{2}} = 9.75 \text{ m}^2/\text{s}$$

The flow behavior of gas phase can be characterized by P_e number

$$P_e = \frac{u_G H}{D_g} = \frac{0.05 \times 50}{9.75} = 0.25$$

The magnitude of P_e indicates the extent of backmixing in the gas phase. The gas phase is essentially flowing as mixed flow.

Example 12.4 A bubble column of diameter 5.0 m is fed with gas velocity of 0.1 m/s. How much interfacial area should be available in the column. The fractional hold up of the gas phase can be estimated using the correlation

$$\phi_G = 0.6(u_G)^{0.7} = 0.6(0.1)^{0.7} = 0.11$$

For most bubble columns the bubble diameter is approximated by 6.0 mm. Thus, the interfacial area

$$a = \frac{6\phi_G}{d_b} = 6\frac{0.11}{6 \times 10^{-3}} \frac{m^2}{m^3} = 110\frac{m^2}{m^3}$$

PROBLEMS

1. What are the operating regimes in a bubble column reactor?
2. When do you prefer a tower bioreactor to a stirred contactor?
3. What is the mechanism of mixing in a bubble column?
4. What is the driving force for liquid circulation in a air-lift reactor?
5. How do you reduce the backmixing effects in a bubble column?
6. What will be the effect of doubling gas velocity in a bubble column on gas-phase holdup?
7. How do you estimate power input in operation of a bubble column?
8. What is the source of shearing action on biocells in a bubble column bioreactor? What are their relative importances?

REFERENCES

1. Hiejnen, J.J.; van't Riet, K. Chem. Eng. J. 1984, 28, B21.
2. Van't Riet, K.; Tramper, J. Basic Bioreactor Design; Marcel Dekker: New York, 1991.
3. Deckwer, W.-D.; Buckhart, R.; Zoll, G. Chem. Eng. Sci. 1974, 29, 2177.
4. Shah, Y.T.; Deckwer, W.-D. In Scale-up in the Chemical Process Industries; Kabel, R., Bisio, A., Eds.; John Wiley: New York, 1981.
5. Towell, G.D.; Ackerman, G.H. Proceedings of the Second International Symposium Chemical Reactor Engineering, 1972; 133–1 pp.
6. Akita, K.; Yoshida, F. Ind Eng Chem Proc Des Dev 1973, 12, 76.
7. Higbie, R. Trans Am Inst Chem Eng 1935, 31, 365.
8. Levich, V.G. Physico-Chemical Hydrodynamics; Prentice-Hall: Engelwood Cliffs, NJ, 1962.
9. Calderbank, P.H.; Mo-Young, M.B. Chem Eng Sci 1961, 16, 39.
10. Akita, K.; Yoshida, F. Ind Eng Chem Proc Des Dev 1974, 13, 84.
11. Kastanek, F. Coll Czechoslav Chem Commun 1977, 42, 2491.
12. Deckwer, W.D.; Karrer-Tien, K.; Schumpe, A.; Serpemen, Y. Biotechnol Bioengg 1982, 24, 461.
13. Verlaan, P.; Tramper, J.; van't Riet, K.; Luyben, K.Ch.A.M Chem Eng J 33, 1986, B43.
14. Sanger, P.; Deckwer, W.D. Chem Eng J 1981.
15. Deckwer, W.D. Chem Eng Sci 1980, 35, 1341.
16. Havenburgh, J.; Joos, P. J Coll Interf Sci 1983, 95, 172.
17. Andrew, S.P.S. Trans I Chem E 1982, 60, 3.
18. Tramper, J.; Williams, J.B.; Joustra, P.; Vlak, J.H. Enzym Microb Technol 1986, 8, 33.

13

Introduction to Downstream Processing

The unit operations that are used in the recovery and purification of the products are described in this chapter. Some of the unit operations described here are used in traditional chemical industries, while others are very specific to biochemical processes. Heat and mass transfer resistances play important role in the design of separation equipments, and the challenge for the design engineers is to maximize the separation efficiencies.

13.1 MASS TRANSFER

Gas–liquid multiphase reactions catalyzed by solid microbes or biocells, as in aerobic process, require the gas to be efficiently transported from the bulk gas phase to the surface of the cell. This gas mass transfer depends on the bubble hydrodynamics, density and the viscosity of the medium, solubility of the gas in the liquid phase, interfacial tension, and type of agitation or mixing. Solubility of oxygen is very poor in hydrocarbon versus in aqueous medium.

The various resistances that are possible during a mass transfer process are (see Fig. 13.1; more details in Chapter 10)

1. Diffusion from bulk of the gas phase to gas–liquid interface.
2. Transport through gas–liquid film
3. Diffusion through bulk of the liquid

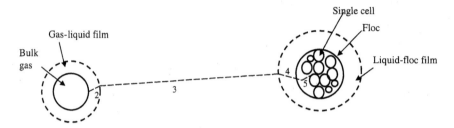

FIGURE 13.1 Various resistances during gas-to-liquid mass transport.

4. Movement through liquid–biocell aggregate and/or floc film
5. Diffusion through aggregate or floc
6. Adsorption on the surface of a single cell

If only one cell is present, then step 5 is absent. The reaction and the mass transfer processes get coupled together giving rise to observed reaction rate, which may be less than the actual reaction rate. Hence innovative reactor designs are chosen to increase gas–liquid mass transport.

The gas and liquid contact in a vessel could be through four different ways:

1. Gas freely rising through a liquid column in the form of a single bubble or bubble swarms.
2. Gas–liquid contact could be enhanced by mechanical agitation. This helps in breaking the gas bubbles into smaller bubbles thereby increasing the gas–liquid interfacial area.
3. Creating artificial circulation patterns in the liquid column using submerged tubes or annular pipes, while the gas is introduced at high velocities.
4. Countercurrent contact, where the gas rises up a tubular column while the liquid flows down. The column is filled with packing material or provided with plates to enhance contact between the two phases.

The rate of gas transport from the gas into a liquid medium is a function of diffusion coefficient of the gas and the thickness of the liquid film through which the gas has to diffuse through, the solubility of the gas in the liquid medium at that pressure, agitation intensity, and the surface tension of the fluid. Several correlations are available in literature (1,2) for estimating mass transfer coefficient as a function of vessel design, agitator geometry and operating conditions. The oxygen transfer rate (OTR), i.e., rate of oxygen transfer from the gas phase to the liquid phase should be greater than the peak oxygen uptake rate (OUR) by the organism. OUR can be determined experimentally in the laboratory. The critical OUR values for organisms lie in the range of 0.003 to 0.5 m mol/l or of the order of 0.1 to 10% of the solubility values for oxygen in organic media or 0.5 to 50% of air saturation values. Although the oxygen solubility appears to be large, in the presence of salts there is a sharp drop in values. Also, OTR becomes critical for systems like penicillin molds, which have OUR in the higher side. Oxygen utilization by the microbes includes cell maintenance, respiratory oxidation for further growth (biosynthesis), and oxidation of substrates into related metabolic end products. If maximum OTR is much larger than maximum OUR, then the reaction is biochemically limited, and the main resistance to increased oxygen consumption is microbial metabolism. If OUR is greater than OTR, then the reaction is gas–liquid mass transfer limited, and good reactor and agitator design should improve conversion.

13.2 HEAT TRANSFER

Heat is added or removed in a biological process to

1. Sterilize liquid reactor feed in a batch kettle.
2. Add heat to increase rate of reaction. For example, since heat generated in anaerobic sludge digestion is insufficient, heat needs to be added.
3. Remove excess heat produced during reaction to prevent denaturation of cells and proteins. For example, heat produced during hydrocarbon oxidation is 200 kcal, while during carbohydrate oxidation it is 80 kcal.
4. Dry cell sludge to remove water.

Bioreactors and fermentors are provided with jacket and/or coils to transfer heat. The rate of heat transfer from the heating–cooling medium to the reactor contents is a function of heat transfer area and temperature difference between the two fluids (more details in Chapter 10). The heat transfer coefficient depends on the physical properties of the two fluids, material of construction of the tubes, accumulated dirt, and agitation intensity and has the units of Btu/ft^2 h °F or W/m^2 °C. Several correlations are found in literature for estimating heat transfer coefficient depending upon the operating conditions and reactor geometry. Since the heat transfer coefficient of steam or condensing vapor is higher than liquids, they are generally used as heat transfer medium. Heat transfer or exchange between the hot and the cold fluids take place in a multitube heat exchanger (Fig. 13.2). A heat exchanger consists of tubes through which one fluid flows while the other flows through the jacket (or shell). The direction of both the fluids could be in the same

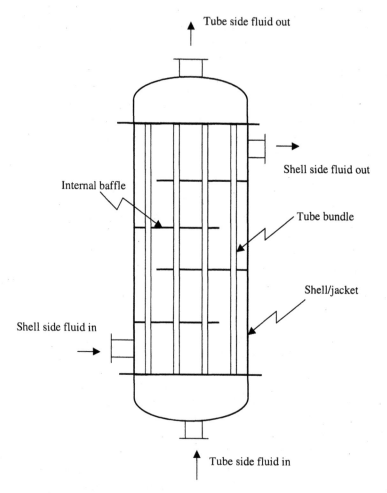

FIGURE 13.2 Sketch of a shell and tube heat exchanger.

direction (cocurrent) or in the opposite directions (countercurrent) to each other. Vaporization of an organic liquid or water is carried in evaporators, while the reverse process (namely, condensing an organic vapor or steam into liquid) in a condenser.

In a bioprocess the heat sources are heat generation from cell growth (Q_{gr}) and due to agitation (Q_{ag}). The heat sinks are heat accumulation (Q_{acc}), transfer to ambient and exchanging fluid (Q_{tra}), heat loss by evaporation

(Q_{eva}), and sensible enthalpy change between inlet and exit streams (Q_{sen}). At steady-state accumulation is zero, and the sum of all heat sources should be equal to all the heat sinks, i.e.,

$$Q_{gr} + Q_{ag} = Q_{tra} + Q_{eva} + Q_{sen} \qquad (13.1)$$

13.3 SEPARATION TECHNIQUES

The viability of a process depends on the ability to purify and concentrate bioproducts in a cost-effective manner on a commercial scale and to meet the required high purity specification. New processes and major improvement of existing processes are needed to attain this goal. Downstream operations involve three steps:

1. Recovery of biomass for reuse or disposal. The processing may involve areas such as cell disruption, post-translational protein processing, protein refolding, protein purification techniques, membrane technology, various types of chromatography including bioaffinity systems, separation techniques using combined force fields, and filtration methods.
2. Recovery of raw material and
3. separation and purification of products.

Steps 2 and 3 may make use of traditional chemical engineering unit operations like filtration, crystallization, distillation, liquid–liquid extraction, and gas–liquid absorption, which are described in the next section (3).

The production of biopharmaceutical drugs such as proteins, peptides, or nucleic acids requires special techniques, equipment, and raw materials. Usually, after fermentation or extraction a multistep protocol has to be followed to finally purify the product. The different types of purification generally used here are sterile liquid and air filtration; clarification and prefiltration (for removal of particulate and microbial contamination); chromatography (for capture, intermediate purification and polishing); tangential flow filtration operations (for protein concentration); virus removal prior to final sterile filtration. Very dilute process streams lead to very high purification costs. The costs for downstream processing is twice the cost of the fermentation itself for low-molecular-weight compounds and could be as high as 10 times the fermentation cost for value-added chemicals and pharmaceutical compounds. Hence downstream processing must not be neglected during bioprocess development. Moreover, the product accumulation during fed batch processes sometimes inhibits cell growth and/or cell productivity.

Thus an integrated downstream processing approach is necessary for product recovery and process optimization.

13.4 TRADITIONAL UNIT OPERATIONS

13.4.1 Crystallization

In crystallization process, solid product is separated out from a supersaturated liquid solution. Supersaturation could be achieved by (1) cooling the mother liquor (as in tank crystallizer), (2) evaporation of solvent (as in evaporative crystallizer), and (3) evaporation combined with adiabatic cooling (as in vacuum crystallizer). Crystallization process can also be used as a purification process, where the impure solid material is dissolved in a solvent and crystallized out of the solvent as a pure material. The impurities remain in the solvent itself.

13.4.2 Drying

Drying is generally carried out in a chamber maintained at high temperature by flowing hot air or nitrogen to remove moisture or solvent, sometimes under vacuum to facilitate the moisture removal process. Spray-drying involves spraying a product solution into a heated chamber through a nozzle. Freeze-drying (lyophilization) is accomplished by freezing the solution and then suddenly applying a vacuum. Water sublimes from solid ice to vapor.

Crystalline solids contain no liquid in the interior, and hence drying takes place at the surface of these particles. Porous material such as catalyst particles has liquid inside the pores and channels, and the rate of drying depends on the diffusion of the liquid through these channels. Nonporous material includes gels, polymers, glue, paste, clay, wood, and leather. Critical moisture content is the amount of moisture that remains in the solid even after drying and the material cannot be dried below this value.

13.4.3 Solid–Liquid Separation (Filtration, Centrifugation, and Sedimentation)

The concentration of insolubles in broth could vary from 0.1% to 60% and the insoluble material may be 1 mm or larger in diameter. Filtration is used as an unit operation to remove solids from a slurry solution. Here, the solid or the liquid could be the desired product. The slurry is passed through a bed of fine particles, cloth or mesh, or perforated plates. Filtration could be carried out by applying pressure to the slurry, so as to force it through the filter medium, or using vacuum at the downstream, so as to suck the clean liquid through the filter medium (Fig. 13.3). The rate of filtrate flow depends on the

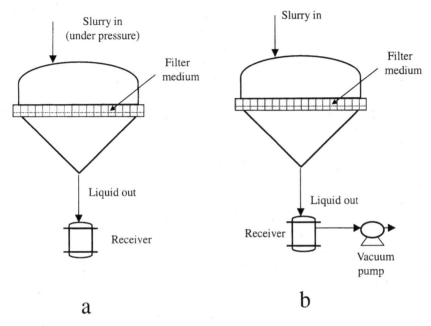

FIGURE 13.3 Filtration assembly (a) pressure and (b) vacuum type.

area of filtration, viscosity of the filtrate, the pressure difference across the filter, the resistance of the filter medium, and the resistance of the cake that deposits on the filter.

Centrifugation and sedimentation make use of the difference in density between solids and the surrounding fluids. In sedimentation the settling force is gravity, while in centrifugation, a centrifugal field is applied to achieve the separation. Centrifugation must be used when filtration does not work too well. At very high solids concentration the settling velocity is also affected by the interaction between the solids.

13.4.4 Distillation

Separation of liquids that have different boiling points is carried out by distillation. Distillation may be performed in two ways, namely, flash distillation and rectification or fractionation. In the former, the liquid mixture is converted into vapor by applying heat and/or vacuum and condensing the ensuing vapor in a receiver without allowing it to return to the still (Fig. 13.4). The condensate will have higher concentration of the lower boiling component, and the reboiler will have higher concentration of the higher boiling

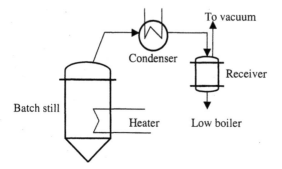

FIGURE 13.4 Batch distillation assembly.

component. This technique is generally used for separating components that boil at widely different temperatures or to recover light hydrocarbons or low-boiling solvents from the product mass. In a batch distillation as the distillation proceeds, the composition of the liquid mixture that is collected at the top and the composition of the liquid that remains in the reboiler keeps changing with time. The moles of the two components (n_A and n_B) in the reboiler are related by the following equation,

$$\left(\frac{n_B}{n_{Bo}}\right) = \left(\frac{n_A}{n_{Ao}}\right)^{1/\alpha_{AB}}$$

(13.2)

where α_{AB} is the relative volatility $= (y_A/y_B)/(x_A/x_B)$. y_A and x_A are the mole fraction of the low boiling component in the gas and the liquid phases, respectively. Subscript o indicates initial moles.

Fractional distillation is employed to separate multiple components whose boiling points are very close. The unit operation is carried out in a system consisting of a tall column placed on top of the still so that the vapor has to travel through out the length of the column before reaching the condenser. A part of the condensed vapor is returned back as a liquid, so that it comes in contact with the rising vapor (see Fig. 13.5). The efficiency of separation depends on the fraction of condensed liquid returned back to the column, to the liquid product removed (known as *reflux ratio*). When the two fluids come in contact in the column, the low-boiling component in the liquid moves to the vapor and the high-boiling component in the vapor moves to the liquid, thereby achieving separation of the components. The column consists of perforated plates, trays, or inert packing so as to improve the vapor–liquid contact. The operation could be carried out in batch or in continuous mode of operation. In the batch mode, the still is charged with the liquid mixture, and the various fractions are collected at the top of the column separately. The

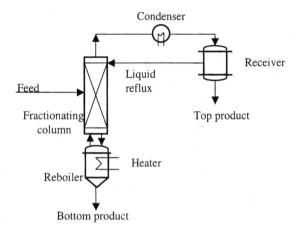

FIGURE 13.5 Continuous fractional distillation assembly.

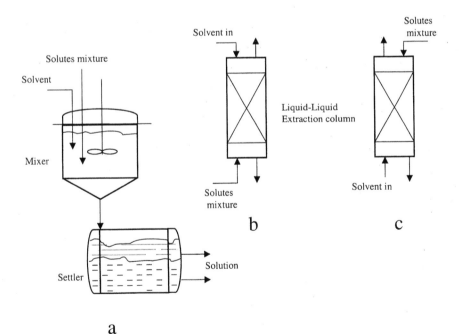

FIGURE 13.6 Liquid–liquid extraction (a) mixer-settler type, (b) for heavier solvent, and (c) for lighter solvent.

highest boiling compound is finally left behind in the bottom still. In the continuous mode, feed is introduced continuously, and the bottom and top products are removed continuously.

13.4.5 Liquid–Liquid Extraction

Here two miscible solutes are separated by a solvent, which preferentially extracts one of them. Close boiling mixtures or solutes that cannot withstand high temperatures in distillation may often be separated by this technique. The extraction method uses the differences in solubilities of the components in the solvent. Ethanol from fermentation broth is recovered by extraction.

Two main designs of extractors are mixer–settler and column type. In the first design, the solvent and the mixture are made to come in contact in a mixer and made to separate out into two layers in a settler (see Fig. 13.6a). In the mixer vessel the solute mixture and the solvent are agitated mechanically. Several mixers–settlers can be placed in series to achieve good extraction. In the column design the solute mixture and the solvent are made to flow through a column in a cocurrent or countercurrent fashion. The column may consist of perforated plates, trays, or packing similar to a distillation column. If the density of solvent is higher than the density of the solute mixture, then con-

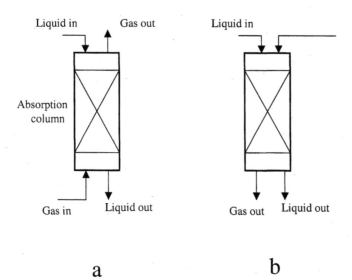

FIGURE 13.7 Gas–liquid absorption column designs (a) counter current and (b) cocurrent.

figuration shown in Fig. (13.6b) is followed and for the reverse situation configuration Fig. 13.6c is followed.

13.4.6 Gas–Liquid Absorption

Soluble gas is absorbed from a gas mixture with a liquid, for example, carbon dioxide from vent gases using monoethanolamine solution. The gas and liquid are brought into contact in a column packed with inert material. The gas–liquid contact is achieved either cocurrent or countercurrent as shown in the Fig. 13.7. The packing material helps in improving the contact between the two fluids. Hence high-surface-area packing helps to increase the efficiency of contact.

13.5 ISOLATION AND RECOVERY TECHNIQUES FOR BIOCELLS

The cell is disrupted for recovery of intracellular products. This is achieved by mechanical or nonmechanical means.

13.5.1 Cell Disruption Through Mechanical Means

The various mechanical disruption techniques are as follows:

1. Bead mills: These are long cylindrical tubes with impellers into which small beads are loaded (up to 80%). The beads are made of tough materials such as titanium carbide or zirconium silicate. As the mill rotates, the beads collide with the biomaterial breaking the cell wall, thereby releasing the products. Product release depends on the microorganism used, cell wall composition, and thickness. Main disadvantage of this system is the rise in temperature, which if not controlled could deactivate the product enzyme.
2. Homogenizer consists of a high-pressure pump that injects the cell suspension through an adjustable orifice discharge valve to achieve pressure of the order of 500 atm. Cell disruption happens by three different mechanisms, namely, impingement on the valve, high liquid shear in the orifice, and sudden pressure drop upon discharge, causing an explosion of the cell. This technique works for most cultures except highly filamentous organisms. Scale-up involves using bigger displacement pumps and discharge valves. Although preferred to bead mills during scale-up, temperature rise again poses a problem.

3. Ultrasonic vibrators disrupt cells using short bursts of ultrasound. This technique is used usually for bacteria and not for yeast (which have a tough cell wall). The heat generated during sonication is also a problem.

Disadvantages of mechanical disruption methods are

1. Since cells break completely, all intracellular materials are released and hence product must be separated from a complex mixture.
2. Released material may increase the viscosity of the solution and may complicate subsequent processing steps.
3. Cell debris, produced by mechanical lysis, often consists of small cell fragments, making the solution difficult to separate.

13.5.2   Nonmechanical Disruption Techniques

These techniques are popular for large-scale applications because of low investment costs and are as follows:

1. Heat shock technique involves applying a high temperature for a very short period of time. This technique is cheap to implement but can be used only if product remains stable during the heat shock treatment.
2. An abrupt change in salt concentration (or solute concentration in general) can create conditions optimal for cell lysis (breakage). This osmotic shock can be used to release protein from the periplasm.
3. Repeated freeze–thaw cycling results in cells bursting (ice crystal formation during this cycling process disrupts cell membranes).
4. Chemical lysis consists of adding detergents (like Triton X-100, which is nonionic in nature) and solvents (including acetone, octanol, DMSO, and benzene) to partially solubilize cell walls. Guanidine-HCl is often added to improve efficiency. The chemicals added must not affect the product or subsequent downstream processing. Chemical permeabilization can also be achieved with antibiotics and thionins.
5. An enzyme is used to lyse bacterial cell walls and recover the desired product. Although this technique is expensive, it consumes less energy, damages the product less, and is very specific. Lysozyme is an enzyme that hydrolyzes the murein present in bacterial cell walls, and cellulase can lyse plant cells. Temperature, pH, and cofactor can influence enzymatic lysis. Additives such as ethylene diamine tetra-acetic acid (EDTA), a chelator of divalent ions (divalent ions are common in cell walls) can also disrupt cell walls.

13.6 PURIFICATION TECHNIQUES IN BIOPROCESS TECHNOLOGY

13.6.1 Membrane Processes

Membranes act as selective barriers to different components, which means it permits some molecules to pass through while retaining some other(s). Hence a feed stream is split into two: a permeate stream with small molecules and a retentate with large molecules. The membrane designs include tubular, hollow fiber, plate and frame, and spiral wound (see Fig. 13.8). The main advantages of membranes are no separate solvent is required to achieve separation, batch or continuous operation at ambient temperature, and easy to scale up. Disadvantages are little resistance to extreme pH and temperature values, cleaning and sterilization is difficult, and problems of membrane fouling.

The driving forces in membrane processes are achieved through different means:

1. Hydrostatic pressure gradient as in microfiltration, ultrafiltration, and reverse osmosis

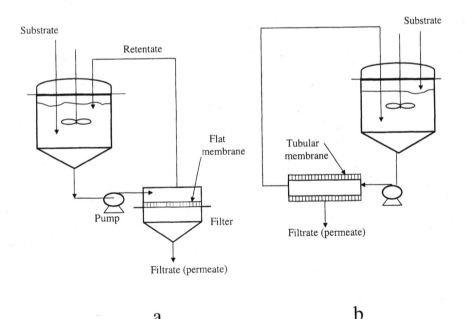

a   b

FIGURE 13.8 Membrane separation (a) flat membrane and (b) hollow tubular membrane.

2. Concentration gradient as in dialysis
3. Electric field gradient as in electrodialysis (purification of small charged molecules like organic acids)
4. Partial pressure gradient as in pervaporation (selective removal of solvents such as ethanol from fermentations)
5. Partition coefficient as in perstraction (extraction of small molecules from aqueous/organic solutions).

Membranes with pore diameter smaller than 0.0001 mm are used in gas separation.

Microfiltration (MF)

Microfiltration is a pressure-driven process that separates micron or submicron particles from the liquid or gaseous feed stream with the help of a membrane having pores of 0.05 to 3 mm. The operating conditions are 5–50 psi (0.3–3.3 bar) pressure and 3–6 m/s cross-flow velocities. Microfiltration is usually carried as a multistage (stages-in-series) operation in industrial scale. Microfiltration is the most open membrane and removes starch, bacteria, molds, yeast, and emulsified oils.

Ultrafiltration (UF)

Ultrafiltration is a low-pressure operation that separates dissolved solutes of 0.005 to 0.1 mm. Ultrafiltration membranes are used in numerous industries for concentration and clarification of large process streams, wherein higher molecular weight compounds are retained while passing solvents and low-molecular-weight compounds.

Nanofiltration (NF)

The NF membrane rejects divalent ions (like Ca^{+2} or SO_4^{-2}) while allowing a majority of monovalent ions (like Na^+ or Cl^-) to pass. Also, organic molecules in the 200–300 molecular weight ranges are rejected. NF is used in the municipal drinking water plants (90% of feed water's hardness ions can be removed) and dairy industry for cheese whey desalting. Since NF is more cost effective than reverse osmosis (RO) in certain applications, new growing markets include RO pretreatment, pharmaceutical concentration, kidney dialysis units, and maple sugar concentration.

Reverse Osmosis

Reverse osmosis membranes concentrate low-molecular-weight organic materials and salts while allowing water and solvents to pass through. During the operation, water flows from the concentrated feed stream to the dilute permeate—a direction that is just the reverse of what would occur naturally during osmosis. High pressures of the order of 35–100 atm are needed here in order to overcome the high osmotic pressures across the membrane. Reverse osmosis is widely used technology for desalinating seawater and reclaiming brackish well water. The reverse osmosis membranes reject organic molecules (molecular weight >150) such as sugars based on its size and shape.

Electrodialysis

Electrodialysis (ED) is the transport of ions through the membranes as a result of the application of direct electric current. If membranes are more permeable to cations than to anions, then anions are retained in the electrolyte solution, thereby achieving its concentration (Fig. 13.9). Similarly membranes, which are more permeable to anions, can be used to concentrate cations. Removal of ionic species from nonionic products can be accomplished by this technique, so that purification is possible. Electrodialysis reversal (EDR) is an electrodialysis process in which the polarity of the electrodes is reversed on a prescribed time cycle, thus reversing the direction of ion movement in a membrane stack.

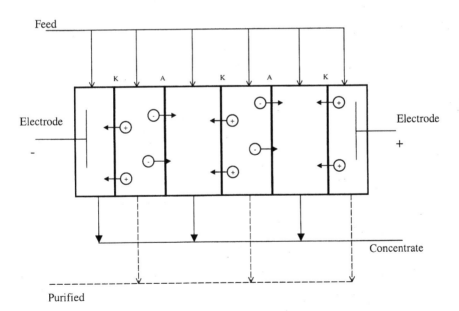

FIGURE 13.9 Electrodialysis unit (K, cation permeable membrane; A, anion permeable membrane).

Pervaporation

Pervaporation (PV) is the separation of liquid mixtures by vaporization of one of the species through a permselective membrane. Permeate first vaporizes, travels through the membrane by diffusion, and condenses into liquid on the permeate side (Fig. 13.10). The driving force here is vacuum applied on the permeate side. PV is an enrichment technique similar to distillation. The heart of the PV is a nonporous membrane, which either exhibits a high permeation rate for water but does not permeate organics, or vice versa, i.e., permeates organics but does not permeate water.

Dialysis

Separation in dialysis is due to concentration gradient and is used for the separation of small solutes from large synthetic or biological macromolecules. The membrane material retains larger molecules due to steric reasons, while the smaller solutes freely diffuse through the membranes, eventually leveling out concentration differences. If the osmotic pressure is different between the two phases, solvent molecules will also diffuse through the membrane, to reach equilibrium. Dialysis is used primarily in the treatment of patients with renal failure and also in the separation of proteins and other macromolecules from salts in pharmaceutical and biochemical applications.

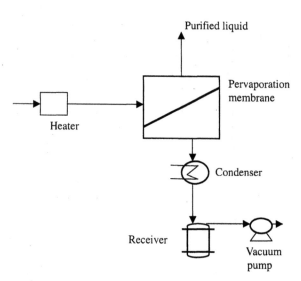

FIGURE 13.10 Pervaporation assembly.

Vortex Flow Filtration

Vortex flow filtration (VFF) consists of a tubular membrane with a high-speed rotating inner annular cylinder. Due to its rotation, vortices are generated on the membrane surface, which improves filtration efficiencies and reduces fouling of the membrane. Hence the feed pump could maintain low flows.

13.6.2 Supercritical Fluid Extraction

Supercritical fluids are materials that remain fluid above their critical temperature and pressure and have the properties of gas as well as liquid. A well-known fluid is supercritical carbon dioxide. In a typical process, supercritical CO_2 is contacted with a feed stream containing the solute (or product) that needs to be extracted. The CO_2 enriched with the product is then taken to a different chamber where the pressure is lowered converting the liquid CO_2 into gas, leaving behind the extracted product as a precipitate. The gaseous CO_2 can be repressurized and recycled.

13.6.3 Adsorption and Chromatography

Adsorption and chromatography techniques involve capturing (adsorbing) the desired product on a solid adsorbent like ion-exchange resins or activated charcoal, and they can be operated in batch, fluidized bed and expanded-bed mode (4). The feed mixture is first contacted with the adsorbent, and the resin is subsequently eluted with a solvent to remove the adsorbed product. The advantage of this technique over liquid extraction is that much smaller volumes of adsorbent are required (vs. organic solvents), while the main problem is fouling of adsorbent by irreversible adsorption by poisons. The adsorbed product needs to be thoroughly desorbed so that it can be reused again.

Chromatography is used for high-resolution purification, and it contains two phases, namely, a stationary phase (adsorbent in a column) and a mobile phase (some suitable solvent). The mobile phase carries the mixture to be separated over the column, and the separation of the mixture into various components is achieved due to the differences in interaction between the stationary phase and the components. Various chromatographic techniques are briefly described below.

Size exclusion chromatography (gel filtration or gel permeation or molecular exclusion) is based on size and shape. Since small molecules enter pores of the adsorbent and hence take longer time to pass through a column, they elute later when compared to large particles, which flow out with the mobile phase. The large molecules interact less with the adsorbent when compared to small molecules.

Ion exchange is based on net charge and is widely used since high-resolution separations can be achieved with this technique. Anion exchangers have positively charged groups (diethylaminoethyl, quarternary amino) as adsorbents that preferentially bind negatively charged molecules from the solution, while cation exchangers have negatively charged groups (sulphonate, carboxymethyl) as adsorbents that preferentially bind positively charged molecules from the solution (Fig. 13.11). Later bound molecules are eluted out by changing the salt concentration of the mobile phase (salt competes with the protein for the resin) or by changing pH (bound molecules lose their charges). One common resin for protein purification is a cation exchange obtained by linking negatively charged carboxymethyl groups to a cellulose background.

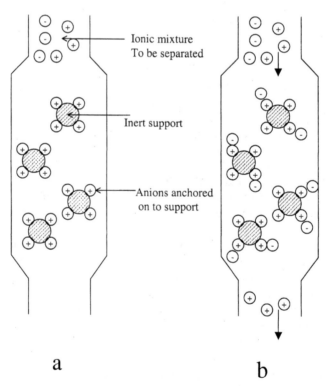

Ionic mixture
To be separated

Inert support

Anions anchored
on to support

a

b

FIGURE 13.11 Ion exchange chromatography (example of an anion exchanger): (a) start up and (b) during operation.

Separation can also be achieved based on hydrophobic interaction between the solute and the adsorbent. Hydrophobic interactions are favorable at high salt concentrations. Elution is achieved by reducing salt or by decreasing polarity of the medium (add ethylene glycol or ethanol).

Affinity chromatographic technique relies on the principle of molecular recognition (Fig. 13.12). An affinity ligand on the adsorbate fishes out a ligate of complementary structure in the solution forming a complex, leaving behind the other unwanted solutes to elute out with the solution. Changing buffer conditions can reverse the binding thereby eluting the adsorbed molecules. Usually extremely specific antibodies are used as ligands.

Immobilized metal-ion affinity technique is based on metal-ion binding in which the amino acid side chains of recombinant proteins are selectively bound to supported transition metals ions like Cu, Ni, Zn, Co, or Fe and later made to elute out by changing pH.

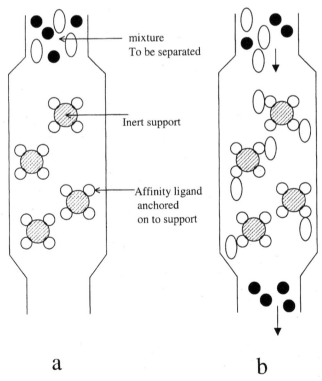

FIGURE 13.12 Affinity chromatography: (a) start up and (b) during operation.

13.6.4 Precipitation

Protein or other biomaterials precipitate out of the fermentation broth when salts such as $(NH_4)_2 SO_4$, or $Na_2 SO_4$ are added in high concentration. This "salting out" happens due to the reduction in the solubility of these biomaterials due to the dissolution of these organic salts.

13.6.5 Electrophoresis

Electrophoresis is the migration of charged molecules such as proteins, nucleic acids, and subcellular-sized particles, namely, viruses and small organelles, in an electrical field based on its size, shape, and charge in a conductive-buffered medium. The side chains of the amino acids of which the proteins are composed contribute to the charges. The charge of the protein depends on the isoelectric pH of the protein and the pH of the surrounding buffer. Since each protein will have different charge, application of an electric field will result in different protein velocity, namely, large molecules migrate slower. Also, greater the charge on the molecule faster is the migration. Cationic species will migrate toward the cathode and anionic toward the anode.

PROBLEMS

1.   A piece of lignin 30 mm in thickness is dried from initial moisture content (X_1) of 25% to final (X_f) of 5% using totally dry air. If the diffusivity of water through lignin $(D) = 8 \times 10^{-6}$ cm^2/s, determine the drying time. The drying time is given by $t = (4 s^2/\pi^2 D) \ln (8X_1/ \pi^2 X_f)$. s is one-half of the slab thickness.

2.   What separation technique should one adopt for separating two biomolecules having sizes 15 and 22 Å and isoelectric pH of 4.6 and 5.2?

3.   If 100 mole of a mixture of 80 mol% ethanol and the rest water is distilled in a batch process, what will be the composition of the mixture in the reboiler after 50% of the ethanol has been removed? What will be the composition of the mixture collected as the overhead product?

4.   Carbon dioxide is removed from air in a continuous absorption column using ethanol amine solution. The airflow rate is 250 kg/h. Air enters at 0.65 g CO_2/kg of air and leaves at 0.2 g CO_2/kg of air. Amine enters with 0.002 g CO_2/kg of amine and leaves at 0.5 g CO_2/kg of amine. Calculate the amine flow rate.

5.   An aqueous methanol solution containing 20% by weight methanol is fed into a continuous distillation column at the rate of 1000 kg/h. If the desire product purity is 99% at the top of the column, what will be the

bottoms flow rate if the aim is to recover at least 80% of the methanol entering the column as the top product?

    6. Initially porous solids dry at constant drying rate. Once the moisture content of the material reaches a critical value, the drying rate falls logarithmically. The equation for drying time is given by the equation $t = [M_s/AR_c]$ $[(X_1 - X_c) + X_c \ln (X_c/X_2)]$. Estimate the drying time to dry 200 kg material (M_s) from initial moisture content (X_1) of 15% to a final moisture content (X_2) of 4%. The exposed area (A) is 20 cm^2, and the maximum rate of drying (R_c) is 0.009 kg/cm^2/s. The critical moisture content (X_c) for this material is 9%.

    7. A shell and tube heat exchanger is used to cool water from 195 °C to 110°C, using cooling water. Estimate the cooling water flow rate if the hot fluid is flowing at 1000 kg/h and the cooling water temperature rises from 30°C to 45°C.

    8. If the maximum solubility of a solute in a solvent is 0.05 kg/kg of solvent. How much solvent is needed to extract 90% of the solute from a polymeric waste of 10,000 kg containing 10% solute?

REFERENCES

1. Bailey, J.E.; Ollis, D.F. *Biochemical engineering fundamentals*; McGraw-Hill Kogakusha ltd: Tokyo, 1986.
2. Mezaki, R.; Mochizuki, M.; Ogawa, K. *Engineering data on mixing*; Elsevier: Amsterdam, 2000.
3. McCabe, W.L.; Smith, J.C. *Unit operations of chemical engineering*; McGraw-Hill Intn. Edn: New York, 1976.
4. Ladisch, M.R. Bioseperations Engineering: Principles, Practice and Economics, Wiley Europe, Apr 2001.

14

Industrial Examples

Successful application of biocatalyst for organic transformations in industries depends on factors that include availability of suitable enzyme, ease of up- and downstream processing, competition with established chemical methods, time required for process development, biocatalyst disposal, possible perception of the technologists and engineers, and government regulations. Biocatalysis is used primarily to carry out two distinct types of transformation, namely, ones in which pairs of enantiomers are resolved from each other and others in which a racemic compound is stereoselectively transformed into a single isomer derivative. Chiro Techs' process for the specialty analgesic(s)-(+)-naproxen where bioresolution is followed by recycle of the unwanted isomer is an example showing the use of biocatalytic route.

The third reason for using biocatalysts is that it can simplify conversions that would otherwise require multiple or difficult synthetic steps. One example is the enzymatic hydrolysis of cyanopyridine to the vitamin niacinamide that Lonza Biotech carries out at Guangzhou, China. Degussa's process for the manufacture of polyglycerine ester is another example where biocatalysis route simplifies the multistep chemical route.

The fourth advantage of biocatalysis is the option of carrying out stereoselective reactions. Production of "unnatural" amino acids is the focus of several companies. For example, Great Lake Fine Chemicals has scaled up production of L-*tert*-leucine using a process developed by NSC Technologies. They already make amino acids like D-phenylalanine and D-tryosine using similar technologies.

Companies are combining biocatalytic step with the conventional chemical process in a multistep synthesis to take advantage of the former. Process for L-tert-leucine by DSM is a typical example of this approach; namely, prepare D, L-*tert*-leucamide through chemical route, followed by the preparation of L-amino acid using the enzyme L-amidase.

Most of the applications for biocatalysts at industrial scale are for low-volume, high-value products, and exception include ethanol and fructose produced by fermentation at more than a million tons per year at less than $1/ kg. Biocatalytic approach offers a few disadvantages, which include

1. Enzyme cost, which makes it suitable for value-added products and not for bulk chemicals. Having said that glucose isomerase is $500/kg, trypsin or lipase is $500/kg, while lactic acid dehydrogenase is $100,000/kg and porcine liver esterase, penicillin amidase, or aspartase are in the range of $15,000 to $10,000 (year 2000). These prices are similar in comparison to asymmetric homogeneous chemical catalysts but are much higher than heterogeneous catalysts. There are a few examples of enzyme being used for bulk chemicals also.

2. Limited number of commercial enzymes available in the market. So one may have to design one's own enzyme for the process under interest.

3. Narrow range of useful transformation that enzymes mediate.

4. Lack of reaction generality. A single enzyme may not perform well even within the same class of reactions.

5. Enzyme instability and sensitivity to environment.

6. Long process development times.

Bioethanol comes from fermenting sugars, which are generated by breaking down starches from corn, potatoes, sugar cane, or wheat. If enzyme costs were less than $0.10/gallon of ethanol, then cost of production from biomass wastes (wood, grass, etc.) would be economical. Major chemical companies like Dow, DuPont, BASF, Degussa, and Celanese are investing heavily to explore opportunities through alliances with smaller firms with specific expertise.

There are five fundamental skills needed in biocatalysis:

1. Screening techniques for finding microorganism that contain the desired enzymes

2. Ability to optimize the organism with directed evolution

3. Capacity to grow the optimized organism through fermentation
4. Immobilization techniques
5. Technology to apply the biocatalyst (either as isolated enzyme or whole cell) in a chemical reaction step

Hence the expertise involves microbiological skills, fermentation capabilities, knowledge of synthetic chemistry, and chemical–biochemical engineering.

14.1 BIOREACTOR

Bioreactor is the heart of the manufacturing operation and the upstream and the downstream processes support this. The raw materials are prepared in the upstream section and the unconverted reactants, products, wastes are separated and purified and recyclable materials such as solvents recovered in the downstream section. A large number of biochemical reactor designs like agitated, sparged, and packed reactors are available, and the selection depends on the reaction kinetics, type of inhibition, mode of operation, aerobic or anaerobic type, whether the enzyme is in the native form or supported, physical properties of the biocell, the chemical and physical properties of the substrates used and the products formed, nature of gases formed, and the amount of heat to be removed or added. The detailed selection and design criteria and different types of bioreactors were described in Chapters 8 to 12. The designs strive toward bringing the substrate, oxygen, and the biocatalyst together in the most optimum way, at the same time supplying or removing heat to maintain high reaction rate. The hardware design also considers maintaining the reactor and the other accessories at aseptic conditions.

Oxygen has a low solubility in organic medium, and it has to be added continuously in aerobic processes for the sustenance of the cells. Agitation prevents cells from settling, brings the substrate and the biocells in close proximity, breaks the large oxygen bubbles, and homogenizes the vessel. Hence for large-scale bioreactors oxygen sparging and agitation is required. The high shear of the agitator forces the cell walls to rupture leading to its death. This contradiction makes oxygen transfer and shear important factors in the reactor design. For example, the cell density in a continuous culture of sf21 insect cells increases with agitation up to 800 rpm and decreases on further increase. When the bubbles disengage at the liquid surface, cells, which are in contact, die due to the disruption of their walls. Cell death does not take place when the cells are rising with the gas bubbles. Hence a tall reactor will have less percentage of cells dieing than a short reactor.

14.2 MANUFACTURE OF AMINO ACIDS

14.2.1 Production of L-Aspartic Acid

L-Aspartic acid is used in medicines and in food additives as artificial sweetener, which can be produced by reacting fumaric acid with ammonia in the presence of aspartase. Tanabe Seiyaku Co. Ltd. first reported the industrial production of L-aspartic acid using E. Coli B immobilized with polyacrylamide gel. The acid is produced (60 t/month) in a multistage packed column reactor, with cooling water tubes to remove the exothermic heat produced during the reaction (Fig. 14.1). The product is crystallized and filtered.

When the immobilization was changed in 1978 to carrageenan (hydrocolloid consisting of high-molecular-weight linearly sulfated polysaccharide), the E. coli activity increased by a factor of 15, leading to increased production (100 t/month). Other strains like EAPc-7 and E. Coli B strain were also found to give better activity and stability for the production of L-aspartic acid in industrial scale.

14.2.2 Production of L-Alanine

L-Alanine, an amino acid used in medicines and food additives, is produced industrially since 1965 by Tanabe Seiyaku Co. from L-aspartic acid in a batch process using an enzyme L-aspartate-β-decarboxylase. L-Aspartic acid was

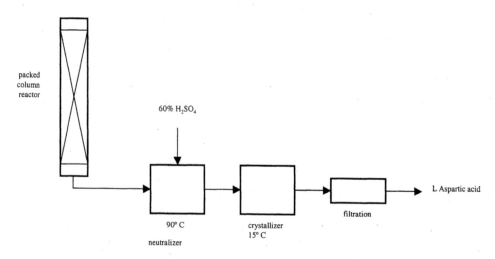

FIGURE 14.1 Schematic of the process for manufacture of L-aspartic acid.

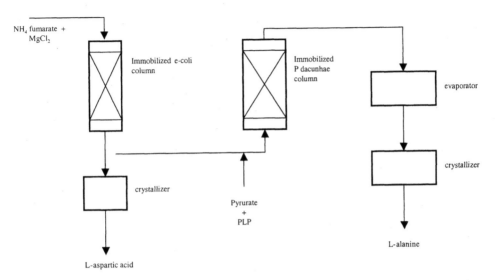

FIGURE 14.2 Simplified process diagram for the manufacture of L-aspartic acid and L-alanine.

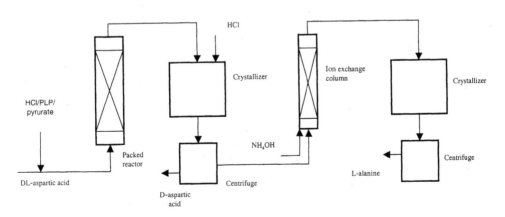

FIGURE 14.3 Schematic for manufacturing D-aspartic acid and L-alanine.

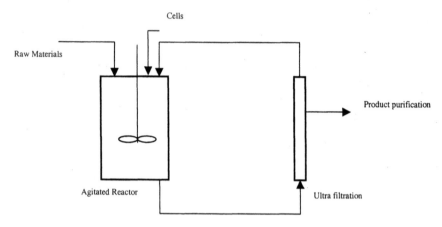

FIGURE 14.4 Schematic for the synthesis of L-malic acid.

produced since 1973 from ammonium fumarate using immobilized *E. Coli* aspartase packed in a column reactor. Continuous production of L-alanine from ammonium fumarate was commercialized in 1982 at a scale of 100 t/month of acid and 10 t/month of L-alanine using 1000 l and 2000 l reactors, respectively. Figure 14.2 shows a schematic of the unit operations.

14.2.3 D-Aspartic Acid

Tanabe Seiyaku Co. has been producing biocatalytically 9.5 t of D-aspartic acid and 5.1 t of L-alanine per month simultaneously from 1988 using a 1000 l pressurized column reactor and a crystallizer for product purification. The enzyme used is *P. dacunhae* cells (L-aspartase-β-decarboxylase) immobilized on a support. The schematic of the unit operations are shown in Fig. 14.3.

14.2.4 L-Aspartic Acid

Mitsubishi Petrochem Co. developed a process for production of L-Aspartic acid using native strain *Brevibacterium flavum* MJ 233. Ultrafiltration was used to separate the cells after the reaction and recycle them back to the reaction vessel.

Similar separation technique is adapted in the production of L-isoleucine from ethanol and α-ketobutyric acid. This process has higher productivity than the fermentation process.

14.2.5 L-Maleic Acid

L-Maleic acid, a food additive, is industrially produced from fumaric acid using lactobacillus traditionally in a batch reactor. Tanabe Seiyaku Co. has been manufacturing since 1980 using immobilized *B. flavum* cells at the rate of 30 t/month with 70% yield. The novelty of the process is the use of microbial cells itself, rather than using the purified enzyme, thereby saving on the cost. A possible disadvantage of this approach is the formation of side products, in this case succinic acid. The native enzyme is recovered using ultrafiltration and recycled back to the reactor. Process schematic is shown in Fig. 14.4.

14.3 MANUFACTURE OF ANTIBIOTICS

14.3.1 6-APA

6-Amino penicillanic acid (APA) is used in the production of semisynthetic penicillin, such as amoxicillin and ampicillin and more than 6200 t of 6-APA is produced annually worldwide. Fifteen years back it was mainly produced by chemical deacylation from penicillin G or V. The process was complicated and hazardous. Today most of it is produced in a recirculation bioreactor with an immobilized *penicillin amidase* enzyme.

The system followed by Asahi chemicals consists of several immobilized enzyme column (Toyo Joto bioreactor system) reactors in parallel of each about 30-l/capacity. Ten percent (10%) penicillin G solution is circulated at a rate of 6000 l/h. The reaction is carried at 30–36°C and at a pH of 8.4. One cycle takes 3 h, and the lifetime of enzyme is more than 360 cycles. To achieve constant output and quality, few columns are replaced periodically. Figure 14.5 shows a process schematic.

Other companies such as Beechams, Pfizer, and Gist-Brocades produce 6-APA using a similar process. Altus Biologics Inc. (Cambridge, Mass.) has developed a new form of penicillin acylase for selectively cleaving the phenyl acetic acid side chain of penicillin G, producing 6-APA. The enzyme is kept active through cross-linked enzyme crystal.

14.3.2 7-ACA

7-Aminocephalosporanic acid (ACA) is an intermediate in the production of medically important semisynthetic cephalosporins. The worldwide annual production is in the range of 1200 t. Toyo Jozo and Asahi Chemicals industry together commercialized a chemical-enzymatic two-step process from cephalosporin C, using immobilized enzyme in a packed column reactor operating at a space velocity of 1.0 and at temperature range of 15–25°C.

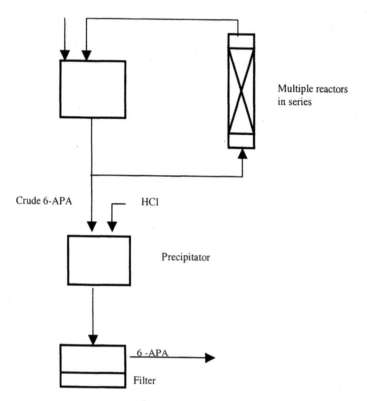

Multiple reactors
in series

Crude 6-APA

HCl

Precipitator

6 -APA

Filter

FIGURE 14.5   Schematic for 6-APA synthesis.

Montedison Group (Italy) also produces 7-ACA using a similar enzymatic route. The amount of waste generated by the enzymatic route is about 0.3 t as against 31 t by the chemical route per ton of 7-ACA production.

14.3.3 7-ADCA

Gist-Brocades, The Netherlands, and Toyo Jozo use supported *cephalosporin C acylase* for the manufacture of 7-amino deacetoxycephalosporanic acid (ADCA) in a packed bed reactor. This is an intermediate for cephalosporin and penicillin. DCM (Heerlen, The Netherlands) has built an enzyme manufacturing plant to be used for the production of several hundred tons of 7-ACA and a 7-ADCA through a fermentation route.

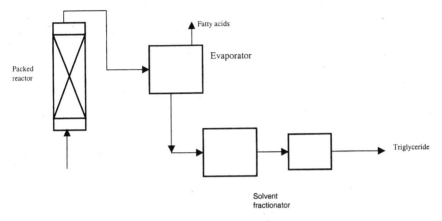

FIGURE 14.6   Schematic for specialty triglyceride synthesis manufacture.

14.4   MANUFACTURE OF TRIGLYCERIDES

Loders Croklaan has an enzymic interesterification plant (Fig. 14.6) at its Wormerveer (Holland) facility for production of specialty triglycerides by use of lipase-catalyzed interesterification in fixed-bed reactor for confectionary use.

A subsidiary of Unilever manufactures Betapol another triglyceride (a formula additive for premature babies) by the interesterification of tripalmitin with oleic acid.

14.4.1   Cocoa Butter Equivalent

Unilever and Fuji oil manufacture cocoa butter equivalent using 1,3-selective lipases replacing palmatic acid with stearic acid. Cocoa butter imparts the "mouth feel" in food products. Unilever subsidiary Quest-loders Croklaan (Netherlands) use palm oil middle fraction or higher oleate sunflower oil as the starting material.

14.5   AMIDES

About 30,000 t/annum of acrylamide is manufactured by Nitto Chemicals (Yokohama, Japan) using *R. rhodochrous* J1. The yield factor is of the order of 7000 g of acrylamide per g of cell at a conversion of >99.9%, while the amount of acrylic acid produced is negligible.

Lonza (Switzerland) in their plant in Guangzhou, China make 3000 t/ annum of nicotine amide (vitamin, animal feed supplement) from 3-cyano pyridine using *R. rhodochrous* J1.

14.6 SPECIALTY CHEMICALS

Enzyme cyclodextrin glycotransferase can be used to convert simple starch molecules into cyclodextrin. Maxygen (Calif) has a process for reducing carboxylic acids into aldehydes.

Thermogen (Chicago, Ill.) uses dehydrogenases to convert ketones to Chiral alcohols. NSC Technology (Mount Prospect, Ill.) has used *d*-trans-aminase to produce D-phenylalanine and D-tyrosine in batch reactor yielding 1 t per batch.

Diversa Corp (San Diego, Calif.) has identified enzymes that can be used to produce a large number of speciality and fine chemicals in the range of 0.5 to 1 million t/yr. It has a partnership with Dow Chemicals to produce 12,000,000 lb/y specialty polymer from epichlorohydrin.

Energy Biosystems (Woodlands, Tex.) has developed oxidases enzyme to convert crude oil into water-soluble molecules so that they can be separated from oil base. They have also patented enzymes to desulphurize crude oil.

Genencor has built whole cell microbial systems to synthesize 2-keto-L-gluconic acid (an intermediate to ascorbic acid) from glucose. Capital costs are estimated to be half that of plant based on chemical process.

Nitto chemical industries in collaboration with Kyoto University have scaled up a process for manufacturing 35,000 t/y of polymer derived from acrylonitrile, eliminating the need for palladium and platinum catalysts. Dupont has licensed Nittos A4-nitrohydroamylase for converting adiponitrile to 5-cyanovaleramide (a herbicide). The enzyme process generates only 0.006 lb of waste when compared to 1.26 lb of waste/lb of product generated by the chemical route.

A business group of BASF, ChiPros, produces annually 2500 t of methoxy isopropyl amine, an intermediate for its corn herbicide through biochemical route. Degussa manufacturers several thousand tons per year of biocatalytic acrylamide in Perm, Russia, for water treatment applications. Its Care Specialities uses biocatalysis to produce fatty-acid-derived easters and ceramides for personal care applications.

Dupont has developed a fermentation technology for 1,3-propanediol (PDO), the intermediate in the manufacture of polytrimethylene terephthalate, from corn sugar. PDO is the key component of DuPont's Sorona 3GT polymer. Cargill Dow's polylactic acid polymer uses lactic acid produced by the fermentation of corn sugar. This process uses 20% to 50% less energy than the conventional plastics.

Japan has tremendous expertise in fermentation because of its unique foods and beverages. Mitsui chemicals produce acrylamide through a bio-catalytic process, while Mitsubishi Rayon manufactures acrylamide through enzymatic hydration, which are four times cheaper than the chemical process. This process is also licensed by SNF (France) for producing flocculent for water treatment.

14.6.1 Approach Through Kinetic Resolution

BASF produces about three dozen amines by enzyme-catalyzed resolution. Racemic amine is treated with ethyl methoxy acetate and a lipase to produce a mixture of (R)-methoxy acetamide and (S)-amine, which are separated by distillation (Fig. 14.7). BASF has a 5,500,000-lb/annum chiral amine plant in Geismar, La. and a 2,200,000-lb/annum plant in Germany.

BASF also has technologies to produce enantiomeric alcohols, resolved by lipase-catalyzed acylation. For example, lipase-catalyzed reaction between

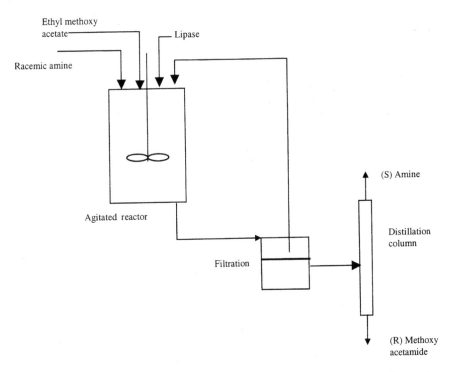

Figure 14.7 Schematic for the synthesis of lipase-catalyzed resolution of racemic amines.

racemic α-phenylethyl alcohol with vinyl propionate yields (S)-alcohol and (R)-propionate, which are separated by distillation (Fig. 14.8).

BASF has technology for producing (R)- and (S)-mandelic acid by kinetically resolving the reaction between benzaldehyde and hydrogen cyanide using the enzyme nitrilase.

DSM's expertise in biocatalysis and synthetic organic methods has led to technologies for resolving nonnatural amino acid (like 2-amino-4-pentenoamide) using aminopeptidase to (S)-acid and (R)-amide. DSM also produces R-glycidol butyrate using porcaine pancreatic lipase by kinetic resolution. Glaxo developed a process to resolve (1*s*,2*s*)-*t*-2-methoxy cyclohexanol, an intermediate for the synthesis of β-lactam antibiotic.

Degussa–Hills (Germany) prepares L-neopentylglycine from 4,4-dimethyl-2-oxopentanoic acid using lencine dehydrogenase and uses acylases enzyme for kinetic resolution of N-acetyl amino acids.

The simulated moving bed (SMB) separation of racemic mixtures into optically pure components is becoming increasingly important in the pro-

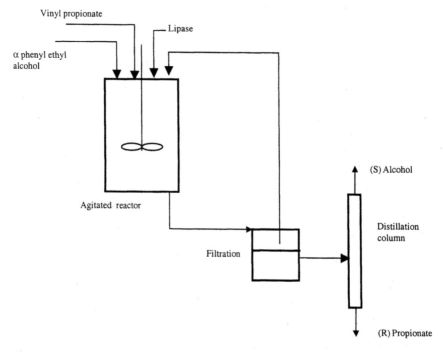

FIGURE 14.8 Schematic for the lipase-catalyzed resolution of enantiomeric alcohols.

duction of pharmaceutical intermediates since there is a countercurrent contact between the mobile fluid phase and the stationary adsorbent phase, leading to fast, cheap, and efficient separation performance. SMB makes use of the chromatography approach for separating racemic mixtures. High selectivity and yields could be obtained by this approach, and also the unit operation is carried out at ambient conditions. Two main disadvantages are that this operation is more complex than biocatalytic approach and productivity although better than biocatalyst is poorer than chemical means. The traditional chemical means involve high temperature operation and medium selectivity but leads to high productivity, yield, and simple downstream operation.

14.7 CHIRAL DRUGS AND INTERMEDIATES

Chiral drugs business in 2001 reached $100 billion, representing one-third of all drugs sales worldwide. The industry's continuing growth is partly because of the discoveries in fundamental biochemistry. Of the top 100 drugs marketed today, 50 are single isomers. Single isomer drugs sale reached $115 billion worldwide in 1999 (32% of the total drug market) and expected to reach $146 billion in 2003. All this activity on chiral drugs has led the fine chemical producers to develop new enantioselective technology to produce chiral intermediates.

Amgen leads in the biopharma industry with more than $3.3 billion in annual revenue and more than $1.7 billion in profit (Oct 2000). Amgen acquired Kinetix, which focuses on inhibition of protein Kinases, enzymes that are key regulators of internal and external cellular communications.

Lilly Research Lab changed its synthesis of anticonvulsant drug from the chemical route to a biocatalytic route, to eliminate 340 l of solvent and 3 kg of chromium waste produced by the former approach. The new synthesis involves biocatalytic reduction of an alcohol to an optically pure alcohol using the yeast zygosacharomyces rouxii. The reaction is carried out in an agitated vessel. The ketone is added to an aqueous phase containing a polymeric resin, buffer, and glucose. Most of the ketone gets absorbed onto the resin, and the yeast reacts with the equilibrium concentration of the ketone in the aqueous phase. The product once again gets absorbed on to the resin surface. This novel triphasic system helps product recovery and keeps the product concentration low in the aqueous phase, preventing enzyme deactivation based on product inhibition.

Monsanto's NSC Technology produces D-phenylalanine and D-tryosine (intermediates for a series of derivative products that have use in HIV treatment, oncology, and cardiovascular drugs) using D-amino acid transaminase enzymes.

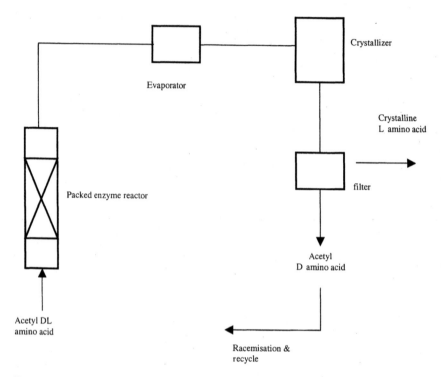

FIGURE 14.9 Schematic for optical resolution of racemic alcohols.

Ajinomoto, a Japanese company, manufactures anti-HIV agent dideoxyinosine through a hybrid process that involves chemical synthesis of dideoxyadenosine from adenosine followed by enzymatic conversion to the desired product. L-Glutamic acid is produced annually at 300,000 t scale by the same company by fermentation.

BASF is producing vitamin B2 through fermentation. Diversia, a San Diego-based company has nitrilases useful for enantioselective one-carbon homologation of aldehydes, preparation of (R)-mandelic acid derivatives at enantiomeric excess (ee) of >99%, preparation of (S)-phenyl lactic acid at 99% ee, and enantioselective preparation of ethyl (R)-3-hydroxy-4-cyanobutyrate an intermediate in the synthesis of the cholesterol-lowering drug Lipitor.

Enzymatic resolution in the pharmaceutical intermediates business include DSM Andeno's (The Netherlands) production of intermediates for the heart drugs Diltiazem and Captopril, Chiro Techs' production of a single isomer form of the analgesic ibuprofen and (-) lactam for the HIV drug abacavir, BASF's production of optically active amines using lipase in organic solvents, and Lonza's conversion of 2.5-dimethyl pyrazine to 5-methyl-2-pyrazine carboxylic acid, an intermediate for the diabetic drug glipzide using a microorganism that grows on p-xylene.

Tanabe Pharma (Osaka, Japan) uses lipase for the production of intermediates for Diltiazem. The reactor consists of immobilized lipase on membrane material. The product is recovered from the solution in a crystallizer, in the form of pure solid. Tanabe Seiyaku and Kyowa Hakko Kogyo (Japan) manufacture. L-Aspartate from fumaric acid using aspartate ammonia lyase.

Daicel produces chiral alcohols, like (S)-2-butanol and agrochemicals intermediates, using biocatalytic route. Intermediates such as amino acids and carboxamides for pharmaceuticals, fine chemicals, food products, and cardiovascular drugs are also being produced using biocatalyst by companies such as Zeneca Specialities, Daicel, Mitsubishi, SmithKline Beecham, etc.

L-Amino acid is used in medicine, food, and animal feed. At present, fermentation and chemical synthetic methods are followed to produce L-amino acid. The chemically synthesized amino acid is a racemic mixture of L and D, which needs to be optically resolved to obtain the pure L form through hydrolysis. Aminoacylase supported on DEAE-Sephadex ion-exchange resin through covalent bond was used for this purpose. A process scheme is shown in Fig. 14.9. The main issues in the manufacturing process are pressure drop across column and slight change in the pH of the native enzyme when supported on ion-exchange resin leading to decrease in its active life. The same technique is used in the manufacture of several D-amino acids.

D-Phenyl glycine and D-4-hydroxy phenyl glycine, the precursors for semisynthetic penicillins, namely, ampicillin and amoxicillin, are currently being produced at a scale of higher than 1000 t/annum using hydantoinase/carbamoylase. NSC Technology has a process for multi-ton production of D-phenylalanine and D-tyrosine using D-transaminase. Kyowa Hakko Kogyo has a process for producing D-alanine using L-glutamate using glutamate racemase and L-glutamate decarboxylase by fermentation. Degussa (Hanau, Germany) has a process for resolving (S)-t-leucine using a leucine dehydrogenase with a turn over number above 10,000.

15

Waste Treatment

Gas, liquid, and solid waste can be treated through biochemical means, either in situ or ex situ. In the ex situ process the contaminated soil or fluid is collected and treated in an external location, whereas in the in situ process the treatment is carried out in the same place. The treatments could be classified as aerobic or anaerobic depending on whether the process requires air or not. Anaerobic processes consume less energy, produce low excess sludge, and maintain enclosure of odor over conventional aerobic process. This technique is also suitable when the organic content of the liquid effluent is high.

The aerobic biodegradation process can be represented by the following equation:

$$C_xH_y + O_2 + (\text{microorganisms/nutrients}) \rightarrow H_2O + CO_2 + \text{biomass}$$

Anaerobic bioprocess can be described by the following equation:

$$C_xH_y + (\text{microorganisms/nutrients}) \rightarrow CO_2 + CH_4 + \text{biomass}$$

Several physical and chemical treatment techniques are also available for treating sludge, soil, water, and air. These techniques are more mature than the biotreatment procedures and are discussed first. This chapter does not deal with the various waste treatment techniques in detail but just introduces them to the reader.

15.1 CHEMICAL TREATMENT

15.1.1 Soil

The in situ physical and chemical treatment methods for contaminated soil, sediment, bedrock, and sludge include chemical oxidation, electrokinetic separation (this technique involves passing an electric current during which the metal ions move toward the cathode and the anions toward the anode; in addition the current creates an acid front at the anode and a base front at the cathode), fracturing (achieved using explosives or pneumatic pressure), soil flushing using solvent and cosolvents, soil vapor extraction, chemical fixation (involves fixing the contaminant to the soil and making it innocuous), and thermal treatment processes like electrical, radio-frequency or electromagnetic heating, and hot air or steam injection. Cleanup times in all cases are generally low when compared to bioprocesses. Except for chemical fixation and cementing (also known as *stabilization* or *solidification*), the operating and maintenance cost of other processes are more than those of biochemical technologies. In situ fracturing and solidification and stabilization processes are also capital intensive.

A large number of ex situ physical and chemical treatment technologies are available which include chemical extraction using acids or solvents, chemical conversion like reduction, oxidation, or dehalogenation, separation using magnetic or physical means, soil washing, solidification, and stabilization (such as bituminization, cementing, sludge stabilization, molten glass formation, etc.). The cleanup times in all these technologies are very short, and they can be generally favored for waste treatment over the biological treatment procedures. Of course, they are capital intensive and also require operation and maintenance cost. Containment technologies include landfill capping using asphalt and concrete caps (landfill capping creates a barrier between the contaminated soil and the environment), vegetative cover, and water harvesting.

15.1.2 Groundwater

In situ physical and chemical treatment methods for groundwater and surface water contamination include air sparging, chemical oxidation, fluid–vapor extraction, thermal treatment techniques such as hot water or steam flushing, and electrical heating. Although the technologies are mature, they need operating and maintenance cost.

Several ex situ physical and chemical treatment methods are well developed and available for treating contaminated groundwater that include adsorption or absorption of the contaminants on activated alumina, clays, carbon, synthetic resins, etc; oxidation using UV or photolysis; air stripping using surfactants for enhancing recovery; ion exchange; and separation techniques like precipitation, coagulation, flocculation, distillation, filtration, freeze drying, reverse osmosis, pervaporation, and freeze crystallization. These technologies are capital intensive and involve operating cost. A few containment approaches are also available that involve building physical barrier walls.

15.1.3 Air

In the case of contaminated air, techniques like membrane separation, water or caustic scrubbing, vapor-phase carbon adsorption, or thermal, ultraviolet, or catalytic oxidation are adopted.

15.2 SOIL BIOREMEDIATION

The toxicity of chemical contaminants can be reduced by transforming, degrading, or fixing it using microorganisms or plants. This activity known as *bioremediation*, is based on the concept that all organisms remove substances from the environment for its own growth and metabolism. For example, bacteria and fungi are very good at degrading complex molecules, and the resultant wastes are generally safe. Fungi can digest complex organic compounds that are normally not degraded by other organisms. Algae and plants are ideal at absorbing nitrogen, phosphorus, sulfur, and many minerals and metals from the environments. Many algae and bacteria produce secretions that attract metals that are toxic in high levels and form complex products, which are harmless. Bioremediation can be carried out in situ or ex situ similar to the chemical treatment procedures.

   The rate at which microorganisms degrade contaminants depends on type of contaminants present, oxygen supply, moisture, nutrient supply, pH, temperature, the availability of the contaminant to the microorganism, the concentration of the contaminants, and the presence of substances toxic or inhibitors to the microorganism.

   Alcanivorax spp. and *Acinetobacter calcoaceticus* may be important contributors to hydrocarbon degradation in oil-impacted marine ecosystems.

   In situ treatment allows soil to be treated without being excavated and transported to the treatment plant. Although it is not expensive, it generally requires longer time periods, and there is less certainty about the uniformity of treatment. Available in situ biological treatment technologies include bioventing, natural attenuation, enhanced biodegradation, and phytoremediation.

   Bioventing is an in situ biodegradation process where oxygen at low flow rates is injected directly into the soil through pipes to the existing microorganisms present in the soil. This operation will aid in the growth of the microorganism and at the same time decrease the contaminant amount. This is a medium-to-long-term solution, ranging from a few months to several years.

In enhanced bioremediation process microorganisms (such as fungi, bacteria, and other microbes) are inoculated from external sources to grow and degrade organic contaminants found in soil or ground water. Nutrients, oxygen, or other materials (other oxygen sources such as hydrogen peroxide) are added to enhance bioremediation and contaminant desorption from subsurface materials. This is a long-term technology that take several years and is not effective in cold climates, since lower temperature retards the rate of the degradation process. Of course, the microorganism may find the new environment hostile for its growth. Bioaugmentation involves the use of microbial cultures that have been specially bred for degradation of specific contaminants or contaminant groups. For example, they may be collected from the site itself and returned to the site later after it has grown in larger number in a controlled laboratory environment. Cometabolism uses microorganisms that are grown on one compound to produce an enzyme that chemically transforms another compound on which they cannot grow.

Phytoremediation is a process that uses plants to remove and destroy contaminants in the soil, and this is done through several means. Enhanced rhizosphere biodegradation takes place in the soil immediately surrounding the plant roots because of the release of nutrients from them to the microorganisms. The roots also loosen up soil leading to easy air ingress. Commonly used flora is poplar tree. In phytoaccumulation the plant roots draw contaminant and accumulate it in their shoots and leaves. In phytodegradation the contaminants are degraded by the plant tissues by the enzymes (like dehalogenase and oxygenase) they produce. In phytostabilization chemicals that are produced by plant helps to immobilize contaminants at the interface of roots and soil.

Electrokinetics involves using electricity to move nutrients, water, and heat into the contaminated soil. Electrodes are placed in the soil and current is passed, which heats the soil while delivering nutrients. Electrolyte pumps move nutrients and water into the soil where they are passed between the anode and the cathode. The heat and additional food source provide the bacteria ideal conditions for growth.

The main advantage of ex situ treatments is that it requires shorter time periods than in situ treatment, and there is more certainty about the uniformity of treatment, however, they require excavation of soils, transportation, etc., leading to increased costs and suitable material handling equipments and possible toxic exposure to workers. Available ex situ biological treatment technologies include biopiles, composting, land farming, and slurry-phase biological treatment.

In biopile treatment excavated soils are mixed with soil nutrients and placed on a treatment area that includes collection systems and aeration. The treatment area will generally be covered or contained with an impermeable liner to minimize contaminants leaching into uncontaminated soil. This is a short-term operation involving a few weeks to few months.

Composting is a biological process in which organic contaminants are converted by microorganisms (under aerobic and anaerobic conditions) to innocuous and/or stabilized by-products. There are three process designs used in composting: aerated static pile composting (in which compost is formed into piles and aerated with blowers or vacuum pumps), mechanically agitated in-vessel composting (in which compost is placed in a reactor vessel where it is simultaneously agitated and aerated), and windrow composting (in which compost is placed in long piles known as *windrows* and periodically mixed and turned using mobile equipment).

Land farming is a bioremediation technology where the contaminated soil, sediments, or sludges are excavated and placed on plastic liners to control leaching of contaminants and periodically turned over to aerate. The rate of degradation is controlled by moisture content, aeration, pH, and other nutrients added externally.

Slurry-phase treatment involves treatment of excavated soil in a bioreactor. The excavated soil is first processed to separate stones and rubble and mixed with predetermined amount of water, so that the slurry contains 10% to 30% solids by weight. Nutrients and other chemicals are added to the bioreactor. Oxygen is bubbled if the process is aerobic in nature. The solids and the liquid separated in clarifiers, pressure filters, vacuum filters, sand drying beds, or centrifuges. Bioreactors are short- to medium-term technologies.

The limitations of bioremediation are

1. Organisms which can thrive on the contaminant need to be developed in the lab.
2. Some environmental modifications need to be done for the organism to thrive in the contaminated soil, which could affect other organisms in that environment.
3. Costs involved are high, and the process is time consuming and labor intensive.

15.3 BIOTREATMENT OF LIQUID WASTE

The main advantage of in situ treatment is that the ground water can be treated without being brought to the surface, resulting in significant cost savings, but the disadvantages are they require longer time periods, and verification of the uniformity of treatment is difficult. In situ bioremedetion consists of the steps like nutrient injection, sampling, and above-ground pumping and air stripping, and finally reinjection. For example chloroethenes can be reduced to ethane using this approach using sodium lactate as the nutrient.

Generally aerobic lagoons or ponds, activated sludge, trickling filter, inverse fluidized bed, etc., are used in aerobic processes. Trickling filter consists of a bed of an inert material on which the microorganisms are attached. A distributor distributes the wastewater over the top of the filter medium, and its flow is cycled on and off at a specified dosing rate to ensure that adequate supply of oxygen is available to the microorganisms.

Cyanide-eating bacteria include *Pseudomonas putida, Pseudomonas pickettii, Pseudomonas paucimobilis, Klebsiella pneumoniae, Klebsiella* sp., and *Alcaligenes xylosoxidans* as well as enzymes such as cyanase. Ammonia and carbon dioxide are the major end products of aerobic and formate and bicarbonate of anaerobic CN metabolism.

Cyclic bioreactor is another concept that is used in aerobic biochemical treatment. A control parameter, which is a representation of the biological process (biometabolism), is monitored, and when it is inferred that the nutrient level is low to support growth, then half of the reactor contents are emptied and replaced with fresh sterile medium. For example, in the treatment of effluent containing phenol, toluene, and xylene, dissolved oxygen concentration is taken as the control parameter, and emptying and filling operation is initiated when the dissolved oxygen concentration reaches a minimum.

A rotating biological contactor is another aerobic reactor for waste liquid treatment and it consists of a series of parallel disks (made of some polymeric material) held close to each other that rotate at 2 to 5 rpm, while half submerged in a semicylindrical trough containing wastewater (Fig. 15.1). Each disk is covered with a biological film that degrades dissolved organic constituents present in the wastewater. As the disk rotates slowly, it carries a film of the wastewater into the air, where oxygen is available for aerobic biological decomposition. The excess biomass produced disengages from the disk and falls into the trough. Several contactors are often operated in series. These are inexpensive and are inexpensive to operate.

Stirred tank, stirred tank with external cell recirculation by sedimentation and settling tank, sludge bed reactors, fixed-bed loop reactors, and fluidized-bed reactors are used for anerobic processes. High-rate anaerobic digestors retain the biomass and require less area. Some of the designs here are up-flow anaerobic sludge-blanket, expanded granular sludge-bed, up-flow or down-flow stationary packed-bed, and fluidized- and/or expanded-bed reactors.

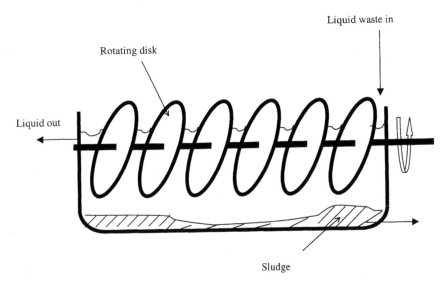

Liquid waste in

Rotating disk

Liquid out

Sludge

FIGURE 15.1 Rotating biological reactor (for aerobic applications).

A residual recalcitrant portion that is resistant to microbial degradation often limits the efficiency of biological effluent treatment systems. Recalcitrant organics are quantified in terms of chemical oxygen demand and color. Anaerobic fluidized bed reactors can handle hazardous waste and inhibitory recalcitrant compositions. A typical anaerobic fluidized-bed design is shown in Fig. 15.2a. The liquid fluidizes the supported biocell. Liquid flow should not be high to disengage the biofilm attached to the inert support.

An inverse fluidized bed is used in aerobic wastewater treatment (Fig. 15.2b), where the solid phase is an inert particle coated with a biofilm, the gas phase is oxygen and/or air and the liquid phase is the wastewater that needs treating. The bed of solids has a density lower than that of the liquid phase, but they are made to be in a fluidized state by the downward flow of the liquid. The gas phase flows countercurrent to the liquid. This mode of operation improves mass transfer rate, reduces attrition rate of solids, and helps to refluidize easily after shutdown. Low concentration synthetic and municipal wastewaters are treated at residence times ranging from 0.6 to 3 h, respectively, in an anaerobic inverse fluidized bed. Sufficient care should be taken during start-up, when the biofilm is forming on the inert support.

An inverse fluidized bed with a draft tube (Fig. 15.2c) is used to degrade phenol. In this design air is introduced at the bottom of the draft tube, thereby creating an up flow in the draft tube and a down flow in the annular

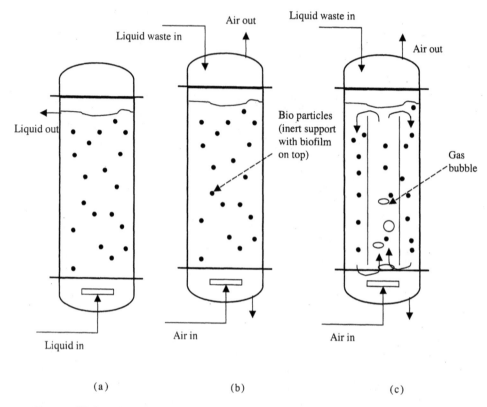

Liquid waste in

Air out

Liquid waste in

Air out

Liquid out

Bio particles
(inert support
with biofilm
on top)

Gas
bubble

Liquid in

Air in

Air in

(a)   (b)   (c)

FIGURE 15.2 Different fluidized bed reactor designs: (a) anaerobic conventional, (b) gas–liquid–solid aerobic inverse fluidized bed, and (c) Same as design (b) but with internal draft tube.

space. Inert particles, which are less dense than water, are fluidized in the annulus section, due to the downward flow of water. When microorganisms are injected, they attach themselves to the inert support over a period of time and grow forming a biofilm.

The effectiveness of the design and opeation of a biological treatment system depends on amount of nutrients available for the organism to grow, dissolved oxygen concentration, food-to-microorganism ratio (this ratio applies to only activated sludge systems; it is a measure of the amount of biomass available to metabolize the influent organic loading to the aeration unit), pH, temperature, cell residence time, hydraulic loading rate (the length of time the organic constituents are in contact with the microorganisms), settling time (time for separating sludge from liquid), and degree of mixing.

Reactor designs that are specifically used in effluent treatment are listed below with example. This does not prevent the use of other reactor designs described in Chapter 8.

Liquid-phase slurry reactor: high-molecular-weight polycyclic aromatic hydrocarbons in soil
Fed batch: decolorization of azo dye
Gas-phase packed bed: perchlorate degradation
Ion-exchange membrane: water denitrification
Biofilm: penta chlorophenol degradation
Packed bed: hexavalent chromium degradation
Inverse fluidized bed: phenol degradation
Membrane: biofiltration of air containing propene
Fluidized bed: anaerobic degradation of distillery waste

15.3.1 Domestic Water Treatment

The domestic wastes generated is generally collected in sewer systems and transported. In the United States alone the wastewater generated per day is of the order of 14×10^9 gal. While about 10% is passed untreated into rivers, streams, and the ocean, the rest receives treatment to improve the quality of the water before it is released for reuse.

The biological oxygen demand (BOD) of drinking water should be less than 1 while that of raw sewage may be of the order of several hundred. BOD is a measure of water quality, and it indicates the amount of oxygen needed (in milligrams per liter or parts per million) by bacteria and other microorganisms to oxidize the organic matter present in a water sample, over a period of 5 d. The simplest method of treatment is to remove undissolved solids, grits through filtration, followed by settling in sedimentation tanks or clarifiers to form sludge, which removes one-third of the BOD. Then the effluent is reacted with oxygen and aerobic microorganisms. This operation breaks down much of the organic matter to harmless substances such as carbon dioxide. Primary and secondary treatment together can remove up to 90% of the BOD (Fig. 15.3). Later chlorination is carried out to remove bacteria. Tertiary treatment is carried out to remove dissolved inorganic nitrogenous and phosphorus salts.

The oxidation tank consists of two sections, an aeration chamber and a settling chamber. Air is bubbled through the first chamber. The residence time of the wastewater here is about 6 to 24 h. The activated sludge developed in the aeration chamber flocculates and settles down. It readily separates from treated wastewater. The settling chamber receives the overflow of the aeration chamber. The sludge settles to the bottom of the clarifier, which is still active,

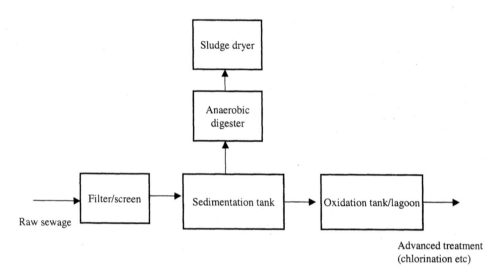

FIGURE 15.3 Water purification system.

and is sent back to the aeration chamber to remove more BOD from the wastewater. This recycled biomass is referred to as *activated sludge*. Occasionally, some of the sludge should be drained to keep the effluent from deteriorating. The cleaner water at the top of the settling chamber overflows through openings at the top of the chamber, which can be treated further or reused.

The sludge treatment involves thickening, by gravity or floatation, to remove as much water as possible, followed by stabilization, which converts the organic solids to more inert forms. Finally, the inert solid is dried.

Anaerobic treatment of sewage waste has several advantages over the conventional aerobic process: (1) produces methane gas, which can be used as a fuel, and (2) is cost effective because the final volume of the waste generated is very less. The digester is similar to the aerobic activated sludge process—often a set of reactors in series with recycle. The up-flow anaerobic reactor allows the waste to flow up from the bottom, and the gas collection chamber is located at the top. In the anaerobic activated sludge process the bioreactor and clarifier are placed in series. Submerged media anaerobic reactors (SMAR) are similar to the up-flow bioreactor: in addition, they have packed bed of rocks which supports bacterial growth. The Fluidized-bed SMAR uses smaller particles as support material, which can be fluidized during operation. In both systems the gas and liquid effluent are separated at the top of the reactor.

15.3.2 Water Purification Using Membranes

Membrane technology finds appliction in process waste treatment and water purification. In the process waste treatment, membranes help to concentrate the waste stream by selectively passing the water and retaining the waste material. As water treatment costs keep increasing membrane technology will play a major role in all areas of water recovery in the near future. Several types of membranes like microfiltration, ultrafiltration, nanofiltration, and reverse osmosis are available for various applications. Pressure needs to be applied to force the fluid through the pores of the membrane, and it increases with decrease in pore size. Microfiltration membranes are porous and operate at 3 to 50 psig. and are ideal for removing suspended particles and bacteria from liquids. They are also used in pharmaceutical industry for cell harvesting and in the brewing industry for the cold sterillization of beer. Ultrafiltration removes oils, colloidal solids, and other soluble pollutants and allows for recycling of industrial waters. This operates at 15 to 200 psig. It works well on waste streams with compositional variability reducing by up to 98% the amount of waste to be treated or discharged. Nanofiltration membranes can retain between 100 and 1000 molecular weight at applied pressure between 75 and 450 psig. Reverse osmosis membranes have very small pores, thereby passing only pure water and rejecting ions from passing through and are used to produce drinking water from sea water. They operate at 200 to 1000 psig.

Reverse osmosis membranes that are commercially available include cellulose acetate, and aromatic polyamides. Ultra filtration polymers available include polysulfone, cellulose–acetate blends, and fluorinated polymer. Microfiltration membranes commercially available are polypropylene, acrylonitrile, nylon, and PTFE. The operating temperature range is generally less than 90°C and over a wide pH range. Commercially available configurations for reverse osmosis and ultrafiltration systems include tubular, larger internal-flow hollow fiber or "spaghetti bundle," plate and frame, and spiral wound. Fouling of membrane due to the presence of sludge and other solid material is an important issue in treatment technology, which has not been fully overcome.

15.4 BIOLOGICAL WASTE AIR TREATMENT

Two different approaches are followed for treating waste air. The first treatment method consists of use of biofilters, trickling filters, or bioscrubbers. In all these setups microorganism is supported on solid carrier and the waste air component is destroyed when it comes in contact with the biomaterial. The contaminated air flows over the bed on which biocultures that degrade the pollutants are supported and during its flow pollutants are transferred from the gas to the biofilm and are degraded. Biofilters and biotrickling filters are successful in the treatment of dilute, high flow waste gas streams containing odors or volatile organic compounds (VOCs). Biofilters are compost beds and work with humid air, whereas biotrickling filters are packed with inert materials and include a mineral nutrient solution trickling downward. VOCs like methyl ethyl ketone and methyl isobutyl ketone can be scrubbed successfully using these methods. Biofilters are not efficient under continuous operation for the removal of high concentration of acid-producing effluents such as chlorinated solvents. Bioscrubbers are ideal for such pollutants, and the degradation of the absorbed pollution takes place either simultaneously or subsequently in another vessel. Apart from biotrickling filter, three-phase air-lift reactors are also used for this purpose, whereas the former could be 50% cheaper than the latter.

The second treatment method (Fig. 15.4) consists of two parts: the first involves the absorption of the waste air components in a solvent (generally water), followed by the destruction of these in the liquid phase using microbial oxidation. Three-phase bubble column or stirred tank reactors or aerated lagoons can be used to treat the liquid containing the absorbed toxins in the waste air.

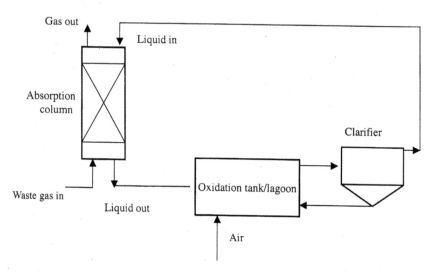

FIGURE 15.4 Waste air treatment technique.

16

Scale-Up of Biochemical Processes

Once a process is established in the laboratory in grams scale it is taken through a pilot plant followed by operations in a "semitech" lab or demonstration unit, where material is handled in kilograms or even in tons scale, before it reaches the manufacturing plant. Scale-up is aimed at finding and collating all the information and data required for the design, construction, and start-up of a new industrial unit. In addition, safety, hazard, and handling of chemicals in large scale are also gauged during this operation.

The science of scale-up for biochemical processes encompasses not only knowledge of biochemistry and biotechnology but several branches of chemistry and engineering including organic and physical chemistry, process chemistry, chemical engineering, and fluid mechanics.

Issues relating to (1) kinetics, (2) thermodynamics, and (3) hydrodynamics emerge during scale-up, and they have an impact on the system performances.

1. Kinetics issue relates to reactor selection, mode of operation, hydrodynamic characterization of the fluids of interest, feed description, product flexibility, and reaction scheme.
2. Thermodynamics relates to densities, enthalpies, solubility, adsorption, heat of formation, preheat requirement, mixing needs, and limitations.

3. Hydrodynamics encompasses flow regime, three-phase holdups, bubble characteristics, solid properties, gas–liquid interfacial phenomena, flow gradient, backmixing, and interphase transport.

Scale-up of biotechnological processes pose extra problems in addition to the problems encountered in chemical processes, namely, maintaining optimum environmental conditions like pH, concentration, mass transport, mixing, shear fields at all scales, maintaining aseptic environment, strict temperature control and regulation, containment, etc. Stability of the biocell, mold and enzyme due to mechanical agitation and collision with air and/or gas bubbles is another issue that is unique to biochemical processes and needs to be addressed. With increasing scale of operation the agitator tip speed increases leading to shearing of the biocell. Also, with increasing scale the average gas bubble size increases leading to greater damage to the biocell.

16.1 LABORATORY VERSUS PLANT

A reaction carried out in a round-bottom flask in the research laboratory at a very high speed of agitation or a shaker flask might give entirely different yields in a large cylindrical-shaped vessel in a plant that is provided with a slowly revolving mixer. A few of the reasons for this variation are due to the difference in degree of agitation, wall effects, and the ratio of surface to the volume of the vessel.

16.1.1 Batch Cycle Time

The overall processing time not only includes the reaction time but also the times for charging of raw materials, heating the reactants, cooling the products, and discharging. These times are generally negligible in the lab but are often longer in the plant, of the order of several hours. This is very relevant in batch fermentors. For example, it takes about 1.5 to 2 h to fill a 10 kl vessel via a 5-cm-diameter pipe. Cooling the contents of a reactor could be of the order of 1 to 2 h. Temperature-sensitive chemicals will pose problems during scale-up because of the extended cooling times leading to product degradation.

16.1.2 Surface Area to Volume

During scale-up the reactor surface-area-to-volume ratio decreases and hence similar rate of heat input and removal cannot be achieved. A volumetric scale-up factor of 10,000 means a reduction in surface-area-to-volume ratio by 21.5. Thus to input or dissipate the same amount of heat, the heat transfer rate must be increased in proportion to the vessel size. This can be achieved by incorporating additional heat transfer area (like internal coils) or higher heating medium temperatures. The latter can lead to increased skin temperature, with the possibility of localized overheating and loss in activity of biocatalyst or formation of tarry or polymeric material.

16.1.3   Fluctuation in Operating Conditions

In the laboratory in gram scale, it is possible to maintain exactly all the operating conditions, i.e., temperature, pressure, addition rate, raw material quantities, etc. Whereas, in the plant strict control may not be always possible, thereby affecting the product yield and selectivity. The ruggedness of the process to variation in operating conditions needs to be studied in the lab by the process chemist before taking it up to larger scales.

16.1.4   Raw Material Purities

In the lab pure or high-grade chemicals are used during process development, whereas manufacturing is generally carried out with industrial grade chemicals. The impurities generally found in the raw material has to be identified and their effects on the process have to be systematically studied. Some of the impurities may be detrimental to the process in the form of poison or inhibitor to the biocatalyst. At times, it may be necessary to carry out pre- or postpurification to remove these impurities. Also quality of the same raw material from different sources may be different, which may affect the reaction chemistry.

16.1.5   Agitation or Mixing Efficiency

Agitation differs with reactor scale. If we want to maintain the same circulation time in the big tank, then we must have higher velocities than on the small scale, because the liquid must travel a greater distance. Hence power per unit volume on the full scale must go up in proportion to the square of the diameter of the tank, which is unrealistic. Thus liquid circulation time in a big tank is generally longer than in a small tank. A good test for the sensitivity of the reaction to bulk mixing time is to perform the reaction in a geometrically scaled-down model of the full-scale vessel with the agitation speed same as on the full scale. The results obtained at these conditions are a good indication of the full-scale performance.

In fermentation studies most screenings are done in shake flasks mounted on rotational, shaking, or rocking devices. There is no relationship between the pumping capacities and shear rates in these apparatuses with the full-scale fermentor provided with a standard agitator. It is impossible to make all the mixing parameters of the full-scale vessel equal to the individual fluid mixing and fluid mechanics variable in a small-scale tank. For example, if the power per unit volume is to be kept constant between a 5-gal pilot plant vessel and a 625-gal plant-scale vessel, then the power has to be increased by a factor of 125, which is impossible to achieve. If one likes to maintain the same agitator tip velocity during scale-up, then one needs to raise the agitator power by a factor of 25.

Scale-up of aerobic bioreactors are carried out based on oxygen transfer rate, since this could be the rate-limiting step. For example, during *A. niger* fermentation in a stirred tank reactor, oxygen transfer is of the order of 10 s in small scale and about 100 s in a large vessel. This rate could be increased by inputting more power to the impeller, applying head pressure or back pressure to increase oxygen solubility, or use of pure oxygen instead of air.

16.2   SAFETY AND HANDLING

Scale-up brings with it problems of handling of toxic, hazardous, obnoxious-smelling chemicals and powders. The handling in a lab is generally done in a fume hood thereby not exposing the chemists to these hazards. In a plant apart from handling of large volumes of chemicals, human exposure every day for prolonged time periods takes up important dimension. Long-term exposure of operators to chemicals need to be ascertained.

Scientific procedures applied to the safety of evaluation of microbial enzymes for food processing include in vivo and in vitro toxicological studies for suspected toxic components in the enzyme preparation. The safety review will go through

1. Safety assessment of the production strain
2. Identity and quality of new materials used in fermentation
3. Processing techniques used in recovery
4. Identity and quality of stabilizers, diluents, and formulation aids
5. Toxicological studies on enzyme concentrates
6. Determination of safety margin

Several differences exist between chemicals and microbes, which need to be considered during the health and safety assessment process. In microbial process multiple exposures could lead to immunity, which is not true in the case of chemical exposure; namely, the person may not get immunity even after several exposures. Microbes reproduce, may assume different strains and may cause primary or secondary infection, which is not the case with chemical exposure. The dose-versus-response relation in the case of chemical exposure is generally linear, whereas it is nonlinear in the case of exposure to microbes.

Potential toxicity of the microorganisms, biocells, or enzymes and toxicity of the degraded cells need to be studied and pathogens have to be identified. If the microbe is well known, nonpathogenic in nature, and needs no containment, then a simple vent gas scrubber needs to be provided to prevent noxious off-gas odors from escaping. Before disposal the microbes are made inactive by heat or chemical means (which involves caustic addition or pH swings). If the microorganism pose a moderate risk to personnel or the environment, then proper containment need to be provided.

16.3 EFFLUENT DISPOSAL

Disposal of large quantities of solid, liquid, and gaseous waste is an important step in a manufacturing scenario. This will involve treating the waste so that it is within the permitted disposal limits of the local authority. The treatment process could be a multistep chemical or biochemical operation involving resources and thereby escalating the production cost.

16.4 STERILITY AND ASCEPTIC CONDITIONS

Unlike a conventional chemical process biocatalytic process requires very sterile conditions. The bioorganisms are susceptible to impurities and toxins. For sterile operation the reactor should have simple geometrical shape, minimum number of flanges, openings, measuring and sampling nozzles, elimination of dead zones, smooth finish, facility to sterilize agitator shaft seal, and provisions for sterilizing the reactor and all components together and separately.

16.5 MIXER AND/OR AGITATOR SCALE-UP

The ratio of the impeller diameter to tank diameter is an important scale-up factor and the nature of agitation influences this ratio. For dispersing a gas into a liquid this ratio is 0.25, for bringing two liquids into intimate contact it is 0.4, and for blending this ratio is greater than or equal to 0.6. Operations that depend on large velocity gradients but low circulation rates are accomplished with high-speed small-diameter impellers, as in the case of gas dispersion in aerobic processes.

Operations, which require high circulation rates are ideally operated with large-diameter slow-moving impellers. In an approximate way it can be said that 0.2 to 1 hp per 1000 gal of thin liquid gives "mild" agitation, 2 to 3 hp per 1000 gal gives "vigorous" agitation, and 4 to 10 hp per 1000 gal gives "intense" agitation. These figures refer to the power that is actually delivered to the liquid and does not include friction and gear losses.

16.5.1 Scale-up Rules

Geometric similarity fixes the ratio of various lengths with the system. A single ratio (R) can be used to define the ratios of all linear dimensions between the large and small scale.

$$R = \frac{D_2}{D_1} = \frac{T_2}{T_1} = \frac{b_2}{b_1} = \frac{Z_2}{Z_1} \tag{16.1}$$

The rpm of the large vessel and the small vessel will be related as

$$N_1 = N_2 \left(\frac{1}{R}\right)^n \tag{16.2}$$

For equal liquid motion or constant Froude number, $n = 1$. This criterion also ensures constant agitator tip speed, which guarantees the velocities leaving the impellers in each case are the same. Impeller tip velocity determines the maximum shear rate, which in turn influences both maximum stable gas bubble size or microbial floc size and damage to viable biocells in biotechnological applications. This criteria also leads to similarity between the gravitational effects in the two vessels.

For equal solid suspension, $n = 3/4$.
For equal mass transfer rates, $n = 2/3$ (Aerobic processes require maintaining constant volumetric mass transfer coefficient (k_{La}) at all scales).
For equal surface motion, $n = 1/2$.
For equal blend time, $n = 0$.
For constant Reynolds number, $n = 2$. This criterion is used as an attempt to obtain hydrodynamic similarity between the two vessels. This scale-up criterion gives the same overall flow pattern but not equality of instantanous velocities.
For constant pumping capacity, $n = 3$.

16.5.2 Power Requirement

Power requirement per unit volume (P/V) under turbulent conditions is given as, $P \propto N^3 D^2$, and at laminar flow conditions, $P \propto N^2$. If power per unit volume is held constant in large and small scales, then agitator speed must change with respect to agitator diameter as

$$N_2 = N_1 \left(\frac{D_1}{D_2}\right)^{2/3} \tag{16.3}$$

During the scale-up of antibiotic fermentors the power per unit volume is kept constant, generally in the range of 1.3–2.2 kW/m3.

16.5.3 Gas–Liquid Mixing

Large tanks tend to have a wider bubble size distribution than smaller tanks. One major principle to be followed for the design of a mixing equipment is that the impeller blade must be two or three times larger than the largest bubble droplet, particle, or fluid "clump" that is of importance to the process. Blend time decreases in the presence of gas bubbles. Thus the percentage of gas holdup is an important factor in scale-up. The linear superficial gas velocity increases on scale-up, and there is usually a greater volume of gas holdup in big tanks than in small tanks. So the larger scale system has longer blend times. Another consideration is that the sizes of the bubbles should not exceed one-half to one-third of the vertical height of the impeller blade in the pilot scale.

The rpm required for completely dispersing a gas in a flat-blade turbine assembly is given by the equation

$$N^2 = Q_g T \frac{g^{0.5}}{0.2D^4} \tag{16.4}$$

where Q_g is the gas flow rate. Of course, the power requirement for agitation decreases with presence of gas. In other words, the motor power decreases as we bubble more gas.

16.5.4 Solid Suspension

Scale-up poses no problem for suspending solids up to about 30% by weight. However, concentrated slurries start behaving like viscous pseudoplastic material. The minimum rpm required for suspending solids is a function of particle diameter,

$$N_{rpm} \propto \frac{d_p^{0.2}}{D^{0.85}} \tag{16.5}$$

and power per unit volume of the liquid is proportional to $D^{-0.55}$.

Scaling up of biotechnology processes which use animal cells pose problems since these tear at increased shear rate. This happens near the tip of the blade (although average shear rate near the impeller zone is lower, the maximum shear rate, which is experienced by the cells at the tip of the agitator, is higher in larger tanks than in smaller tanks). Many approaches including encapsulating the organism in or on micro particles or conditioning the animal cells to withstand required shear rate are being attempted.

Both shear rate and circulation rate affect the solid–liquid mass transfer rate. Larger particles tend to slip behind the liquid motion, while smaller particles tend to follow the flow pattern. In the case of aerobic biochemical processes the living organism should have access to dissolved oxygen throughout the tank. Hence, top to bottom blending is especially crucial in fermentation reactors having tank height to diameter ratios of the order of 2:1 to 4:1.

16.5.5 Liquid–Liquid Emulsion

The average emulsion droplet size, d is proportional to $N^{-2.56} D^{-4.17} T^{1.88}$ and to achieve uniform dispersion, agitator rpm N should be proportional to $D^{-2.15}$.

16.5.6 Blending

Typical mixing times in small scales could be about 10 s, while the same in the large plant scale could be of the order of 100 s. Blending time t is proportional to TH/ND^2, and in geometrically similar tanks t is proportional to $1/N$.

16.6 HEAT TRANSFER SCALE-UP

In scaling up of many homogeneous stirred systems the rate of heat transfer controls the design of heat transfer equipment, i.e., $h \propto u^m d^{m-1}$, where h is the heat transfer coefficient and u is the fluid circulation velocity. m is generally of the order of 0.6714.

A scale-up based on equal heat transfer coefficient is desirable when pilot plant studies indicate that the heat transfer resistance increases slowly over a period of time, due to the formation of polymeric or tarry deposits along the heat transfer area, caused by the temperature-sensitive nature of the material:

$$\frac{U_2}{U_1} = \left(\frac{D_2}{D_1}\right)^{(1-m)/m} \quad \text{and} \quad \left(\frac{N_2}{N_1}\right) = \left(\frac{D_2}{D_1}\right)^{(1-2m)/m} \tag{16.6}$$

The power per unit volume in such a situation will scale up as

$$\frac{(P/V)_2}{(P/V)_1} = \frac{D_2^{(3-4m)/m}}{D_1^{(3-4m)/m}} \tag{16.7}$$

For $m = 0.67$, power per unit volume will be scaled up as one-half power of agitator diameter.

For highly temperature sensitive material it is often essential to assure that same rate of heat transfer is maintained in the large scale unit to avoid material degradation, i.e., $h_2/h_1 = D_2/D_1$, and for geometrically similar vessels,

$$\frac{N_2}{N_1} = \left(\frac{D_2}{D_1}\right)^{(2-2m)/m} \tag{16.8}$$

When constant velocity (i.e., constant tip speed in agitated vessel) is desired, then $h_2/h_1 = (D_2/D_1)^{m-1}$. Dynamic similarity requires equal tip Reynolds number for turbulent flow which guarantees similar maps of local to average rates of energy dissipation. This leads to

$$\frac{N_2}{N_1} = \left(\frac{D_1}{D_2}\right)^2 \tag{16.9}$$

16.7 FROTHING AND FOAMING

Presence of dissolved salts and viscosity of medium inhibits bubble coalescence. Intense agitation also lead to trapping of air leading to generation of foam. Temperature, gas flow, and pH have effect on froth formation. The liquid content in foam is very less (generally of the order of 10–40%). Foaming leads to decrease in effective reactor volume and product loss when overflow occurs. Fragile biocells can be damaged when foams collapse. Foaming liquids are also noncoalescing; therefore, they will have favorable mass transfer characteristics.

Surface tension gradients persist during a relatively long time in a surface to which protein is adsorbed, leading to highly stable protein foams. So very stable foams are observed during fermentation. Here antifoam liquids like oils, fatty acids, esters, polyglycols, and siloxanes are used to break foam.

16.8 SCALE-UP OF CHEMICAL REACTOR

16.8.1 Fixed-Bed Reactor

Fixed-bed reactors in the laboratory and pilot plant are operated under plug flow conditions. But this is not so in the commercial plant where temperature and concentration gradients across the tube diameter are never uniform leading to different effective heat conductivity. Reactor throughput increases as the square of diameter of the tube for constant fluid velocity. However, increase in diameter during scale-up leads to highest temperature in the tube center for exothermic reactions and lowest for endothermic reactions. Radial temperature gradients and hot spots cause loss of selectivity and activity. In pilot plant small catalyst particles are preferred so as to minimize mass transfer effects to develop rate equations. But in commercial plants, the particle size is kept high to reduce pressure drop across the bed, so the reaction rate will be limited due to intrapellet heat and mass transfer. Hence the activity or effectiveness of larger pellets has to be predicted apriori.

For the scale-up of a fixed-bed catalytic reactor, in which a first-order reaction is taking place under isothermal conditions, the correct scale-up demands that catalyst particle size and bed depth (or height) be kept constant, while the cross-section is multiplied by the scale-up factor.

16.8.2 Fluidized-Bed Reactor

Fluidized-bed reactors lessen the problems inherent in the scale-up of packed-bed reactors by smooth temperature and concentration gradients due to circulation of catalyst particles. However, the effect of the physical properties of the catalyst and the fluid velocity on bed activity, bed expansion, and catalyst attrition need to be studied thoroughly first. The scale-up of fluidized-bed reactor involves not only understanding the increase in the size of the reactor (bed expansion) but also the interaction of the fluid solid in the larger scale. This design is ideally suited if the particles exhibit gel-type behavior forming clumps or agglomeration.

16.8.3 Stirred Tank Reactor

In small-scale stirred tank reactors are operated under well-mixed isothermal conditions, but in the large-scale one may observe dead unmixed zones. When the *aspect ratio* of the reactors in the small and the large scale is the same, the increase in heat transfer area will be only a fraction of its volumetric capacity, leading to slow heating and cooling. (*Aspect ratio* is the ratio of reactor cylindrical length to its diameter.) For geometrically similar vessels to achieve same reaction yields, same mixing times, and constant turn over time in both the scales, the scale-up rule states that

$$\frac{P_c}{P_s} = \left(\frac{T_c}{T_s}\right)^5 \quad \text{and} \quad \frac{P_c}{P_s} = \left(\frac{q_c}{q_s}\right)^{5/3} \tag{16.10}$$

where subscript c and s represent large and small scales.

16.8.4 Gas–Liquid–Solid Reactors

The fundamental assumption in a slurry reactor is that solid is homogeneously dispersed in liquid. The most critical element in developing a scale-up methodology is to understand the hydrodynamic issues, liquid circulation pattern, and slip between the solid and the liquid. Bubble size depends on the dynamic equilibrium between the coalescence and breakup through out the column. These processes depend on the physical properties, operating condition, and localized hydrodynamic behavior. The presence of solids would affect the bubble coalescence and breakup. It is observed that large particles would promote bubble breakup. Solid free gas holdup increases with increasing reactor diameter, and liquid holdup would correspondingly decrease. Actual liquid residence time depends on the liquid holdup and is the principal scale-up parameter for predicting product yield. Bubble shape, concentration, and resultant rise velocities are also important for reactor performance.

Minimum rpm required for lifting the solids in a slurry rector is,

$$N \alpha D^{-n} \qquad n = 0.67 - 0.85 \tag{16.11}$$

To obtain a degree of homogeneity, the solid particles have to be not only lifted from the bottom but also have to be carried throughout the volume of the vessel in sufficient quantities, which leads to a relation,

$$N^2 \frac{D^2}{d_p} = \text{constant} \tag{16.12}$$

The stirrer blade thickness (b) also has an effect on solid and

$$\frac{ND^{11/12}}{b^{1/4} d_p^{1/6}} = \text{constant} \tag{16.13}$$

It is observed that lower stirrer speeds are needed to suspend solids in larger tanks if geometric similarity is maintained. For the same impeller size and clearance ratios, the critical speed is almost the same for turbine and flat paddle agitators, but the turbine draws twice as much power as the paddle and 15 to 20 times as much power as the propeller to achieve complete suspension of solids.

16.9 INNOVATION IN SCALE-UP

Penicillin production requires the growth of large quantities of the mold, which is aerobic and sensitive. Although it grows well on the surface of a thin layer of nutrient in a petri dish or a flask, in large vessels the mold forms a thick blanket of fibers that smother the organisms. A submerged fermentation technology was developed, where the mold is grown directly in the nutrient medium leading to more efficient contact between the mold and its liquid food source. This innovation led to the scale-up of penicillin production from flasks to deep tanks of thousands of gallons in the year 1943. Foaming due to bubbling of air into the broth was overcome by using antifoaming agents.

Innovative strategies adopted during scale-up of a few biochemical processes are listed in Table 16.1.

PROBLEMS

1. Estimate revolutions per minute (rpm) of stirrer in the large scale to achieve each of the following conditions (diameter of vessel in large scale is 5 m, diameter in small scale is 0.5 m, and rpm = 220):(a) equal solid

TABLE 16.1 Innovative Strategies Adopted During Scale-up of Selected Biochemical Processes

Manufacture		Reactor type	Observations
Ligninolytic enzyme production	Semi-solid-state cultivation (growth on solid material with small quantity of free liquid) Noninert material like corn cob (acts as both a support and nutrient)	A 1 l static bioreactor with air diffuser Tray bioreactor	Maximum activity and poor selectivity Highest selectivity because of thin layer of solids and good contact with oxygen.
Bioethanol	Fermentation of corn step	External loop, liquid-lift reactor (Oleic acid used for circulating flow and also absorb ethanol produced. Spinning sparger for oleic acid entry.)	Oleic acid increased extraction of ethanol.
Daidzeim and genistein	From immobilized soya bean culture (supported in calcium alginate spheres)	Magnetic fluidized-bed continuous reactor (Axial magnetic field helps to fluidize beads.)	
Recombinant mammalian protein (HIV-1 gp 120)	*Vaccinia* virus T7 supported on protein alginate beads	A 2.2-l bioreactor equipped with basket and a vertical mixing system	

Toluene extraction from toluene-p-xylene mixture	Pseudomonoas putida culture	A 4-l cyclic bioreactor (When nutrient levels are low, 5% of reactor contents are removed and replaced with sterile medium.)	About 3 times higher toluene removal rates were achieved when compared to steady-state operation.
Wine distillery wastewater treatment		A 3.5-l Continuous flow fluidized-bed anerobic reactor (open pore sintered glass beads for cell immobilization).	Anerobic process when compared to aerobic consumes less energy, no excess sludge production, and absence of odor.
Lactic acid production	Rhizopus oryzae supported on mineral oil + polyethylene glycol	A 3-l air-lift bioreactor.	Cottonlike morphology of the floc leads to high yield.
Fermentation (viscous systems)	Pseudoplastic polysaccharide fermentation broth	A 6-l centrifugal field aerobic fermentor.	Air flow into the rotor of the centrifuge and disperses as fine bubbles. Very high gas–liquid contact.
7-ADCA from hydrolysis of cephalosporin G (Na salt)	Immobilized penicillin amidase (cross-link of soluble enzyme with glutaraldehye)	A 80-l multichamber tower reactor. Particles held in each chamber between wire mesh.	Design takes in advantages of packed bed and agitated vessel.

suspension, (b) equal mass transfer rates, (c) equal surface motion, (d) equal blend time, and (e) constant Reynold's number (assume density and viscosity of that of water)?

2. Which criterion is beneficial during scale-up based on heat transfer—equal heat transfer coefficient, equal heat transfer or power per unit volume?

3. If 0.1-mm-diameter particles can be lifted and dispersed well in a vessel of 0.25 m diameter at an rpm of 310, what should be the rpm for just suspending the solid in a vessel of 5 m diameter and the rpm required to disperse the particles uniformly?

4. Compare slurry reactor, packed-bed reactor, and fluidized-bed reactor for gas–liquid–solid reactions. Which gives the best gas-to-liquid mass transfer coefficient?

5. What should be the agitator rpm for a 3-m-diameter continuous-stirred tank reactor to achieve the same mixing pattern as in a 0.25-m-diameter vessel? The residence time in both the vessels has to be maintained constant (60 min).

6. If the power requirement to agitate a fluid in a vessel of 8 m diameter and 1.4 m height is 5 hP, what will the power required be if the vessel is scaled up to a size of 2 m diameter and 3 m height to achieve the same heat transfer coefficient in both the scales?

7. What will be the power requirement for the above system if the scale-up criteria is based on equal heat transfer rate?

8. What will be the increase in power to get the same emulsion droplet size if the tank diameter and height are scaled up by a factor of 2?

9. What is the change in rpm to completely disperse a gas in a liquid if the tank diameter is scaled up by a factor of 3? Assume that the volumetric flow of gas per unit liquid volume in both cases is the same.

10. Redo the calculations of Problem 9 if volumetric gas flow is assumed to be inversely proportional to the mass transfer coefficient, i.e., Q_g large scale/Q_g small scale $= k_L$ small scale/k_L large scale. $k_L \propto N^{3/4}D^{5/4}$.

NOMENCLATURE

D	agitator dia
K_L	mass transfer coefficient
N	rpm of the agitator
P	agitator power
T	tank diameter
Q_g	gas flowrate
U	overall heat transfer coefficient
V	liquid volume

Z	tank height
b	agitator blade thickness
d_p	particle diameter
g	acceleration due to gravity
h	heat transfer coefficient
q	vessel throughput
t	blending time
u	fluid velocity
ρ and μ	density and viscosity of the fluid
Reynolds number	$D^2 N \rho / \mu$
Froude number	u^2 / gD

REFERENCES

1. Euzen, J.P.; Trambouze, P.; Wauquier, J.P. Scale-up Methodology for Chemical Processes. Gulf: Houston, TX, 1993.
2. Oldshue, J.Y. *Fluid Mixing Technology*; McGraw-Hill: New York, 1983.
3. Nienow, A.W. Chem. Eng. Sci. 1974, *29*, 1043.
4. Mezaki, R.; Mochizuki, M.; Ogawa, K. *Engineering Data on Mixing*; Elsevier: Amsterdam, 2000.

Index

9 780367 394431